Introduction to RF Design Using EM Simulators

DISCLAIMER OF WARRANTY

The technical descriptions, procedures, and computer programs in this book have been developed with the greatest of care and they have been useful to the author in a broad range of applications; however, they are provided as is, without warranty of any kind. Artech House, Inc. and the author and editors of the book titled *Introduction to RF Design Using EM Simulators* make no warranties, expressed or implied, that the equations, programs, and procedures in this book or its associated software are free of error, or are consistent with any particular standard of merchantability, or will meet your requirements for any particular application. They should not be relied upon for solving a problem whose incorrect solution could result in injury to a person or loss of property. Any use of the programs or procedures in such a manner is at the user's own risk. The editors, author, and publisher disclaim all liability for direct, incidental, or consequent damages resulting from use of the programs or procedures in this book or the associated software.

For a complete listing of titles in the
Artech House Microwave Library,
turn to the back of this book.

Introduction to RF Design Using EM Simulators

Hiroaki Kogure

Yoshie Kogure

James C. Rautio

ARTECH HOUSE

BOSTON | LONDON

artechhouse.com

Library of Congress Cataloging-in-Publication Data
A catalog record for this book is available from the U.S. Library of Congress.

British Library Cataloguing in Publication Data
A catalogue record for this book is available from the British Library.

Cover design by Adam Renvoize

ISBN 13: 978-1-60807-155-5

(KAITEI) DENJIKAI SHIMYURETA DE MANABU KOSHUHA NO SEKAI
Copyright © 2010 Hiroaki KOGURE and Yoshie KOGUREAll rights reserved.
Original Japanese edition published in 2010 by CQ Publishing Co., Ltd.
English translation rights arranged with CQ Publishing Co., Ltd. through Japan UNI Agency, Inc., Tokyo

All rights reserved. Printed and bound in the United States of America. No part of this book may be reproduced or utilized in any form or by any means, electronic or mechanical, including photocopying, recording, or by any information storage and retrieval system, without permission in writing from the publisher.

All terms mentioned in this book that are known to be trademarks or service marks have been appropriately capitalized. Artech House cannot attest to the accuracy of this information. Use of a term in this book should not be regarded as affecting the validity of any trademark or service mark.

10 9 8 7 6 5 4 3 2 1

Contents

Preface

My collaboration with the Kogures goes back to 1985. At that time, the IBM PC had just been introduced, and I had written an antenna analysis program for amateur radio operators that Aki and Yoshie promoted and sold in Japan. We have been working together, and with computers, ever since.

It is truly amazing what has happened with computers over those 25 years. As described in this book, a typical moderately high-end desktop computer (which is what the 4.77-MHz IBM PC was back then) has increased in speed by about 300 million times as applied to matrix inversion. It is an incredible time to be involved in technology.

With the huge increases in speed, we have also seen huge increases in the numbers of engineers who need to solve microwave problems. The Kogures have been addressing this need for the last quarter-century in Japan by giving innumerable courses to help bring engineers new to our field up to speed as quickly as possible. While these students were eager and motivated, and very skilled in other areas, their knowledge of microwaves was often nearly nonexistent.

This book, along with the companion antenna book, is the culmination of all those years of training Japan's new-found microwave engineers. It is intended for anyone who finds themselves in the same situation as the Kogures' Japanese students—lots of motivation, but little knowledge in this highly specialized field. It is for those who will go on to immerse themselves in equation-filled college text books but would like a quick, easy read to get the big picture, to get a fundamental understanding of what is happening before dealing with massive amounts of technical detail. It is also for those who deal with microwave designers, who would like to have a deeper understanding of exactly what is going on, rather than just trying to memorize a few impressive-sounding technical terms.

The Kogures performed the initial translation of this book, and then I provided some polish to the prose and added a few points here and there. The original content of the book is due to work performed by all three authors. This book has been a labor of love for a quarter-century. We hope you find it useful.

The DVD included with this book includes a copy of Sonnet Lite and various files used for the examples in the book. You can do the examples later, or follow along and do the examples as they appear. If you need help, just load the appropriate file and figure out what happened.

James C. Rautio
North Syracuse, New York
March 2011

1

Electricity with Good Manners

1.1 From Source to Load, Nicely

The purpose of a transmission line is to transmit the desired signal from the source to the load with maximum efficiency and fidelity. We first saw this in our grade school science experiment with a light bulb and a battery. We learned that electrons come out of the negative terminal of the battery. They run through the inside of the wire to the light bulb, and then they return to the positive battery terminal through the second wire.

Electricity flowing efficiently through the electric wire is "electricity with good manners." However, this is sometimes not achieved with today's extreme requirements. Problems have arisen.

In this chapter, we review the primary role of transmission lines and start to explore the reason why we can, and do, have problems.

1.1.1 Parallel Lines Model

Two wires are used for the experiment with a light bulb and a battery. Figure 1.1 shows a simple model that consists of only the wires. This example uses a three-dimensional electromagnetic simulator, XFdtd by Remcom USA, based on the analytic technique of finite difference time domain (FDTD) (see Chapter 9 for further description).

1.1.2 Model of a Light bulb and a Battery

For our first example, we view the result of electromagnetic simulation of our grade-school experiment. Three elements that compose an electric circuit are

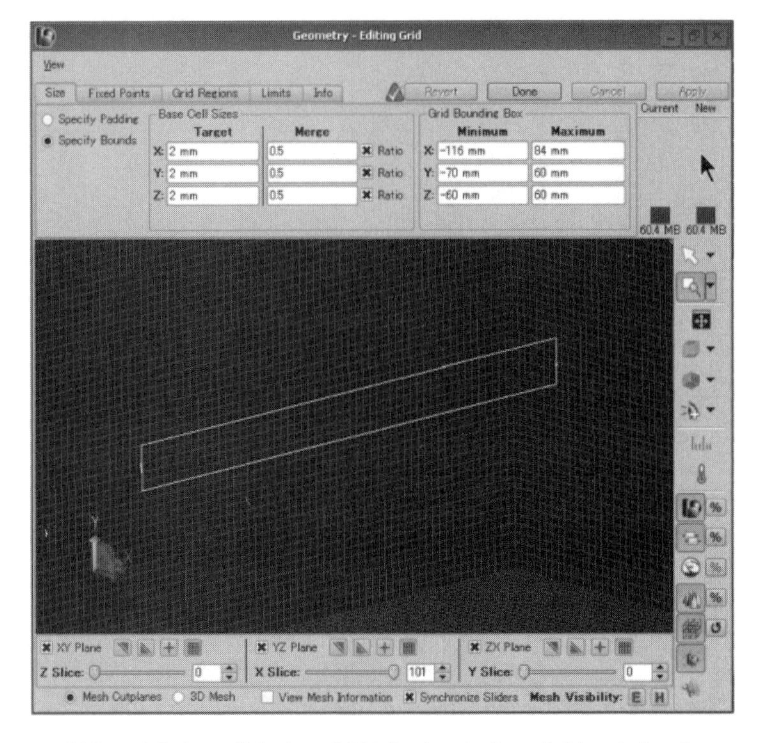

Figure 1.1 CAD model of parallel wires. A 50Ω load at the front (left) end absorbs power from the power source at the far (right) end, port 1.

the power supply, the load, and the transmission line that connects them together. In this case, the battery is the power source, the light bulb is load, and the wiring is the transmission line. Figure 1.2 shows the parallel lines with the power source at the far (right) end and the 50Ω load at the near (left) end. The vertical section between the source and the load displays the electric field generated by the voltage on the two wires. The electric field can be viewed as the electrical tension or stress due to the voltage on the two wires. And, indeed, if the tension becomes too strong, things will break!

The flow of current is the movement of electrons. Imagine the loop of current starting from the power source of Figure 1.2, running along the lower line to the load resistance, passing through it, and then returning in the upper line. (The direction of the current, for the purpose of performing calculations, is defined to be the direction opposite of the direction of electron movement. This is because physicists arbitrarily assigned the direction of positive current long before the electron was discovered.)

The electric field is strong between the upper and lower lines in Figure 1.2. The electric field becomes weaker as we move out from the lines. Note that

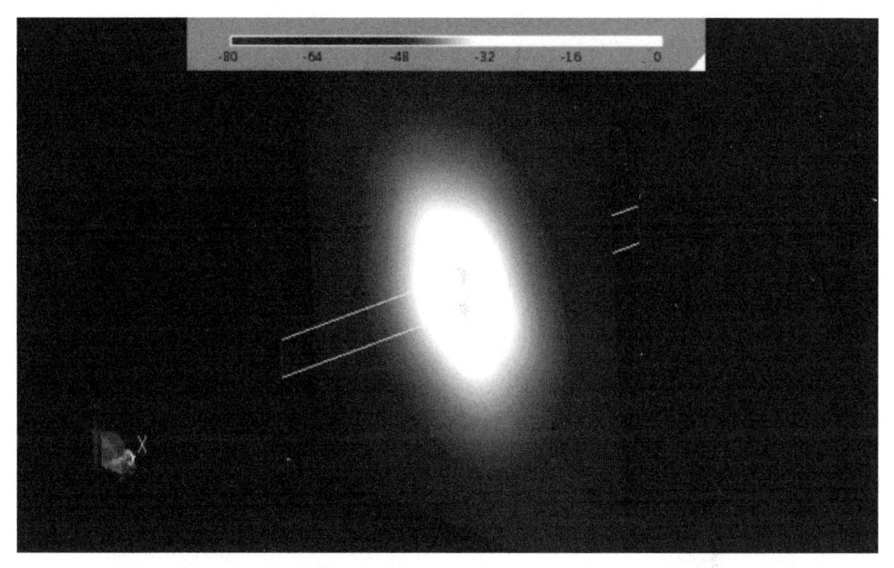

Figure 1.2 Electric field in a vertical cross section due to the voltage between two parallel lines.

the current on each wire flows in the opposite direction of the other wire. The current flows into, and out of, the region with the strongest electric field.

Although we do not have an actual light bulb and battery in Figure 1.2, what we do have is an electric circuit generated with a computer-aided design (CAD) tool, which we use to graphically draw a "model."

1.1.3 Electric Field in the Vicinity of the Wires

The electric field is the potential gradient caused by the voltage of the power supply (electric potential difference) in the space around lines. While voltage is a magnitude only, electric field has both a magnitude and a direction. It is a vector. In Figure 1.2, we just show the magnitude of the electric field.

Figure 1.3 shows the complete electric field vector across the same cross section as Figure 1.2. Small cones show the direction of the electric field. The size of the cones indicates the magnitude of the electric field. Notice that the electric field extends a considerable distance into the surrounding space.

If we draw lines that connect the small cones of Figure 1.3 head to tail, we see the electric lines of force. If we were to then place an electron anywhere in the electric field, it will move along the electric lines of force. In Figure 1.2, the electric field appears to be strong only in the vicinity of the wires. However, after changing the scale of the display to enhance where the electric field is low level, we see that it covers a large volume, as shown in Figure 1.3.

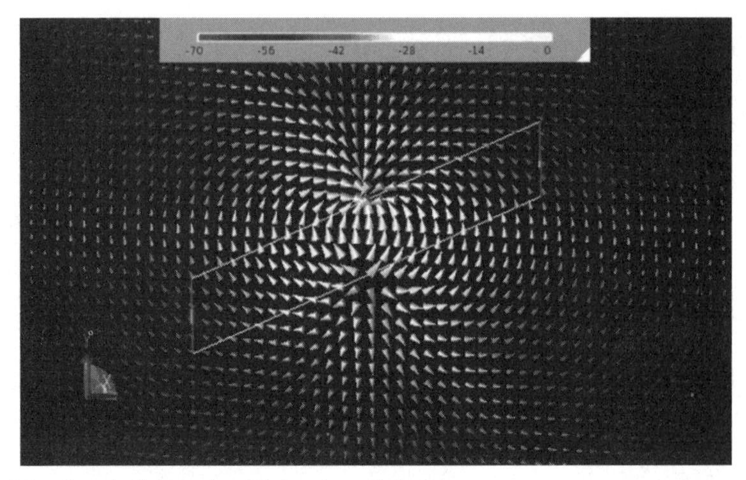

Figure 1.3 Electric field in the vicinity of parallel wires.

1.1.4 Electric Field Due to AC Power

In the previous example, we used direct current. What about alternating current? If we were to look at the electric field distribution at 50 Hz, for example, a display of the electric field at a single instant in time would be the same as Figure 1.3.

Figure 1.4 shows the magnetic field vector at 50 Hz at a single instant in time. If we connect these cones head to tail, we draw the magnetic lines of force. If we were to place a compass in the magnetic field, the compass needle would try to point in the direction of the magnetic lines of force. Note that the magnetic lines of force all form complete loops.

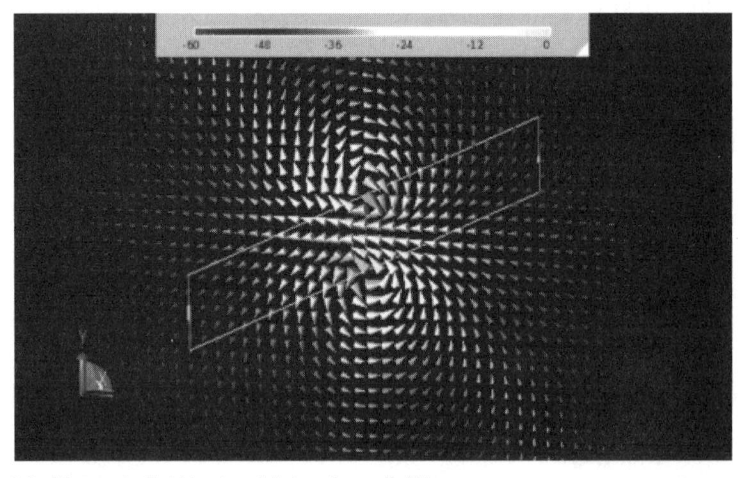

Figure 1.4 Magnetic field in the vicinity of parallel lines.

The loop direction around the upper line is clockwise, and the lower side is counterclockwise. We can use Ampere's right-handed screw rule to determine the direction of the current. Curl the fingers of your right hand in the direction of the magnetic lines of force. Your thumb now points in the direction of the current. Remember that positive current was arbitrarily defined hundreds of years ago to flow in a direction that happens to be opposite to the direction that the electrons actually flow.

1.2 Transmission Lines with Widespread Fields

The concept of the transmission line often appears in textbooks concerning microwaves and communications. The theory of the transmission line started with the telegrapher's equation in the nineteenth century and is due to Oliver Heaviside.

According to the *Dictionary of Electrical Engineering* (Gihodo, 1962), the transmission line is "in the wide sense, all electric circuits by which electric power is sent and received, the transmission system being in the middle." Perhaps the first major transmission line was historically constructed in the United States to carry a telegraph signal (via Morse code) in 1840s.

1.2.1 Two Parallel Wires, the Basis of the Transmission Line

Two parallel wires were announced by the Austrian physicist Ernst Lecher in 1888. Called a Lecher wire, it was chiefly used to transmit microwaves, as seen in Figure 1.5. When viewing the electromagnetic field simulation result in Figure 1.2 and Figure 1.4, we see that the electric and the magnetic fields spread out over a large area. Moreover, when we closely compare Figure 1.3 and Figure 1.4, we see that the electric field vector and the magnetic field vector are everywhere mutually orthogonal.

1.3 Transmission Lines with Confined Fields

In high-power microwave communication, a transmission line consisting of a waveguide tube is sometimes used, as shown in Figure 1.6. Regardless of whether the tube cross section is a square or a circle, it is basically a metallic tube or pipe. Waveguide tubes are useful for transmitting high-frequency power. Since a metallic tube is a solid, one-piece construction, we cannot view it as a voltage applied between two parallel wires.

In this case, electromagnetic energy is transmitted by confining the electric and magnetic field inside the waveguide tube. This is actually a splendid transmission line. As long as the connecting screws are firmly tightened at all

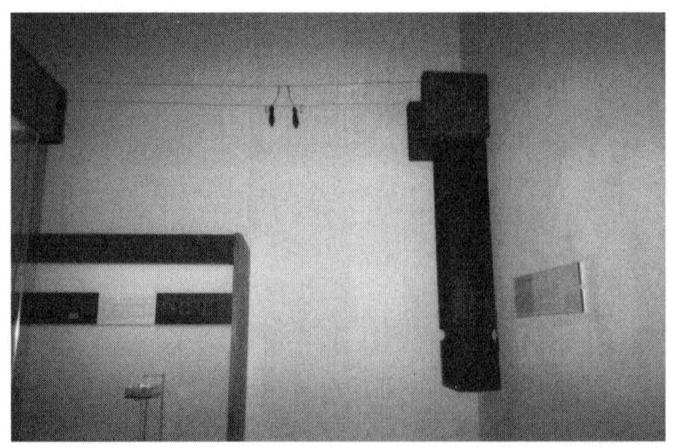

Figure 1.5 Lecher wire. (Photo taken in Deutsches Museum by the author.)

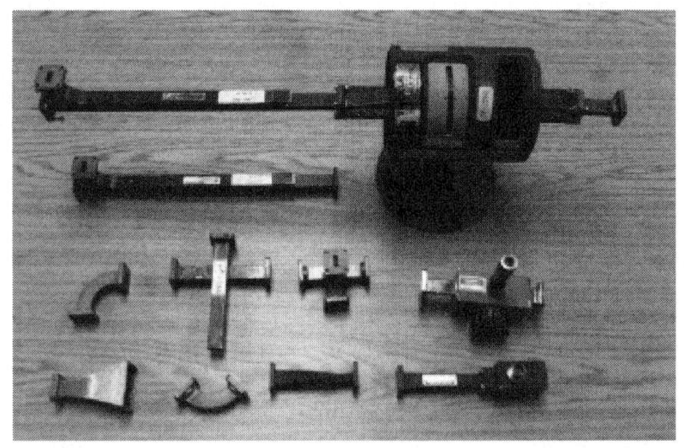

Figure 1.6 Examples of waveguide tubes. (Components courtesy of Ed Foley, Southwest Microwave.)

waveguide joints, all the fields stay inside the tube. There are no fields spreading out over a large area, as we have with the parallel wire transmission line.

1.3.1 Model of a Waveguide Tube

Figure 1.7 is a model of a simple rectangular waveguide tube. It is simulated with the electromagnetic field simulator MicroStripes, which uses the transmission line matrix (TLM) method of CST in Germany (further description is provided in Chapter 9).

Both ends are matched to the characteristic impedance of the waveguide. This means that there is a seamless, smooth transition of power from the source to the waveguide, and then to the load. Thus, when the electromagnetic wave being transmitted inside the tube gets to the end, no portion is reflected back. The excitation is set at the far end, and the electromagnetic energy travels only forward to the front. This is called a traveling wave.

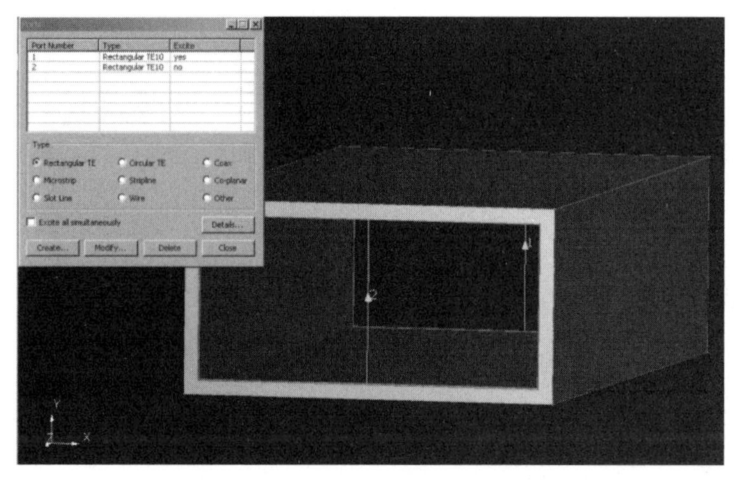

Figure 1.7 Model of a rectangular waveguide tube, excited at the rear side.

Figure 1.8 shows the electric field in a rectangular waveguide tube, and Figure 1.9 shows the magnetic field, both at the same single instant in time. They form a beautiful pattern. Note that a portion of the waveguide wall is removed from the image, so that we can easily see the fields inside the waveguide. Notice, also, that once more the electric and magnetic fields are always at right angles to each other.

1.3.2 Observation of Electric Fields, Magnetic Fields, and Currents

It is understood that the direction of the electric field in Figure 1.8 is vertical from the bottom to the top. During one AC cycle of the microwave energy, the direction reverses for one half of the cycle.

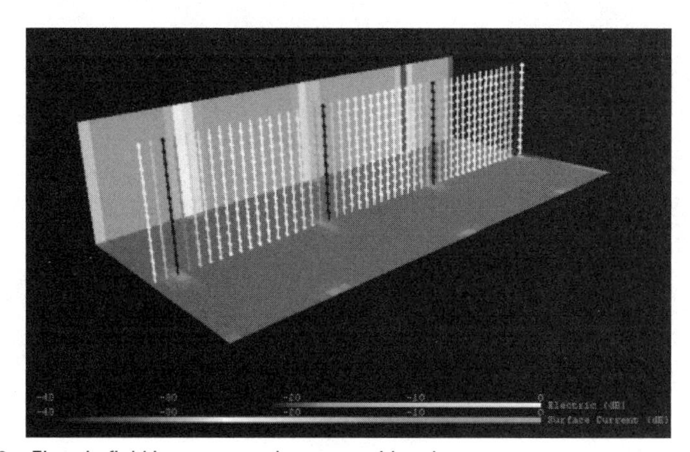

Figure 1.8 Electric field in a rectangular waveguide tube.

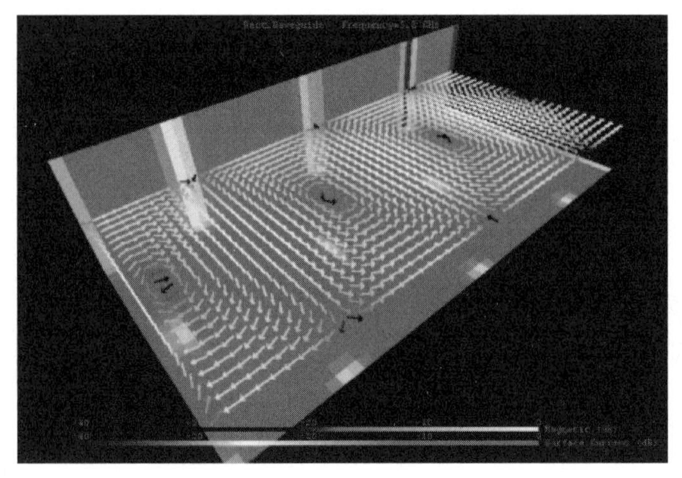

Figure 1.9 Magnetic field in a rectangular waveguide tube.

Next, when viewing the magnetic field sliced in a horizontal plane, as shown in Figure 1.9, we see that many closed loops occur. It is clear that the clockwise and the counterclockwise curls appear alternately, and a beautiful pattern is seen.

In Figure 1.10 we see the surface current distribution in the waveguide. The areas of strong and weak current correspond to the pattern of the electric and magnetic fields. Using these visualizations, the beautiful pattern of the electromagnetic energy being transmitted in the waveguide is easy to imagine.

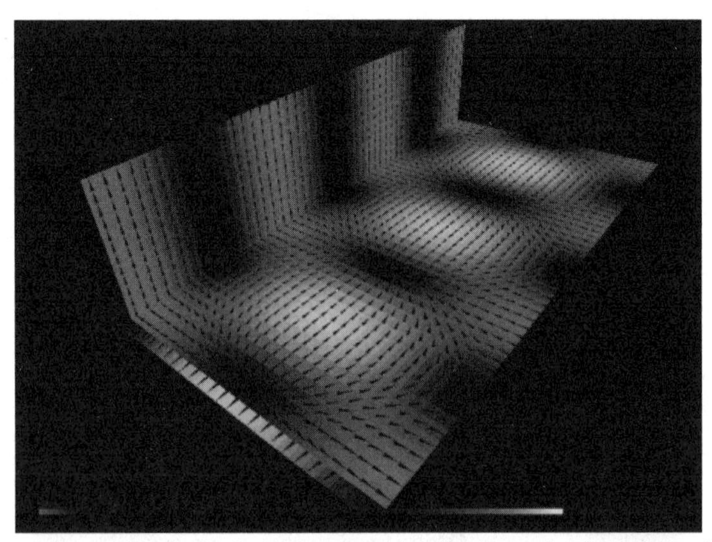

Figure 1.10 Current distribution inside a waveguide tube. Acute triangles indicate the direction of the electric current.

1.3.3 Tubes Can Be Transmission Lines

Two parallel wires and waveguide tubes, as we described earlier, can both be transmission lines. At first, it might seem strange that an empty tube could be a transmission line. However, optical fiber is similar to an empty tube in that the electromagnetic energy is confined to the inside of the tube without radiation. Therefore, optical fiber is also a transmission line. Other transmission lines also use dielectrics to confine and guide the electromagnetic wave. The transmission line concept is widely used.

1.4 What Is a Microstrip Line?

The most common transmission line used at microwave frequencies is the microstrip line. A microstrip line is a flat strip of conductor placed on a thin dielectric substrate. Since this is not a tube-waveguide, we must have a second wire to complete the circuit. The second wire, which we call the ground plane, covers the entire underside of the dielectric substrate, (Figure 1.11).

As the electric field and magnetic field vectors do not extend beneath the ground plane, there is some electromagnetic shielding of any circuits below the ground plane. One example of a simple simulation model is shown in Figure 1.12.

1.4.1 Microstrip Is Great for Printed Circuit Boards

Microstrip lines are often used for the common printed circuit board (PCB), also known as a printed wiring board (PWB). This is because of problems seen in the past with PCBs as the clock frequency of microprocessors continued to rise.

As PCB wiring was originally thought to be predominantly inductive and often did not have a clearly defined ground return (the second wire going back to the source), the characteristic impedance (see Section 1.5) cannot be controlled or even kept constant along the length of the line. As a result, for example, overshoot, undershoot, and ringing is seen in the rise and fall time of a square wave (Figure 1.13).

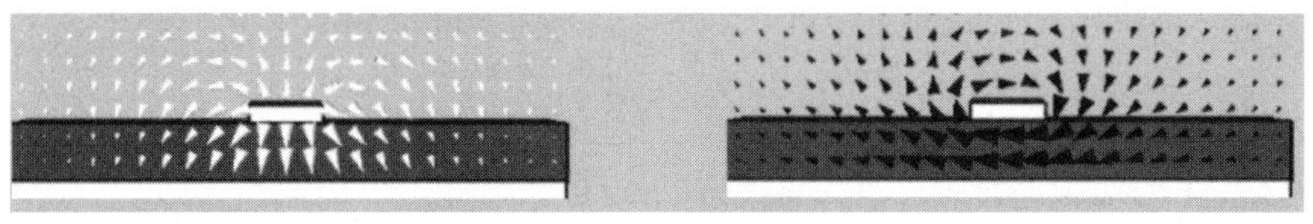

Figure 1.11 Electric field vectors and magnetic field vectors in the vicinity of a microstrip line.

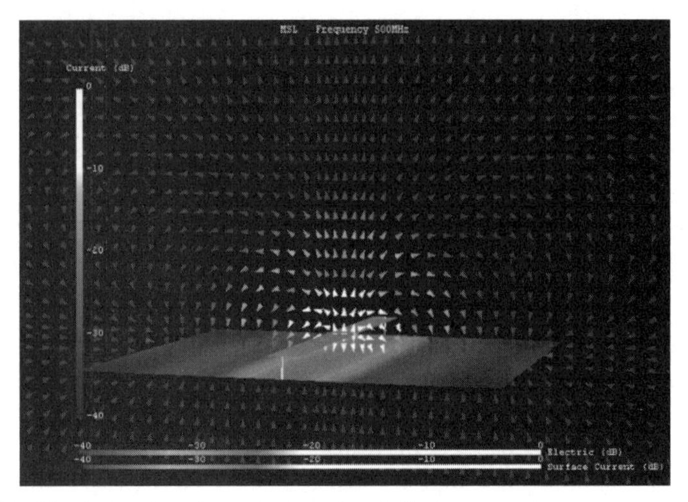

Figure 1.12 Simulation of a microstrip line, electric field vectors shown.

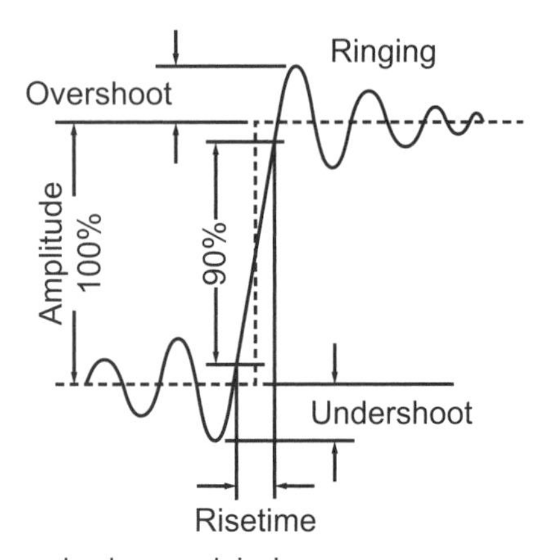

Figure 1.13 Overshoot, undershoot, and ringing.

On the other hand, the characteristic impedance of a microstrip line is mainly determined by the line width w, the thickness of dielectric substrate h, and relative permittivity ε_r, so it is possible to precisely match the impedance of a device. Now designers are using, among other types of transmission lines, microstrip lines by adding a well-defined ground plane to allow carefully controlled characteristic impedance. See, for example, the distributed-lumped circuit shown in Figure 1.14.

1.4.2 Fundamentals of the Microstrip Line

Figure 1.15 shows the result of simulation using Sonnet Suites of Sonnet Software, Inc. Viewed from above, we see that the current distribution is not uni-

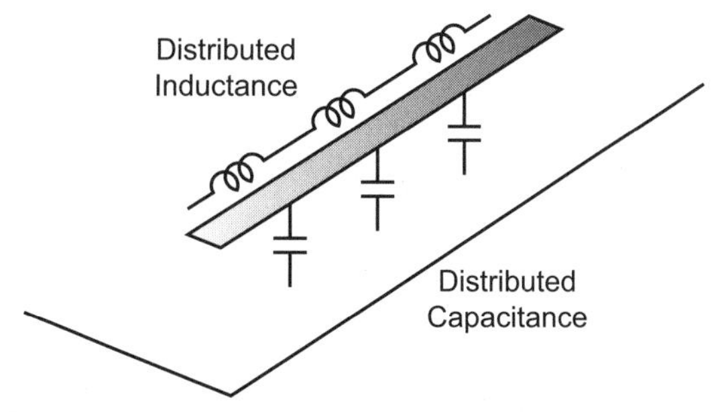

Figure 1.14 A microstrip line can be viewed as a distributed-lumped circuit.

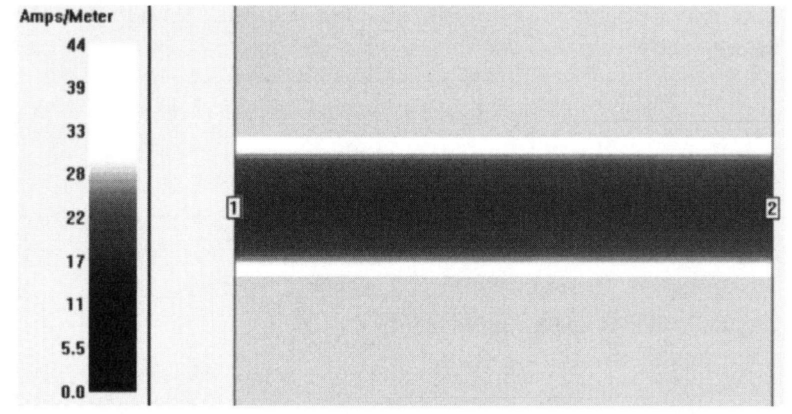

Figure 1.15 High current is seen along both edges of the line, viewed from above.

form across the width of the line. It flows most strongly along both edges of the line.

This is called the edge singularity. We have very high current traveling along the edge, and very high electric fields extending out in all transverse directions from the edge. This is called the edge effect. Except at very low frequency, this is always seen in microstrip lines.

In the previous section, we saw that the electric lines of force extend out from the two parallel wires into space. The direction of the current in the two lines is opposite of one another, and the loops formed by the magnetic lines of force wrap around the two wires in opposite directions.

If the electromagnetic fields are too loosely bound to a transmission line, we might get radiation. The transmission line acts like an antenna. If the two wires above are close to each other, the radiated wave of each wire tends to cancel the radiated wave from the other wire. The total field at a large distance is then very low. We might say that this is "electricity with good manners."

However, if we make the space between the lines wider or the current on one line is not exactly equal and opposite to the current on the other line (for a variety of possible reasons), electromagnetic energy might be radiated into space. Because this can cause noise to be radiated from the wiring (hold your cell phone near your computer sometime), this is "electricity with bad manners," as described in the next chapter.

Thus, we see that it is important that the electric and magnetic lines of force be confined to reduce radiation. If we keep the microstrip line close to the ground plane, like a capacitor, a microstrip line can used for a transmission line on a printed circuit board.

Viewing Figure 1.12, we see that a strong electric field is present between the signal line and the ground plane. The electric field tends to be kept under the line if the line is straight. However, when, for example, a bend in the line is encountered, the electric lines of force can bulge to the outside and we might get radiation.

1.5 Characteristics of Microstrip Lines

In order to do engineering, we need numbers. Let's see what numbers we can get.

1.5.1 What Is Characteristic Impedance?

When the electricity (an electromagnetic wave) is traveling along an infinitely long and uniform transmission line in only one direction, the ratio of the voltage to the current is a constant that is independent of the location. This value is directly related to the ratio of the inductance to the capacitance per unit length, and it takes on a value specific to the transmission line. This value has the dimension of voltage divided by current, ohms (Ω). We call it the characteristic impedance.

Figure 1.16 shows the appearance of fields from which the characteristic impedance of a microstrip line is accurately determined using this definition. The ground conductor is not displayed, and the dielectric substrate is also omitted in the figure, in which we view the line by looking up from the ground side. In addition, because it is a symmetric structure, the solution space is cut in half, with the half not shown assumed to be the mirror image of the half that is shown.

In Figure 1.16, we get the value of voltage and current by the integrating the electric field and surface current, respectively, from the result of the electromagnetic field simulator, MicroStripes (version 3.0). To integrate the electric field from the ground plane to the microstrip line, we click on the corresponding blocks in space sequentially (six small blocks piled up). Next, to integrate the surface current, we define a path surrounding the microstrip conductor for integration.

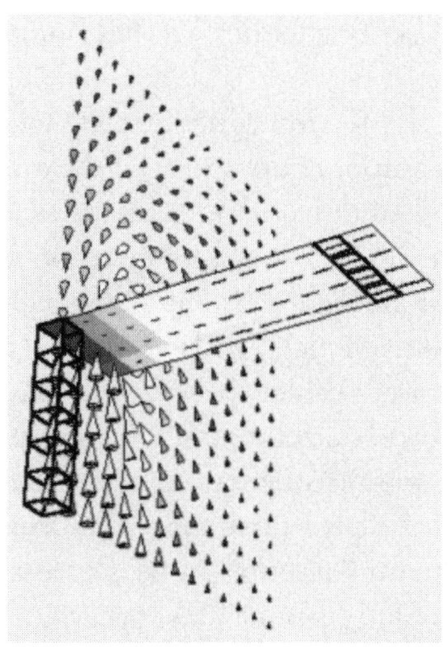

Figure 1.16 Characteristic impedance of a microstrip line can be determined from the current flowing on it and the voltage from the signal line to ground.

When you read the results of each integral, the voltage is found to be 8.390E-14V and the surface current is 9.594E-14A. Then, the characteristic impedance of the microstrip line is 44Ω, the ratio of the two values. The value of the characteristic impedance obtained with electromagnetic field simulator Sonnet is 43Ω, essentially the same answer.

1.5.2 Electromagnetic Field of a Transmission Line

So far in this chapter, we have discussed typical examples of the following three transmission lines:

1. Parallel wires;
2. Microstrip lines;
3. Rectangular waveguide tube.

As for 3, the electromagnetic field (electric and magnetic fields) is completely shielded. In contrast, 1 and 2 leak (very weakly, if correctly designed) the electromagnetic field into space. Looking at the the electric field and the magnetic field in detail, we see they are both always mutually orthogonal. In fact, this is also seen in an electromagnetic wave transmitted in so-called free space.

1.5.3 TEM Mode

When a bell is struck, or a violin is played, the sound we hear is a sum of natural modes. The same is true with the electromagnetic field. In fact, in the transmission line examples we described earlier, we have only one natural mode that propagates along the line.

We see a very special electromagnetic mode when both the electric and magnetic fields have no component pointing in the direction in which the wave is traveling. All the electric and magnetic fields are exactly at right angles to the direction of propagation. In addition, as is true for all electromagnetic fields, the electric and magnetic fields are also at right angles to each other. This is called a transverse electromagnetic (TEM) mode.

Commonly used coaxial cables work well all the way down to zero (DC) frequency. This is described as a zero cutoff frequency mode. Not all transmission lines have a zero cutoff frequency. A transmission line essentially stops working when it is below the cutoff frequency. The mode with the lowest cutoff frequency is known as the fundamental mode. For lossless coax, the fundamental, zero cutoff frequency mode is pure TEM because all of the electric field goes between the center conductor and the shield and all the magnetic field loops around the center conductor. Both are exactly perpendicular to the length of the coax.

However, if there is conductor loss or dielectric loss, and there is enough loss that we cannot ignore it, coax is no longer strictly TEM. For example, resistance in the metal causes a small voltage to form along the length of the coax conductors. This voltage generates an electric field that is pointed exactly along the length of the coax. When this extra electric field is very small, we often refer

Microwaves and Microstrip

When I (Hiroaki Kogure) started studying microwaves, I had thought at first that the wavelength should be micron (μm) because it was a "micro"-wave. However, for such short wavelengths, we find that we are dealing with terahertz (THz) submillimeter waves or even higher. I quickly learned that "micro" is used only subjectively to mean any fairly short wavelength—in this case, super high frequency (SHF) with wavelengths from 1 to 10 cm (3 GHz to 30 GHz).

Note that we sometimes call 1 GHz or higher a quasi-microwave. However, 1.2 GHz for the radio amateur is described as "microwave" by the sectional committee of Japan Amateur Radio League.

Meanwhile, we have more questions: Why are "micro"-strip dimensions on the order of "mm"? Perhaps it is because it is used as a transmission line of "micro"-waves? Or perhaps it is just because it is small, and the name microstrip sounds nice.

to the mode as being quasi-TEM, as described next. The attenuation of modes due to loss tends to increase with frequency.

The electromagnetic field on two parallel wires can also be considered a TEM mode as long as the frequency is not too high. At higher frequencies, the attenuation can become large and the two wires must be kept extremely close together in order to avoid radiation. Thus, parallel wires are not often used for microwave applications.

1.5.4 Quasi-TEM Mode

Because the upper half of a microstrip line is filled with air, the electric field and the magnetic field run through two different media, air and dielectric. Whenever this happens, a pure TEM mode is not possible. However, if the frequency is not too high, it is called a quasi-TEM mode. A strongly non-TEM mode is strongly dependent on frequency. This is called a dispersive mode. A quasi-TEM mode is only weakly dispersive. By assuming we are very close to a TEM mode, it is convenient to use simple transmission line theory, and a range of numerical analysis techniques can be applied directly. This allows us to easily apply microstrip lines to high-speed and high-frequency miniaturized circuits.

It is also suitable for applications like microwave integrated circuit (MIC) and monolithic microwave integrated circuit (MMIC). MIC circuits are often built on ceramic substrates, like alumina. MMICs are often built on semiconductor substrates, like silicon and gallium arsenide.

In addition, microstrip lines can be used in multilayer printed circuit boards, which are now often used. As these applications grow, the use of electromagnetic field simulation has spread rapidly with our circuits always shrinking and becoming even more complicated.

A few transmission line types are summarized in Table 1.1.

1.6 Confirming Results of This Chapter by Simulation

The key points in each chapter of this book can be confirmed using an electromagnetic field simulator. Here, we use Sonnet Lite, which is nice because it is free. Please refer to the Appendix for the installation procedure.

1.6.1 Modeling a Microstrip Line

For our first example, we draw a microstrip line (MSL), as shown in Figure 1.17.

After starting Sonnet Lite, press the Edit Project button of the Task Bar, and select New Geometry. An initial screen, Figure 1.18, is displayed. This plane is the surface of the substrate viewed from above. All circuit metal and

Name	Mode	EM Field Distribution	Characterisitic Impedance
Parallel Wire (Lecher Line)	TEM		$Z_0 = 120 \ln\left(\frac{2d}{a}\right)$ a = Diameter d = Separation
Wave-guide Tube	TEmn TMmn	TE10 TM01	$Z_0 = \frac{120\pi}{\sqrt{1-\left(\frac{\lambda}{2a}\right)^2}}$ a = Length, long side λ = Wavelength in air
Micro-strip Line	Quasi-TEM		$Z_0 = 30 \ln\left[1+\frac{4h}{W}\left\{\frac{8h}{W}+\sqrt{\left(\frac{8h}{W}\right)^2+\pi^2}\right\}\right]$ W = Line width h = Sub. thickness
Coax	TEM TEmn TMmn	TEM TE11	$Z_0 = \frac{1}{2\pi}\sqrt{\frac{\mu}{\varepsilon}} \ln\left(\frac{b}{a}\right)$ a = Inner diameter b = Outer diameter
Free Space Plane Wave	TEM	Propagation →	$Z_0 = \sqrt{\frac{\mu_0}{\varepsilon_0}} \approx 120\pi \approx 377\ \Omega$

Table 1.1 Typical Transmission Lines and Their Characteristics

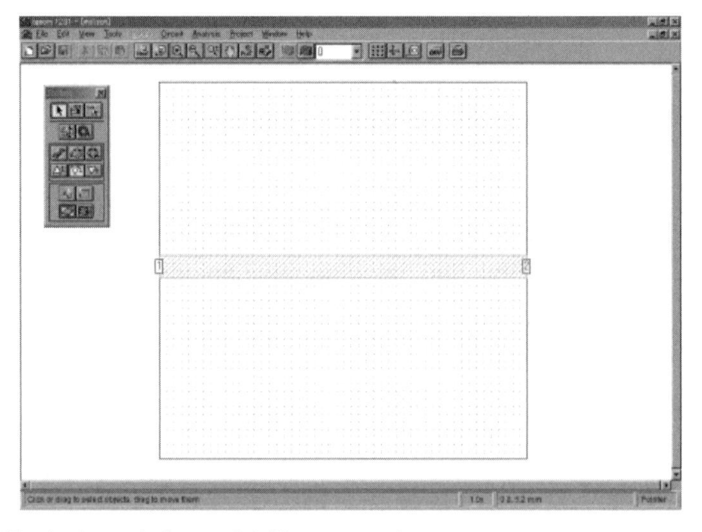

Figure 1.17 A microstrip line model. File name: msl.son.

wiring is drawn here. The fine mesh of points on the substrate show the (user-specified) minimum drawing dimension. The smallest possible rectangle is determined by four points and is called a cell.

Next, select Circuit > Units..., shown in Figure 1.19. Set the Length unit to mm and the Frequency unit to GHz. Next, select Circuit > Box..., shown in Figure 1.20, and set the cell size in both the *x* and *y* directions to 0.1 mm. The next row is the dimension of the substrate. Set both dimensions to 5 mm. The number of cells (Num. Cells) is determined automatically.

When drawing a rectangle on the surface of the substrate, first click the button in the lower right of the Tool Box in Figure 1.21. Next, place the point of the mouse cursor on the point of the cell, drag with the left button held

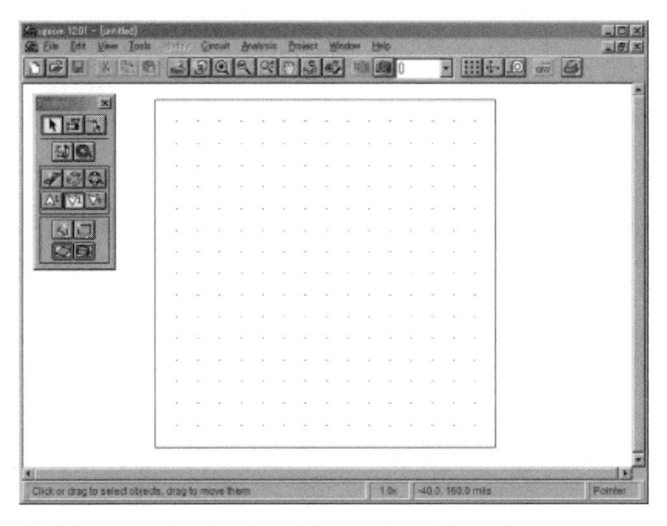

Figure 1.18 Initial screen showing the substrate viewed from above.

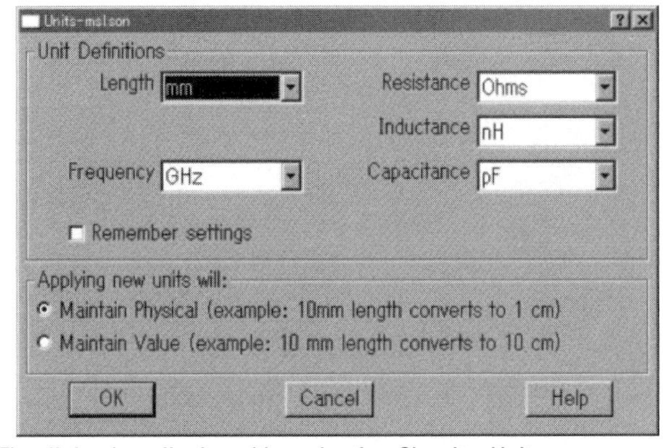

Figure 1.19 The dialog box displayed by selecting Circuit > Units...

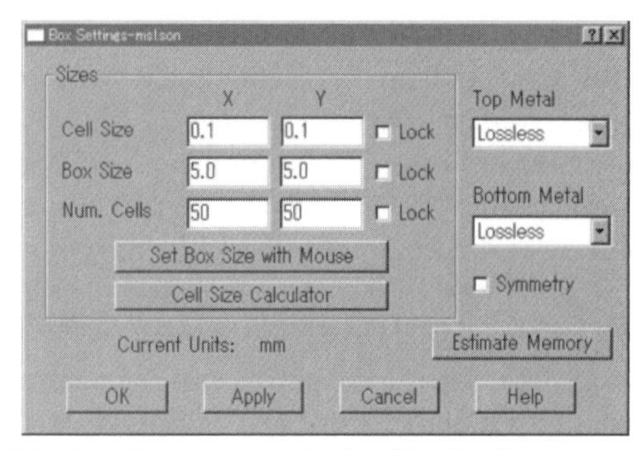

Figure 1.20 The dialog box displayed by selecting Circuit > Box...

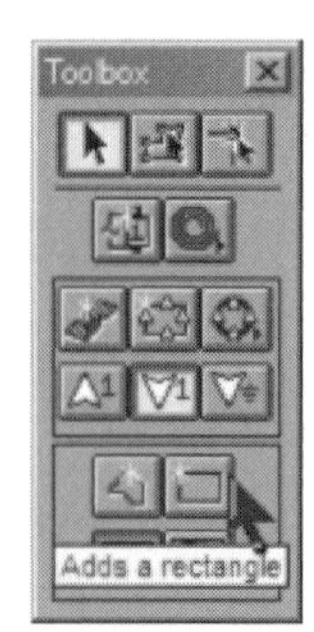

Figure 1.21 The Sonnet Tool Box.

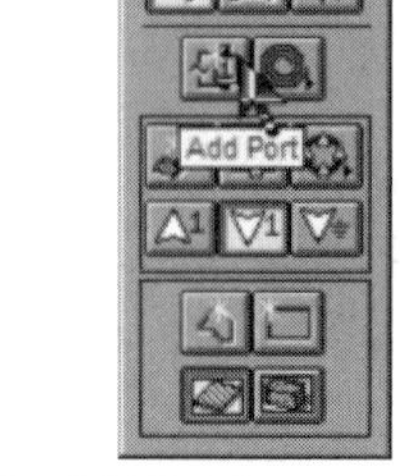

Figure 1.22 Add Port button.

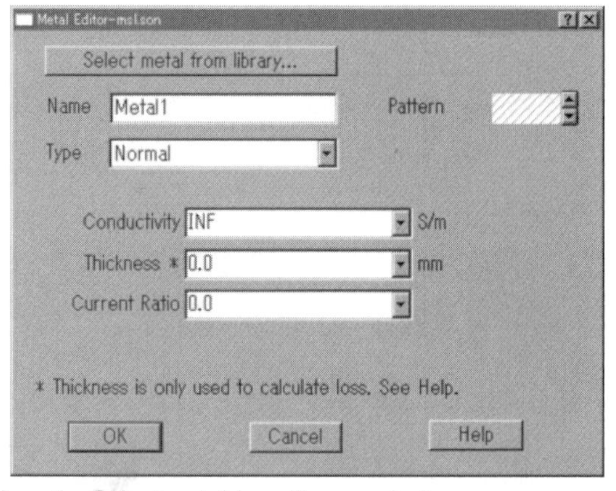

Figure 1.23 Click on the Select metal from library... button.

down, and release the button when the rectangle is the right size. The rectangle will appear with red slashes (lossless conductor).

Draw a rectangle 0.3 mm (three cells) in width and 5 mm in length at the center of the substrate in this way. As shown in Figure 1.17, both edges of the

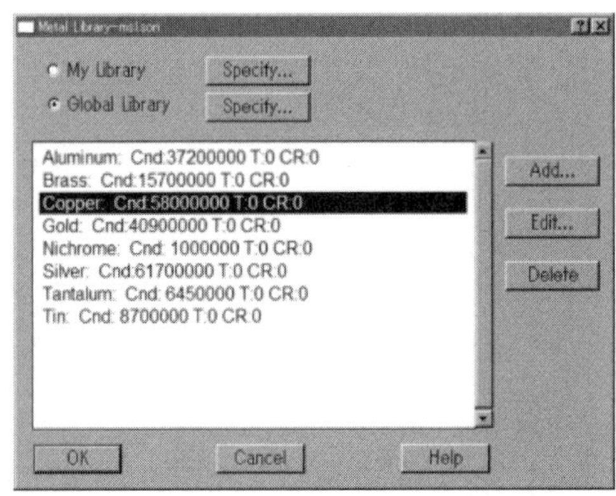

Figure 1.24 List of conducting materials present in the Sonnet global library.

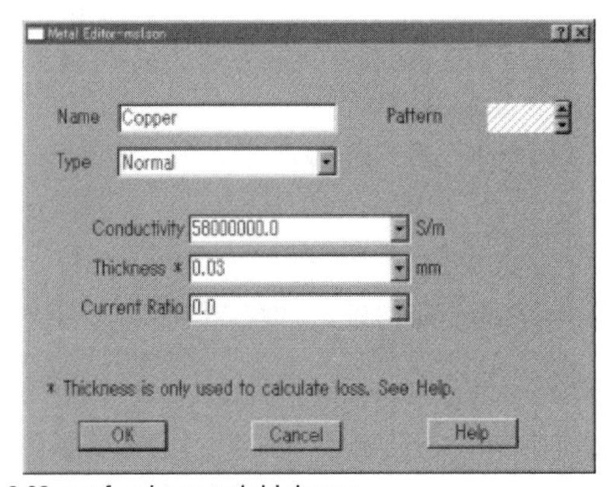

Figure 1.25 Input 0.03 mm for the metal thickness.

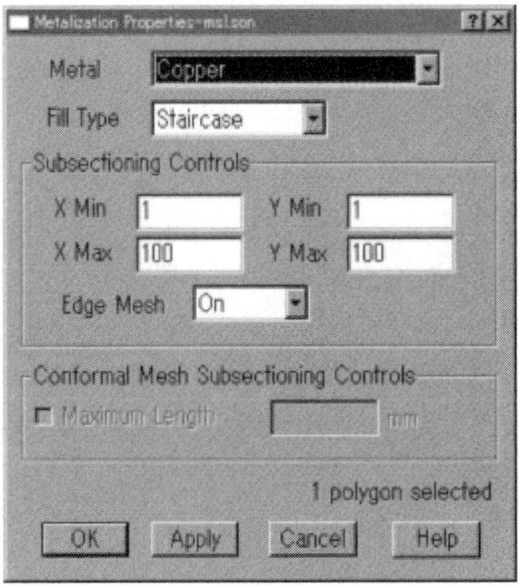

Figure 1.26 Change the metal from lossless to copper.

line touch the edge of the substrate. After clicking the button shown in Figure 1.22, click on the edge to set port 1 and port 2. Imagine building the circuit and placing connectors at those two locations.

Next, the material of the metal is specified. Click on Circuit > Metal Types..., then click on Select metal from library... button shown in Figure 1.23, and click on the Add... button. Select the Global Library and the list shown in Figure 1.24 appears. Select Copper and click OK. Input 0.03 mm for the Thickness, as shown in Figure 1.25. You are now allowed to specify copper in your circuits.

Double-click the mouse on the cross-hatched rectangle we just drew, and the dialog box in Figure 1.26 is displayed. Change the Metal from Lossless to Copper.

Select Circuit > Dielectric Layers..., as shown in Figure 1.27. The dielectric of our substrate is defined here. Here, we see the substrate stackup. In this case, there are only two dielectric layers in the stackup. The boundaries between

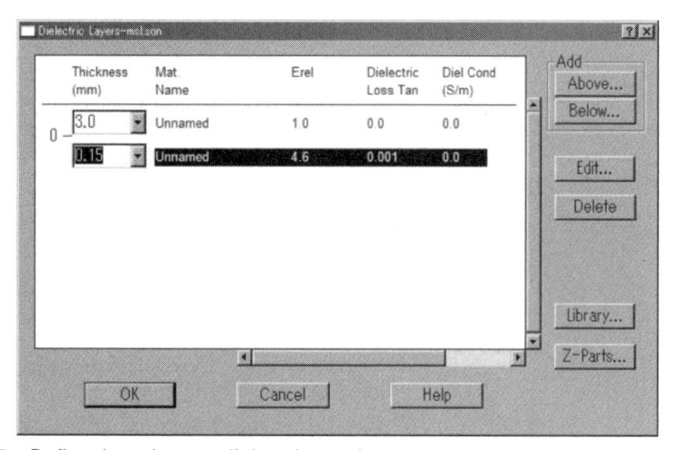

Figure 1.27 Define the substrate dielectric stackup.

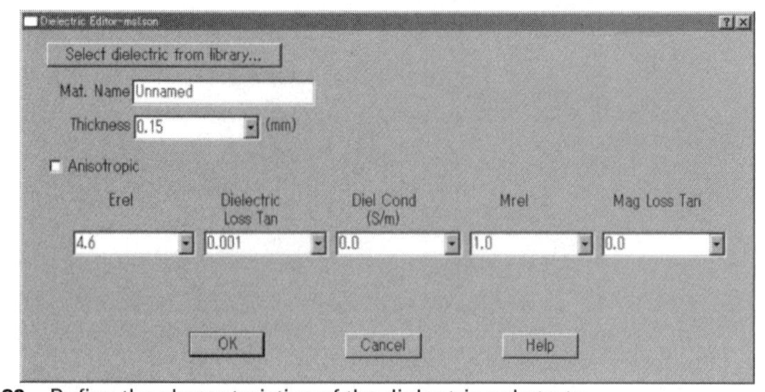

Figure 1.28 Define the characteristics of the dielectric substrate.

layers are referred to as levels. With two layers, we have only one level, level 0. Set the substrate thickness to 0.15 mm for the lower substrate and 3 mm for the upper dielectric layer (which will be air).

To define the substrate dielectric constant, also known as the relative permittivity, double click on the lower dielectric layer in Figure 1.27. Then, input the relative permittivity (Erel) and tanδ (Dielectric Loss Tan), Figure 1.28. Since the upper "dielectric" is air, we will leave its dielectric constant set at 1.0.

Sonnet uses the method of moments in a shielded environment (i.e., the circuit is inside a perfectly conducting box). We can allow radiation from the circuit by setting the top cover (Top Metal) to Free Space. The ground plane

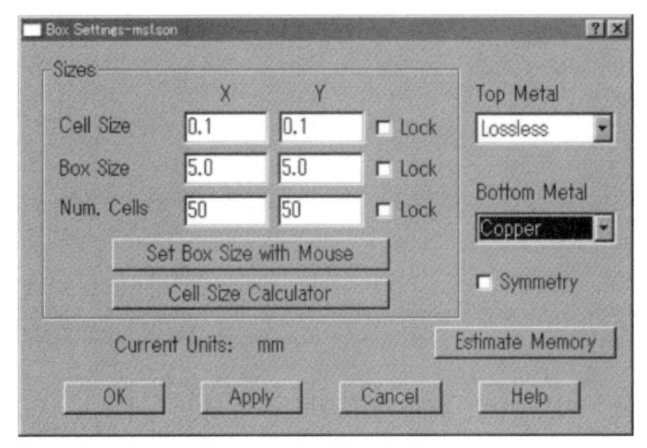

Figure 1.29 Dialog box displayed by selecting Circuit > Box...

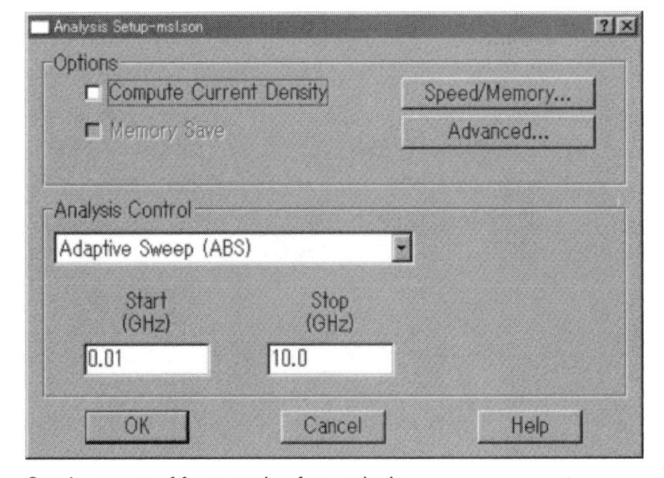

Figure 1.30 Set the range of frequencies for analysis.

of our MSL can be set to Copper by setting the box bottom (Bottom Metal). Please do this by selecting Circuit > Box…, as shown in Figure 1.29.

Don't forget, every now and then, to select File > Save.

1.6.2 Understanding Our MSL—Obtaining S-Parameters

Next, we tell Sonnet what range of frequencies is desired for analysis, among other analysis setup details. Select Analysis > Setup…, as shown in Figure 1.30, and set a range of 0.01 GHz to 10 GHz. Adaptive Band Sweep (ABS) in Analysis Control is the default setting. This is usually the most efficient, requiring actual EM analysis at only a few frequencies, regardless of the smoothness or complexity of the circuit response. There are many other choices, like Linear Frequency Sweep, which analyzes the circuit at user-specified equal intervals. In order to later display the surface current on the line, check Compute Current Density on the top left.

Click on the Speed/Memory... button at the upper right, and the slider bar in Figure 1.31 appears. This adjusts the calculation accuracy and the amount of the memory use. If the result of estimating the amount of memory (click the Memory: Estimate… button) is under 16 MB, it is possible to analyze the circuit using Sonnet Lite. If your circuit is over the limit, you might be able to run within the 16-MB limit when the slider bar is moved to the center or even to right edge, though accuracy is reduced.

Let's analyze this model. Select Project > Analyze, and the window in Figure 1.32 appears, showing the progress of the analysis. When it is complete, select Project > View Response > Add to Graph, and the graph shown in Figure 1.33 is displayed. This is one of the S-parameters, S_{11}. The "S" stands for scattering. Imagine an electromagnetic wave incident on our circuit. Some is scattered (reflected) back. Some is scattered forward. What we see here is the magnitude of the reflection coefficient at port 1.

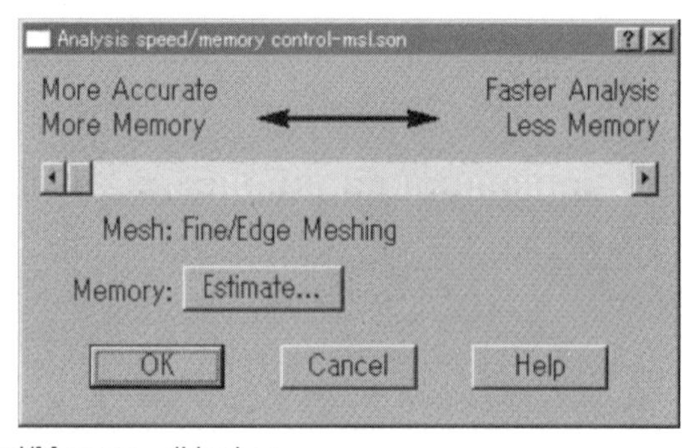

Figure 1.31 Speed/Memory... slider bar.

When it is displayed using decibel (dB) like this, it is sometimes called the return loss. There is always less power reflected from a passive circuit than is incident. Strictly speaking, return loss should be positive dB, because it is a positive loss. However, in practice, you will find return loss often listed with negative dB. Just keep in mind that passive devices do not generate power.

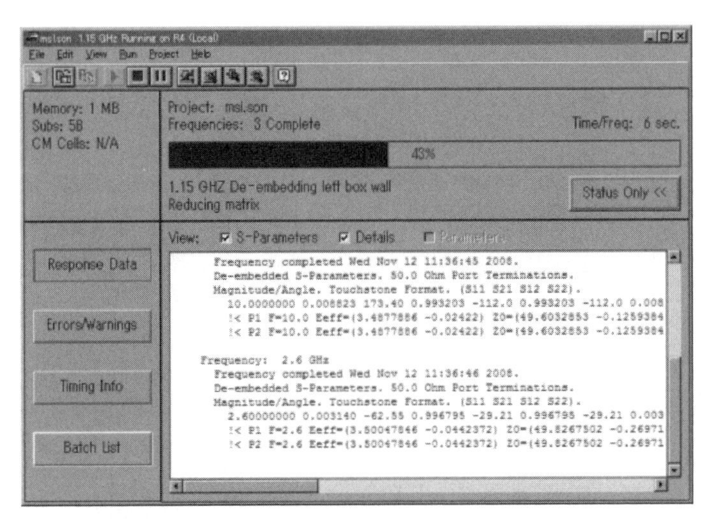

Figure 1.32 Project > Analyze window.

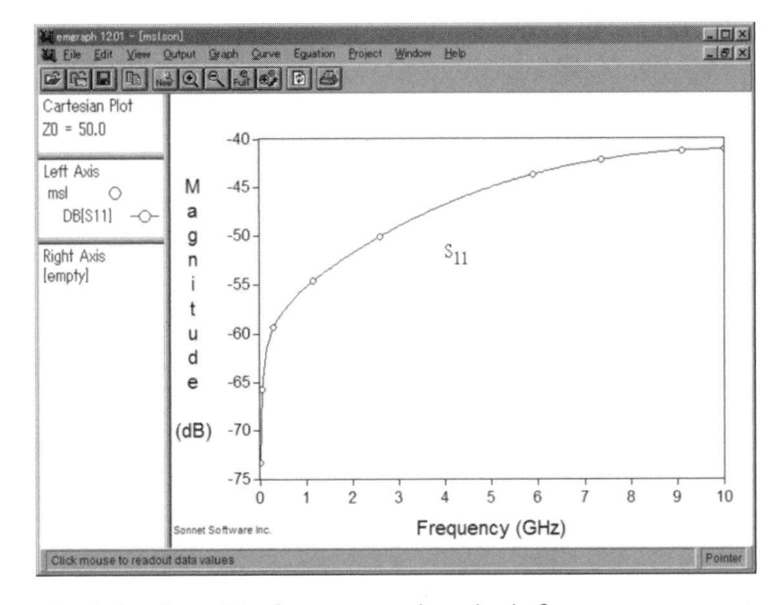

Figure 1.33 A plot of one of the S-parameters of our circuit, S_{11}.

In Figure 1.34, when DB[S21] is moved to the right to Selected items and the OK button is clicked, S_{21} (how much power is transferred from port 1 to port 2) is displayed, as seen in Figure 1.35.

1.6.3 Characteristic Impedance of the MSL

In the line starting with "!<"in Figure 1.32, "Z0=" indicates the characteristic impedance of the line. If you want to plot the characteristic impedance, specify

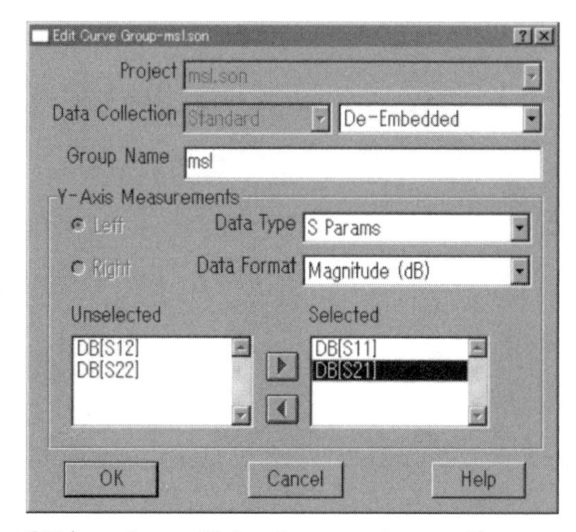

Figure 1.34 Select S21 (transfer coefficient from port 1 to port 2).

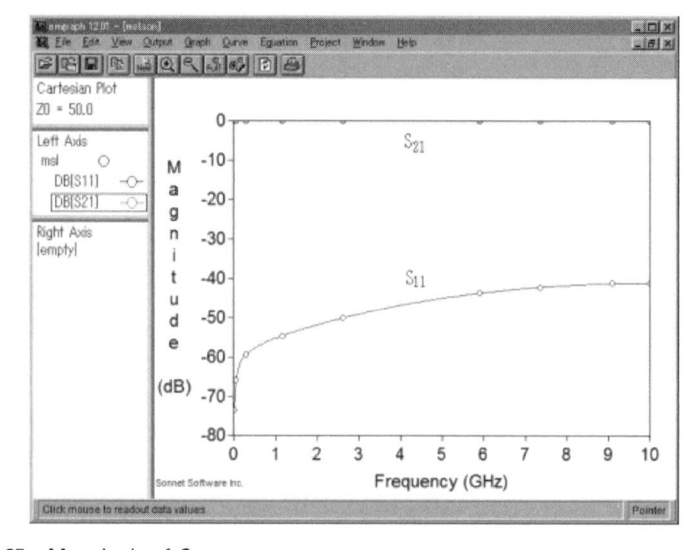

Figure 1.35 Magnitude of S_{21}.

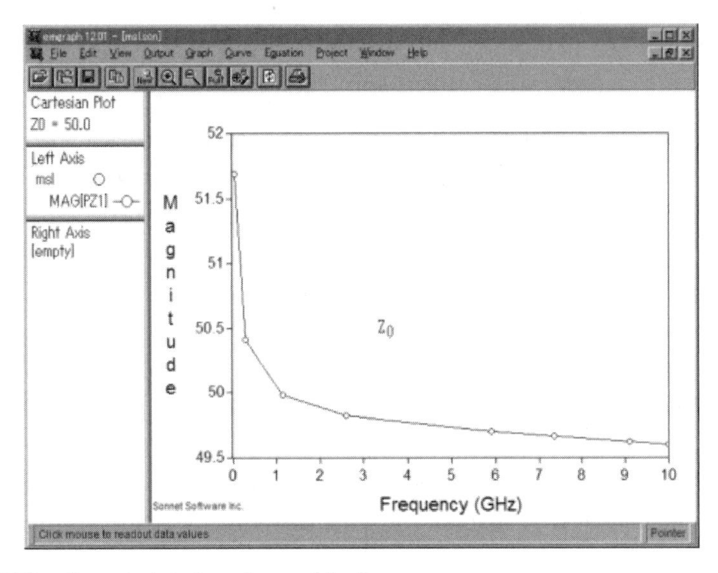

Figure 1.36 Characteristic impedance of the line.

the Data Type of Port Z0, as shown in Figure 1.34, and a plot of the characteristic impedance, seen in Figure 1.36, is obtained.

The definition of the characteristic impedance is the ratio of the voltage to the current of the line as determined by the structure of the transmission line. The voltage of the MSL is the potential difference between the line and the ground. This is obtained from the electric field vector, as shown in Figure 1.11. The electric field is the gradient (rate of change) of the electric potential. It is widely distributed in the surrounding space.

Moreover, the current of the MSL flows through the line and back through the ground plane underneath it. The magnetic field is generated in loops surrounding the flow of the current. The direction of the loops is determined by Ampere's right-hand screw rule (see Figure 1.11).

Thus, the electric field is related to the voltage, and the magnetic field is related to the current. It makes sense that the characteristic impedance also corresponds to the ratio of the electric field to the magnetic field. Because the distribution of the electric and magnetic fields changes when adjusting the width of transmission line, the characteristic impedance changes, too.

An important role of a transmission line is to carry the energy of the input signal efficiently. When the impedance of the power source that feeds the line and the characteristic impedance of the line are different, perhaps due to the discontinuity formed when they are connected together, energy can be reflected. As we show in the next few chapters, impedance matching can be very important. See, for example, Chapter 3.

1.7 Summary

1. Electric and magnetic fields are widely distributed in the space around two parallel wires.
2. Electromagnetic wave energy is transmitted as the energy of the electric and magnetic field around the transmission lines.
3. When our transmission line is a waveguide tube, the electric and magnetic fields travel inside the tube, and radiation to the outside is minimized.
4. The microstrip line can be used to wire PCBs and multilayer substrates when high-frequency or high-speed operation is required.
5. The waves traveling on a transmission line can be viewed as a sum of modes. In typical application there is only one mode propagating on a transmission line.
6. Transmission lines are in wide use.

2

Electricity with Bad Manners

2.1 What Could Possibly Go Wrong?

In Chapter 1, we examined several transmission lines. We found that electromagnetic fields expand around the two parallel wires and the microstrip line. This is because the voltage applied between the lines and the loop current that flows in the circuit generate electric and magnetic fields in the space around the transmission lines. This is how they propagate electromagnetic energy.

In this chapter, we examine transmission lines more closely. As for multilayer substrates, we consider how current flows on the surface of the transmission line conductor, ground plane, and (DC) voltage distribution plane. We examine these matters using substrates based on familiar examples. We search for the reason why electricity with good manners can transform into electricity with bad manners, which couples with other transmission lines and radiates into space causing a variety of problems.

2.1.1 Transmission Line Bend

Figure 2.1 is a microstrip line (MSL) that has a right angle bend in the middle. It is modeled using Sonnet Lite. Both ends are terminated by a 50Ω resistor, and the number 1 at the left edge refers to the input terminal (port 1).

Wiring patterns on printed circuit boards often have right angle bends and other discontinuities. When the signal frequency is low, it does not matter. However, when the clock frequency of a personal computer is several gigahertz and the circuitry is carrying a high bandwidth signal, like digital video, such a discontinuity can be a big problem.

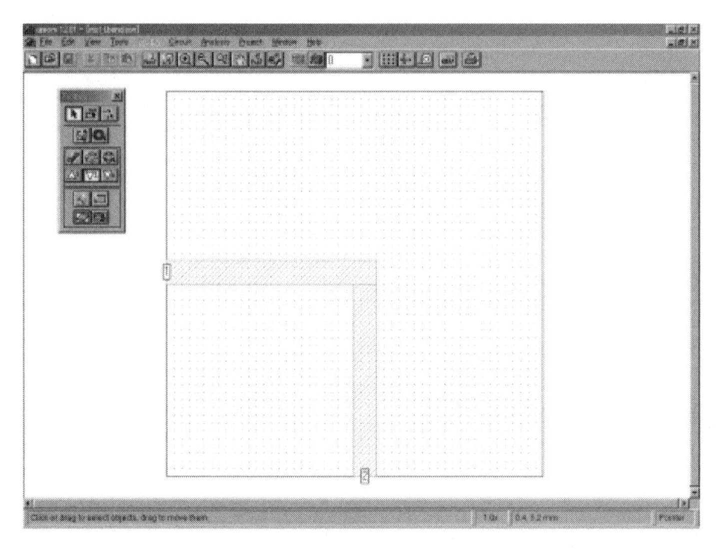

Figure 2.1 Microstrip line with a right-angle bend.

2.1.2 Looking at the Current Distribution

Figure 2.2 shows the current distribution of the MSL in Figure 2.1. As described in Chapter 1, we see that current does not distribute uniformly. The strongest current flows along the two sharp edges of the line. This is called the edge singularity.

However, at the right-angle bend, current does not flow simply along both edges. Rather, it distributes asymmetrically. Comparing the inside path with the outside path of the bend, strong current flows around the inside corner. The inside corner path is shorter, and it is easier for the current to get around the corner by taking this path. Along the outside path, current does not flow at all at the very corner of the bend.

Sonnet is an electromagnetic field simulator based on the method of moments and by default assumes the conductor has zero thickness. FDTD and TLM methods using 3-D CAD also by default handle only the surface without calculating the fields inside metal.

Real metal has thickness, and it has fields and currents inside. The interface between the metal and air or metal and the substrate entails what electromagnetic experts call boundary conditions. There are various ways to handle this boundary condition. For example, a thick conductor can be allowed and all the current is assumed to be flowing only on the surface. This is what FDTD and TLM tools often do.

Method of moments tools can do this, too, but the usual default is to just assume zero thickness and assume that all the current flows on a surface. When

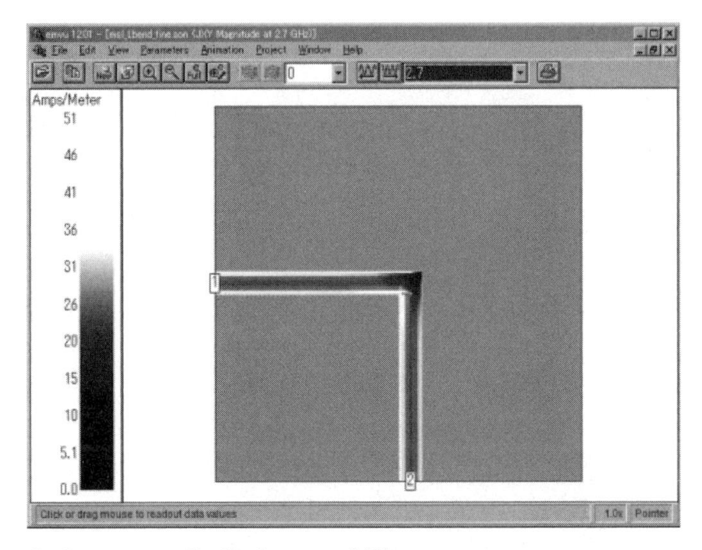

Figure 2.2 Surface current distribution on an MSL.

the correct boundary condition is used, both of these approaches can give good answers.

A full solution, solving for the volume currents throughout the entire volume of the metal, can be carried out by any of these tools, but it can take quite a long time to do so and is rarely justified.

2.1.3 Viewing Electric and Magnetic Fields

How does electric current generate the electromagnetic field around this bend? We demonstrate how to analyze the right-angle bend in the exercise simulation at the end of this chapter. Figure 2.3 shows the electric field on a cross section around the bend. There is uniformly strong electric field between line and ground. We see that the electric field (electric lines of force) flows in an arc running off into space in the vicinity of the bend (results from XFdtd).

How about the magnetic field? Figure 2.4 shows the magnetic field of the same surface. We see that the magnetic field lines are perpendicular to the electric field in the area along the right half of the line. As in other places, when looking at the line around the bend, we see the loop pattern extends beyond the line and ground. These results are for 1 GHz. We also analyzed the bend at 5 GHz. The shapes of the fields are essentially identical (not shown).

In Figure 2.5, we set the space around the substrate wider and display the field intensity at 10 GHz. On the right side of the substrate, we can see that the electric field goes to the backside of the ground plane. Depending on the oper-

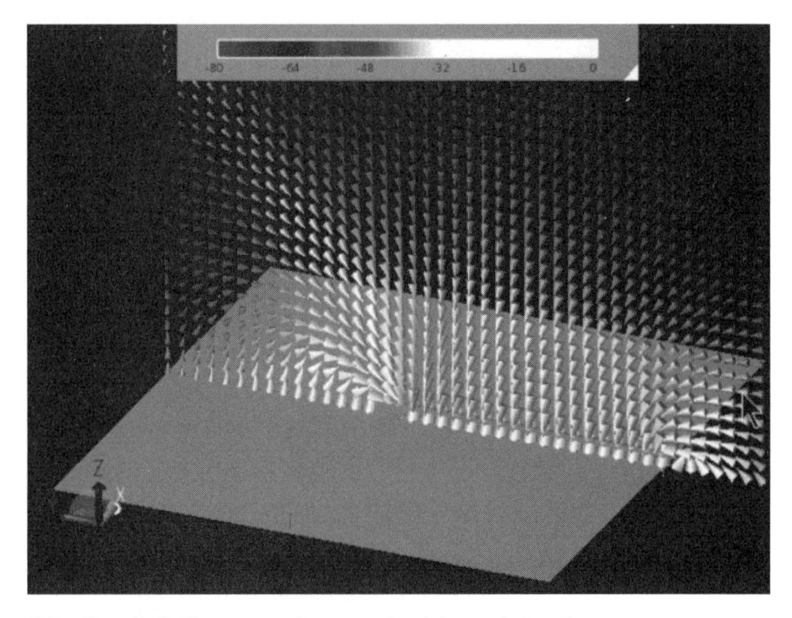

Figure 2.3 Electric field cross section near the right-angle bend.

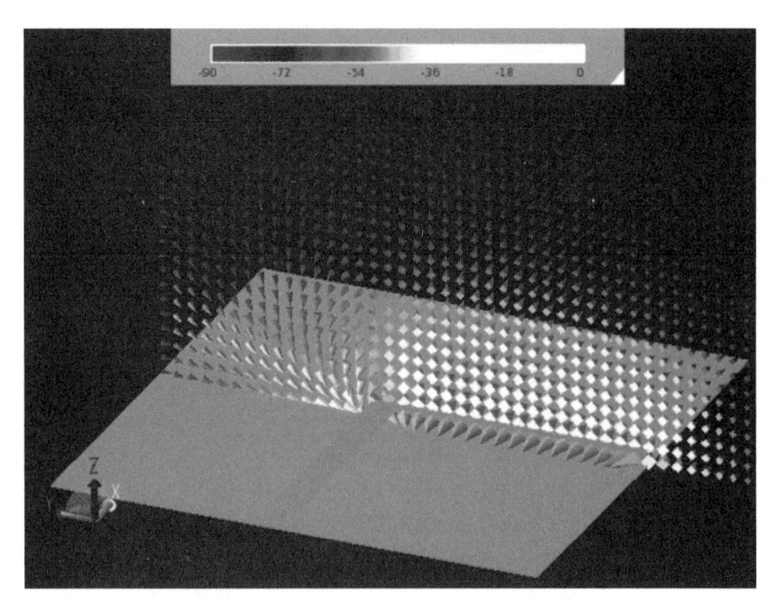

Figure 2.4 Magnetic field cross section near the right-angle bend.

ating frequency, the shielding effect of a finite-sized ground plane is sometimes insufficient.

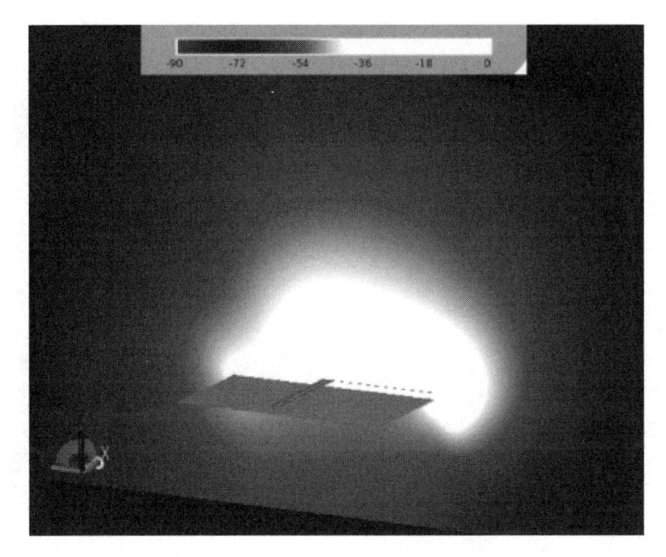

Figure 2.5 Electric field on the cross section at 10 GHz.

2.2 Radiation from Substrates

Now we examine how electromagnetic waves are radiated from substrates. In order to simulate this, we set the far zone sensor. We might think of it as a virtual area that surrounds the radiating object and senses all electric and magnetic fields on that surface. This is a feature provided originally to evaluate the characteristics of an antenna; however, we can also use it to see if our circuit (like the right-angle bend) is acting as though it were an antenna.

This virtual area, called the equivalent surface, is a rectangular solid surrounded by six surfaces. The software solves for the fields on the equivalent surface and then replaces the fields with equivalent surface currents. The radiation is determined from the equivalent surface currents.

2.2.1 Higher Frequencies Usually Mean More Radiation

Figure 2.6 shows the far-field radiation pattern indicating the direction of the electromagnetic wave at 1 GHz. Figure 2.7 shows the result at 10 GHz. Both models are floating in free space and have no obstacles around them.

When comparing only the far-field radiation patterns of the two, we do not see a large difference except for the directions of reduced radiation. However there is a difference in the total amount of energy radiated into free space.

In each case, we use a 1V sine wave to excite port 1, on the left end of the substrate. Although the value of radiant power itself is small, the excitation

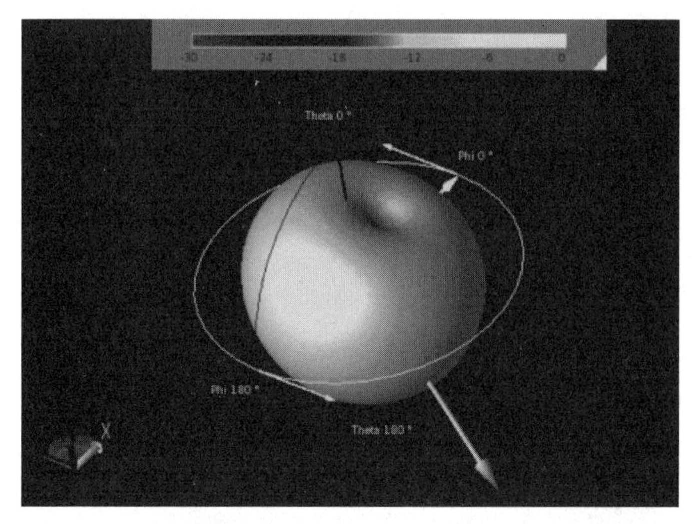

Figure 2.6 Far-field radiation pattern at 1 GHz.

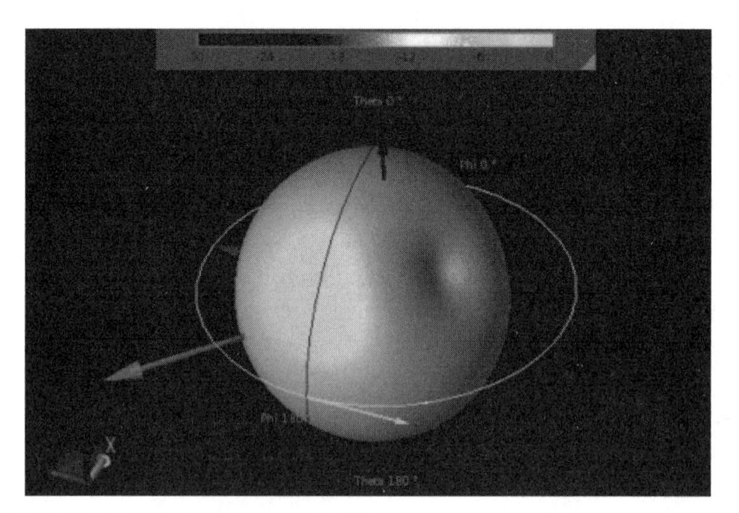

Figure 2.7 Far-field radiation pattern at 10 GHz.

conditions are the same, so we can compare results. We have 33 nW at 1 GHz and 1.4 μW at 10 GHz.

With current, I, flowing on a wire antenna, the electric field magnitude E observed in free space more than half a wavelength away from the antenna is proportional to dI/dt (time rate of change of I). We cannot determine this result easily using only the results of Figures 2.5 and 2.6. However, in general we can say that the higher the frequency (the larger the time rate of change of current), the more strongly that electromagnetic energy is radiated into free space.

2.3 Larger Reflection Coefficient Means...

How much of the signal applied to the input port is transmitted to output port? Also, how much of the input signal gets reflected from the input port? With this information, we can verify whether a transmission line is working properly.

We need to quantitatively evaluate how much of the input signal gets reflected. This is where we use the reflection coefficient. In terms of S-parameters, which we will examine closely in Chapter 3, we are interested in "S_{11}."

2.3.1 The Right-Angle Bend Has Problems

Figure 2.8 is a model of an MSL that has a right-angle bend. The dimension of the substrate is 30 mm × 30 mm, which is larger than the MSL in the previous section.

Figure 2.9 shows the results of reflection coefficient, S_{11}, and transmission coefficient, S_{21}, of this MSL. An S_{11} magnitude of 1.0 is total reflection. When the magnitude is close to 0, then almost no power is reflected from the input port. For this case, we see that S_{11} goes up and down repeatedly. At few specific frequencies, the reflection is extremely low.

In an ideal transmission line, whose characteristic impedance has been carefully selected, there is very low reflection even at high frequency. In this case, nearly all the electric energy arrives at the output port. In practical application, however, even a straight MSL has frequency dependency (dispersion). Add

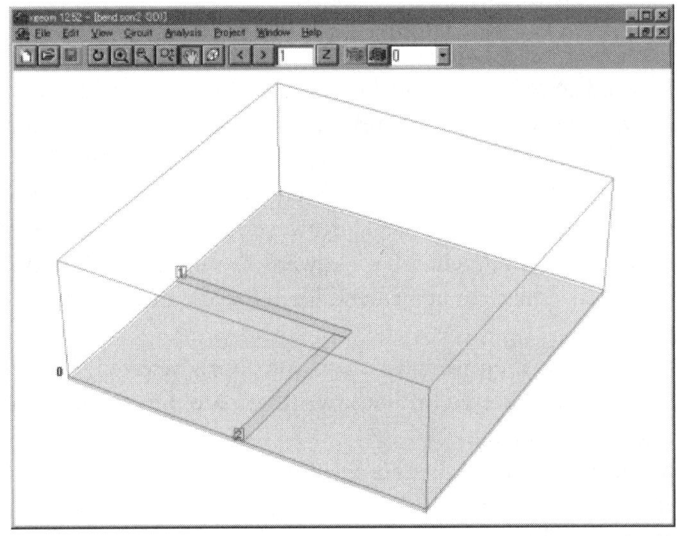

Figure 2.8 Right-angle MSL bend. File: bend.son. Line width: 1 mm, dielectric thickness: 0.3 mm, relative permittivity: 4.8, and tanδ: 0.001.

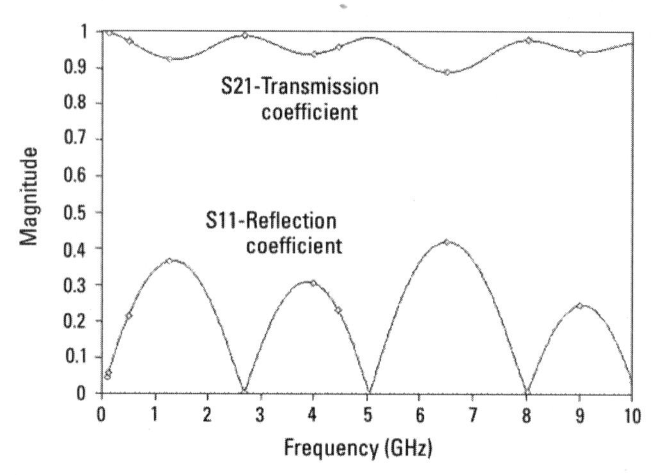

Figure 2.9 S_{11} and S_{21} of an MSL right angle bend, plotted in voltage magnitude (not dB).

in a few discontinuities, like right-angle bends, and things get even worse. So, what happens at a frequency where the reflection coefficient is large?

Figure 2.10(a) shows the surface current distribution of a line at 4 GHz. Starting at the input, port 1, we see a high-current region, followed by a low-current region. At 9 GHz, Figure 2.10(b), we see that the alternating low-/high-current regions are closer together. The distance between two lows, or two highs, is one half wavelength.

2.3.2 Standing Waves Come from Reflection

Let's get a feel for what is happening by becoming a wave in the sea of electrons that form a good conductor. We will take a trip along a transmission line. First, we are pushed and pulled into the input port. Looking at the transmission line ahead of us, it seems that the dielectric substrate goes on forever. Then, we start to roll along the line, everything is just fine. Suddenly, we see a sharp curve! A portion of our wave squeezes through. However, our entire wave did not make it, and a small wave is launched back toward the input. That portion of our wave was reflected. The right angle bend introduced a discontinuity and reflection occurred.

What happens when we have a bad (or poorly selected) load at the end of a perfectly straight line? To find out, we terminate the end of the line with a 100Ω resistor and simulate.

Figure 2.11 is a plot of S_{11} and S_{21} of this MSL. Figures 2.12(a, b) are the surface current distributions at 4 GHz and 9 GHz. Both show similar repeating patterns of high and low current. The current at each region of high current

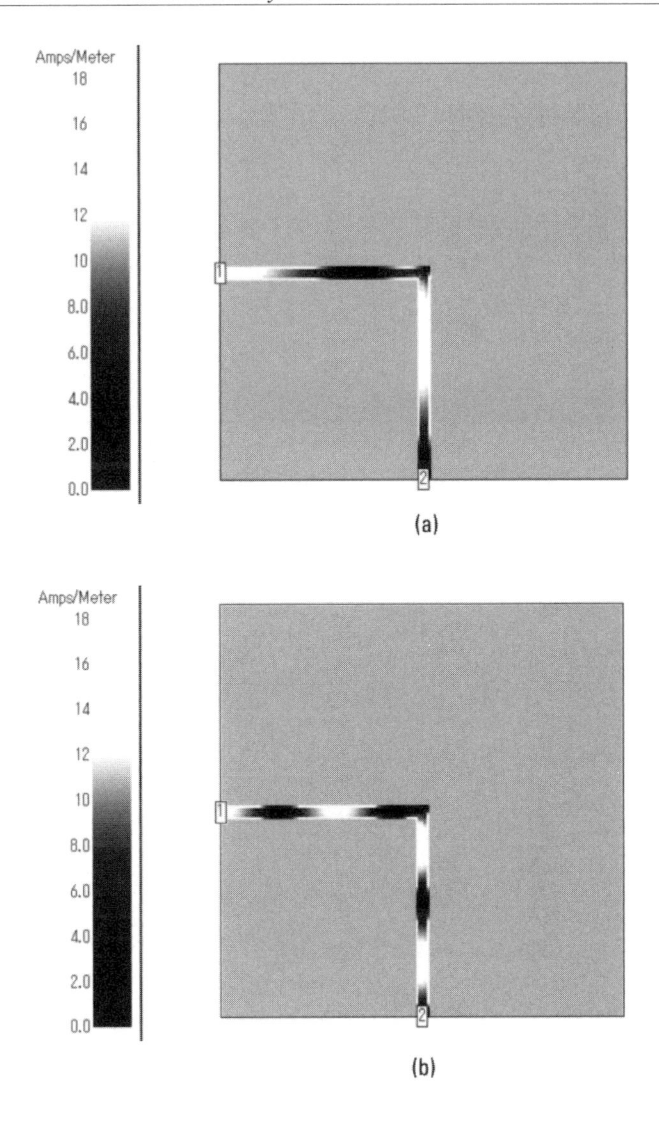

Figure 2.10 (a) Surface current distribution at 4 GHz; and (b) surface current distribution at 9 GHz.

flows in the opposite direction of adjacent high-current regions (remember, the sine wave curve starts out positive, goes through zero, and then goes negative).

2.3.3 How Is a Standing Wave Generated?

In Figure 2.12, the incident wave at Port 1 on the left edge travels to the right. When the wave encounters the 100Ω resistor terminating port 2, a portion of the wave is reflected and travels back to the right. In this case, when we have

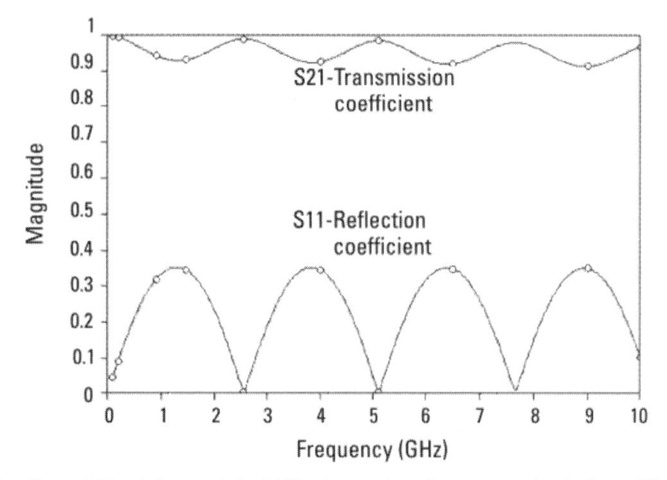

Figure 2.11 S_{11} and S_{21} of the straight MSL, plotted in voltage magnitude (not dB).

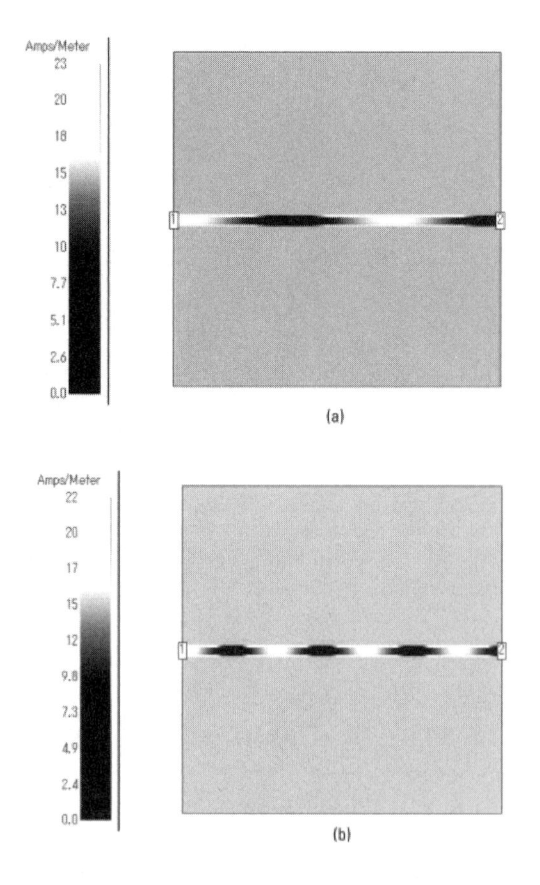

Figure 2.12 (a) Surface current at 4 GHz; and (b) surface current at 9 GHz.

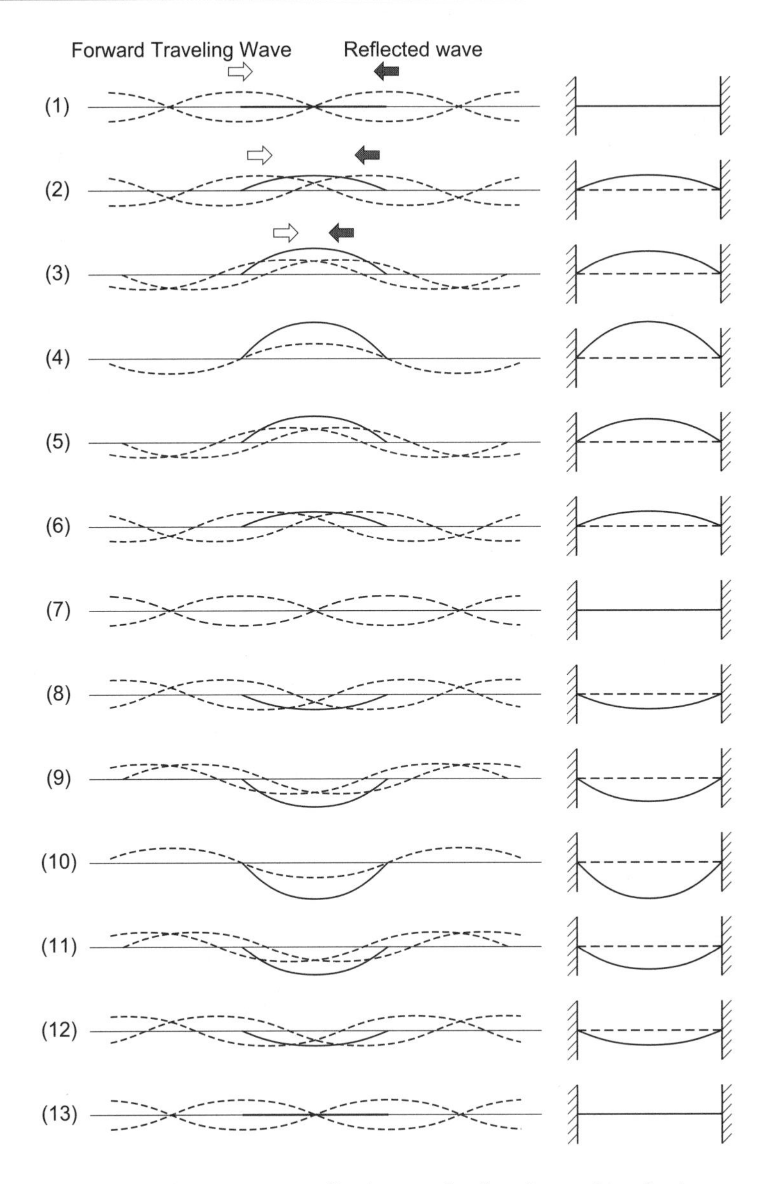

Figure 2.13 Two traveling waves, traveling in opposite directions, add and subtract creating a standing wave.

waves traveling in both directions on a line, the sum of both is a standing wave, as shown in Figure 2.13.

The two traveling waves are drawn on the left with dotted lines. Time instances are drawn from (1) to (12). During each time interval, each wave travels one-twelfth of a wavelength further along. The total effect is simply the sum of the two traveling waves, the solid line. In (1), the two traveling waves are exactly opposite phase and they exactly cancel. In (2), they no longer completely cancel, and the net result is the start of a standing wave.

The solid lines of (1) to (12) are redrawn, for clarity, on the right. We can see that the standing wave is the same as a string fixed at both ends vibrating up and down, as when we strum a guitar. A standing wave is always generated when we have two waves traveling in opposite directions on a transmission line. If you are familiar with the dipole antenna, we have a situation just like the vibrating string. There is a strong standing wave on the dipole. When we have high current in the dipole antenna standing wave, then the radiation becomes large. Could this be a problem?

2.4 Meander Lines

Figure 2.14 is a line that winds its way along, trying to get a maximum amount of length into a minimum area. This is a meander line. This structure is used to adjust the length of a line, for example, making all the lines in a digital bus the same electrical length. This is because it is best if all the digital bits on a high-speed bus arrive at the same time. The meander line also finds use as an inductor for monolithic microwave integrated circuits (MMICs). Because it is so important, let's examine how the current flows.

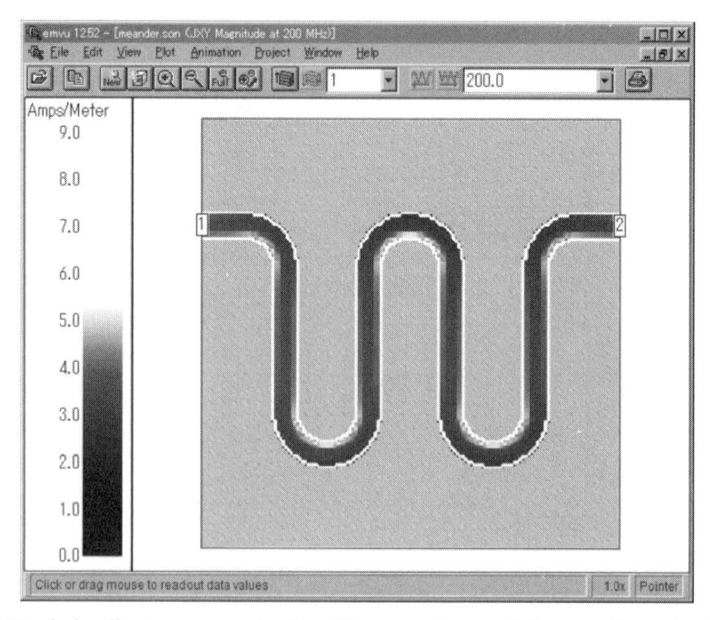

Figure 2.14 Indy effect on a meander line. File: meander.son. Substrate dimension: 30 mm × 30 mm, line width: 2 mm, at 200 MHz.

Figure 2.14 is the simulation result from Sonnet. The current that comes from the left input port (Port 1) flows strongly on the both sides in the straight regions of the line. On the curves, we see that the strong current always flows on the inside of the curve, trying to take the shortest possible distance. Sonnet Software calls this the Indy effect, inspired by the Indy 500 formula car race.

2.4.1 Electric Field Representation

Figure 2.15 displays the electric field distribution 1.5 mm above the substrate surface. The electric field tangential to the surface of the substrate is plotted. We see that low-current regions of the meander line have strong electric field. Likewise, high-current regions have weak electric field. High voltage (electric field) corresponds to low current, and low voltage corresponds to high current. This is characteristic of standing waves.

Sonnet uses a sense layer to examine the electric field. A sense layer is a film of high impedance (e.g., Xdc might be set to 1.0 MΩ/□). With very high impedance, the sense layer has no effect on the circuit. However, the tangential electric field causes a very small current to flow. That current is proportional to the electric field (Ohm's law). Then, to see the electric field, we just plot the current. To determine the electric field, just multiply the magnitude of the current by the impedance of the sense layer.

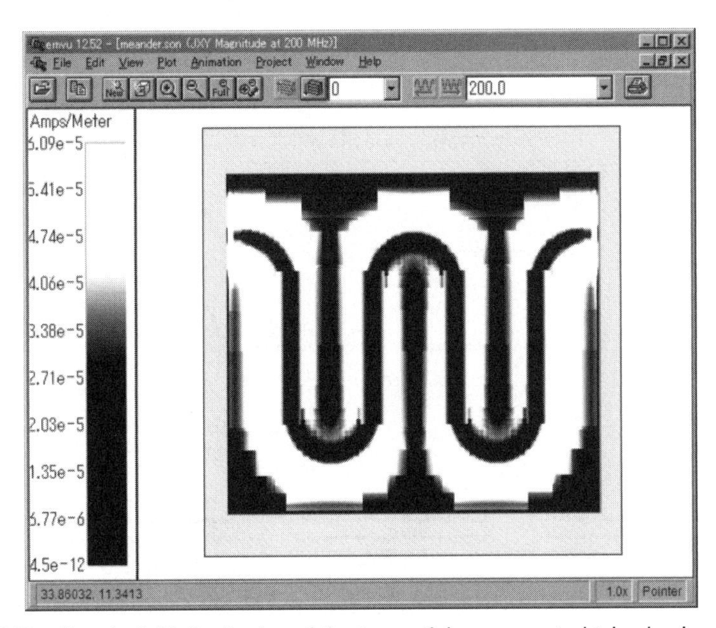

Figure 2.15 Electric field distribution of the tangential component obtained using a sense layer in Sonnet.

2.5 What Causes Bad Manners?

Bad things happen. Let's start exploring why this is.

2.5.1 The Normal Mode and the Common Mode

We will make a model that puts a transmission line on a multilayer substrate and examine the relationship between the line layout and radiation from the substrate, as shown in Figure 2.16. There are two finite area sheets of copper beneath the transmission line. The upper one represents a ground plane. The lower one represents a Vcc (DC power) distribution plane.

1. Normal/differential/balanced/odd mode: This transmission line is a loop with one wire going out and another one coming back. We call this the normal mode. It is also often called the differential, balanced, or odd mode. As a differential mode, the current going out on one line must equal the current coming back on the other. Operated in this way, this transmission line is often called a differential pair. If there is an alternate path for return current (say, a portion of the loop current can flow through adjacent circuitry in an unexpected way, somewhere

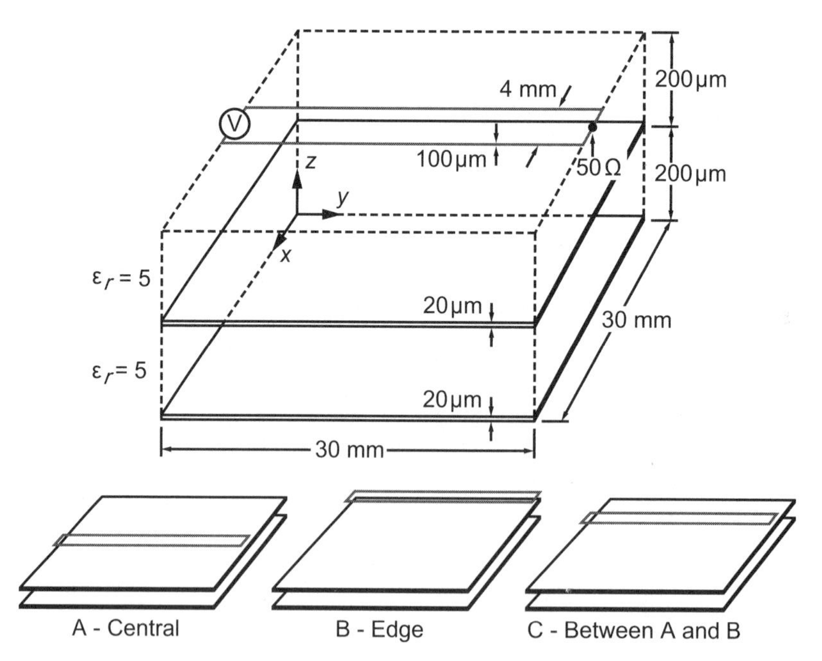

Figure 2.16 A transmission line on a multilayer printed circuit. Three different possible line locations are examined.

along the loop), now there can be problems. Also, if the loop area is made too large, it will start to act like a loop antenna, and, depending on the frequency, it might radiate electromagnetic waves.

2. Common/even mode: When the current flows on both lines in the same direction, then the only way for that current to return to the source is over the ground plane at the bottom. We call this the common mode current. It is also called the even mode. In Figure 2.16, the loop is floating—it is not physically connected in any way to the ground—and thus there is no way for current to return over the ground plane. Well, almost no way. We explore the common mode current in the next section.

2.5.2 Location of the Loop and Common Mode Current

Let's run simulations with the transmission line in three different locations.

- Location A: center of dielectric substrate;
- Location B: along the edge of dielectric substrate;
- Location C: between A and B.

We calculate the common mode current by looking at the current difference between the two lines. The differential mode has currents equal and opposite. The common mode has both currents equal and in the same direction. In general, the actual current will be part common mode and part differential mode. Thus, because we have ports that excite only the differential mode, any difference in current on the two lines is due to the common mode being excited.

We check this for the three locations indicated earlier and in Figure 2.16. We find that the closer the transmission line is to the substrate edge, the larger the common mode current, as shown in Figure 2.17. The vertical axis indicates the difference in current between the two lines (this is the common mode current), and the horizontal axis indicates the position along the line.

Recall that there is no direct connection between the ground plane and the transmission line. How does the common mode ground return current get into the ground plane? It does so using a combination of inductive and capacitive coupling, depending on the frequency and the specific geometry. Notice also that the common mode current is zero at the ends of the transmission line. This is because the ports we are using force all the current at the ports to be differential mode current. Thus, we have a differential mode line that is coupling to a common mode line.

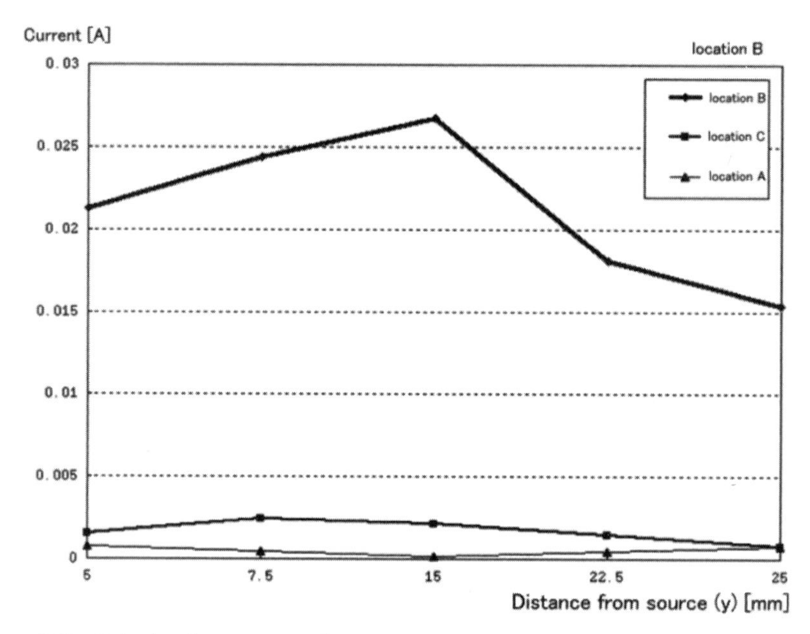

Figure 2.17 Calculated common mode currents at three different locations.

2.5.3 Effects of Common Mode Current

Because the common mode current induces radiation (among other problems) and the common mode current coupling at position A is smallest, it is best to put the transmission line in the center of the ground plane. Figure 2.18 shows the magnetic field on the *x-z* plane when the transmission line is on the edge of the substrate. We see that the magnetic flux density is high, and the current concentrates at the edges of both the ground plane and the Vcc (power supply) plane.

2.5.4 Relation Between Common Mode Current and Radiation

When a portion of the current flowing on the two lines in our differential pair is in the same direction, that portion is common mode current and it contributes to radiation.

Normally, a coaxial cable does not radiate because its outer conductor is a shield. When we connect coax to two parallel lines directly, there is no problem as long as the common mode is not excited on the parallel lines. If there is some common mode, then the common mode portion must flow on the outside of the coax shield. This is because the current flowing on the coax center conductor must be exactly equal and opposite to the current flowing on the inside of

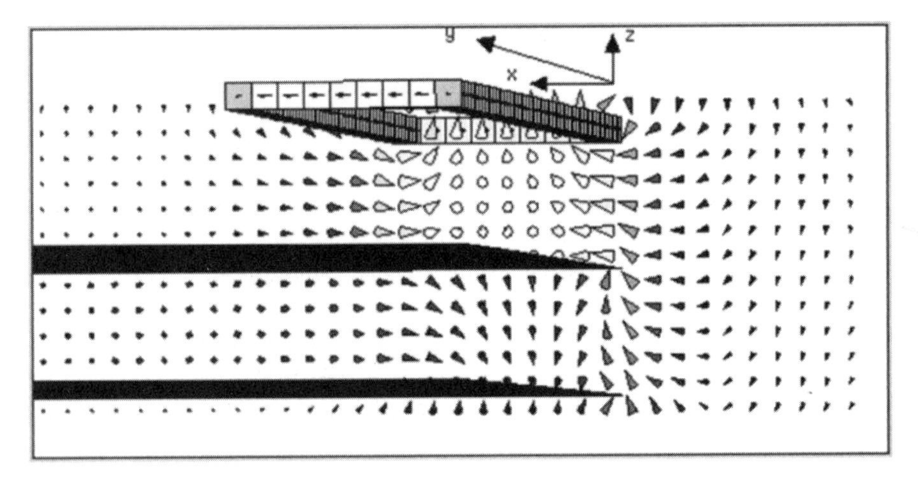

Figure 2.18 Magnetic field on the *x-z* plane.

the coax shield. If we have current also flowing on the outside of the coax shield, the coax will start to radiate.

Current flowing along a line in space, like the current on the outside of a coax shield, causes radiation. This can be explained by looking at a small dipole with a positive and negative charge oscillating back and forth, as shown in Figure 2.19. Sequences of these small dipoles can be thought of as current. The current is generated by the charges changing with time. The time rate of change of current generates changing, and radiating, electric and magnetic fields in free space. The field intensity is proportional to the time rate of change of the current, so in general a higher frequency has stronger radiation, as described in Section 2.2.

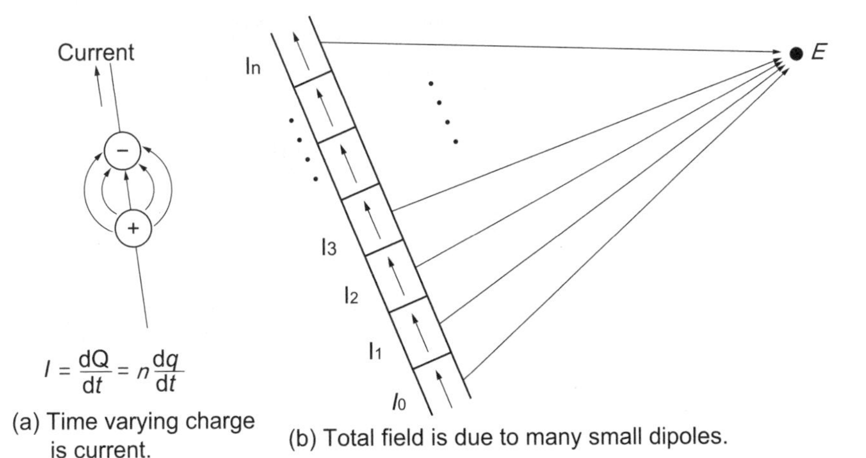

Figure 2.19 Common mode current and the electric field in free space.

2.6 Considering Multilayer Substrates

Figure 2.20 displays the ground conductor current density distribution at 10 GHz with the transmission line at location C (Figure 2.16). Notice that the current distribution right below the transmission line location looks like an image of the transmission line. Notice that we can even see a small standing wave in the ground plane current. There are nulls every half wavelength. In addition we know that every peak in the standing wave current flows in a direction (not shown in the image) opposite to the adjacent peaks.

Figure 2.21 displays the current distribution, again at 10 GHz, but this time with the transmission line at location B. Figure 2.22 displays the current distribution on the Vcc level. The patterns are similar. Both figures show the current distribution on the parallel plates formed by the ground plane and the Vcc plane. Because they show periodic patterns, we can see that there are a number of resonating areas where there is high current.

2.7 Where Does the Radiation Go?

Typically, the radiation from a circuit tends to be strongest broadside to the direction of current flow. Figure 2.23 is the radiation pattern at 8 GHz due to current flowing on the transmission line plus the radiation due to all the

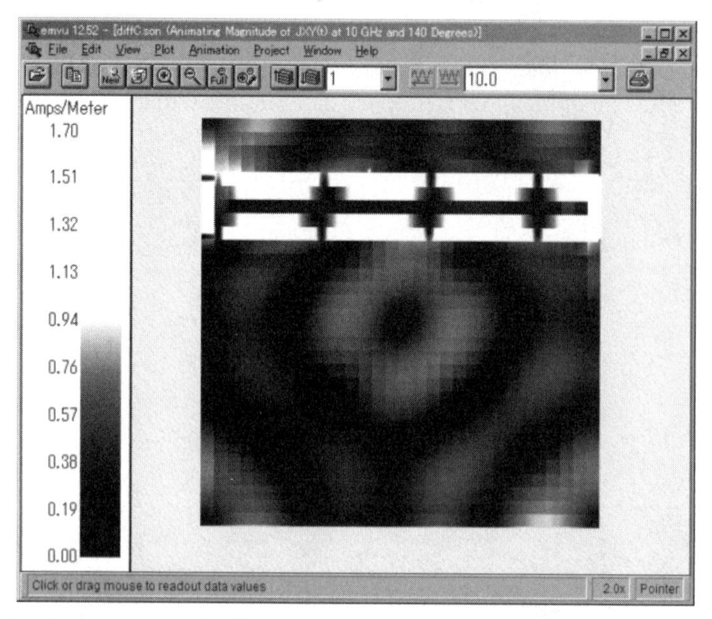

Figure 2.20 Current density distribution at 10 GHz, on the ground plane with the transmission line over position C. File: diffC.son.

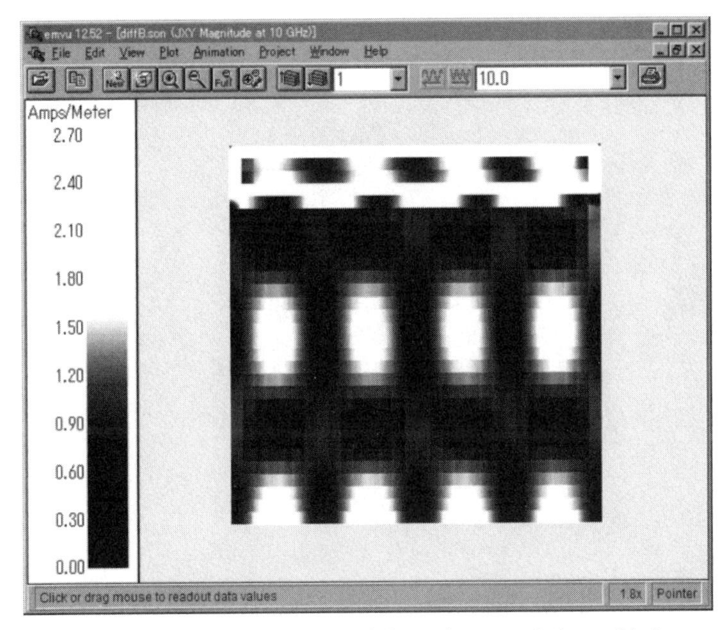

Figure 2.21 Current density distribution at 10 GHz, on the ground plane with the transmission line over position B. File: diffB.son.

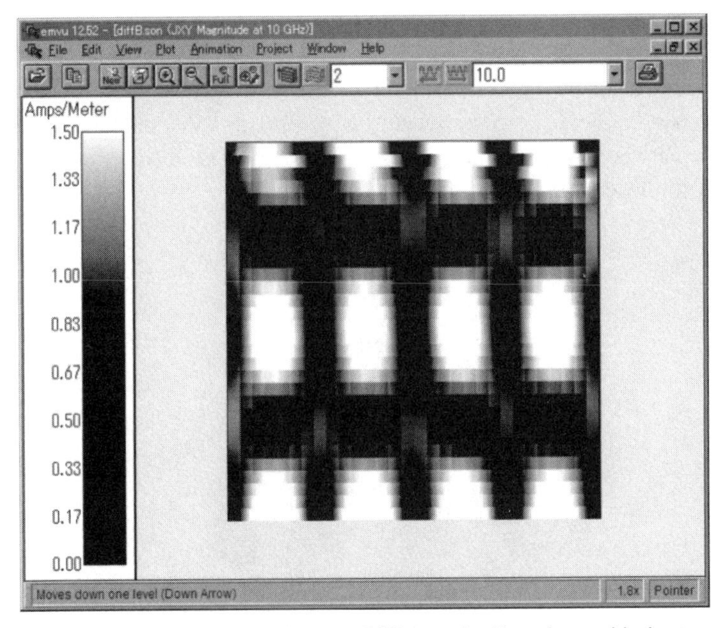

Figure 2.22 Current density distribution at 10 GHz, on the Vcc plane with the transmission line over position B. File: diffB.son.

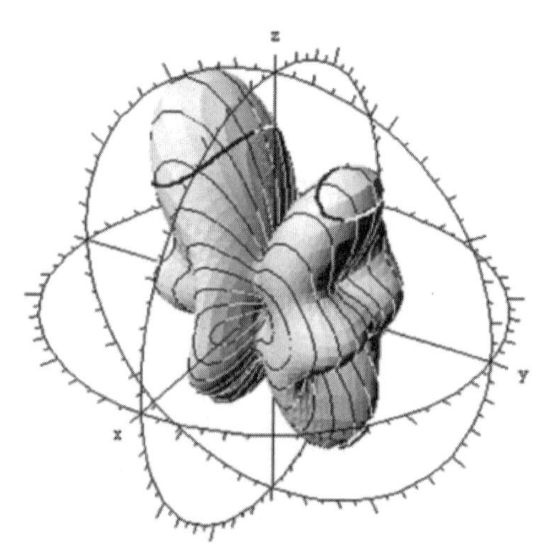

Figure 2.23 Radiation pattern at 8 GHz.

induced currents on both ground planes. At this frequency, we can see that the radiating pattern is complicated.

Figure 2.24 displays the radiation pattern at 500 MHz. In this case, the transmission line is shorter than the wavelength, and because of this the radiation pattern must be smooth and simple.

The radiation at 19.4 GHz, shown in Figure 2.25, has a section removed so we can see more detail in several planes. At this high frequency, the radiation is complicated. The distance between the ground and Vcc plane is large enough that they start acting like an aperture antenna (satellite dishes and horn antennas are other examples of aperture antennas).

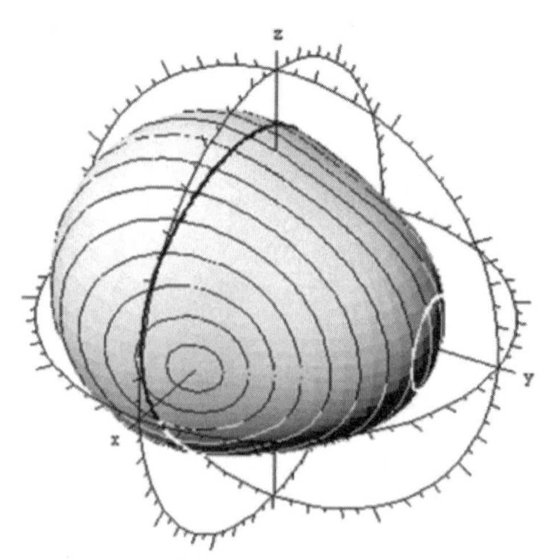

Figure 2.24 Radiation pattern at 500 MHz.

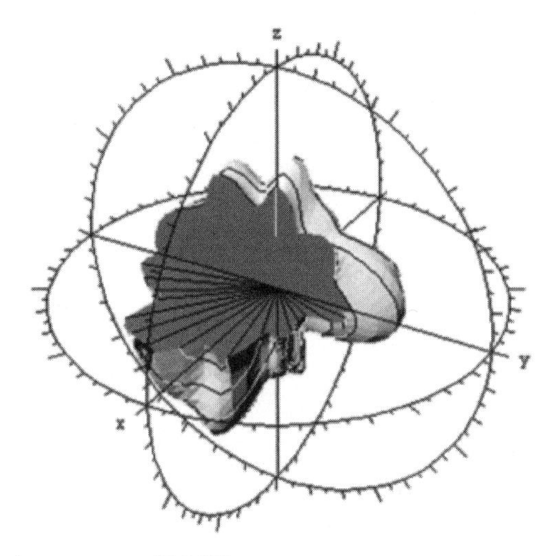

Figure 2.25 Radiation pattern at 19.4 GHz.

Typically, we have no interest in frequencies as high as 20 GHz. However, if we have a larger substrate, we might see similar behavior even at low frequencies.

2.8 Maxwell Predicted Displacement Current

Our entire field is based on Maxwell's equations. Let's learn about this quiet giant, upon whose shoulders we stand.

2.8.1 Maxwell's Achievements

James Clerk Maxwell (1831–1879) is a scientist who was born in Edinburgh, Scotland. He predicted the existence of electromagnetic waves in 1865. His equations were later put in their modern form (using vector calculus, which Maxwell did not have) and wrapped up in four rules of electricity and magnetism that we now call Maxwell's equations.

He unified the previously known laws of electricity and magnetism. To this, he then added the fateful concept of a displacement current. His theory was ready to take on the universe.

2.8.2 Maxwell's Hypothesis

Figure 2.26 shows the case where alternating current flows on a parallel plate capacitor. First looking at the electric leads, which connect the battery (an AC power supply, in this case) to the capacitor, the electrons in the metallic con-

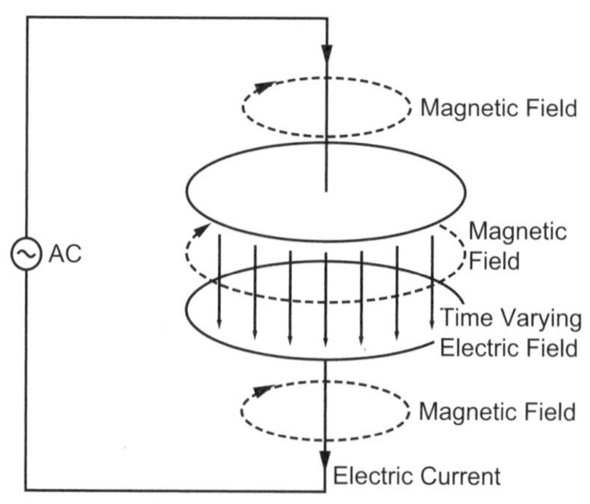

Figure 2.26 Alternating current flows through a parallel plate capacitor.

ductor are forced to move by the AC power supply. This movement is electric conduction current. The flowing current causes magnetic field to loop around the wire.

Next, how about the capacitor? Because the space between plates is empty, there can be no movement of electrons and current cannot possibly flow. With no current, there will also be no magnetic field. Through a stroke of genius, Maxwell hypothesized that the magnetic field does not need an actual physical current to create it. Instead, it can also be generated by a time-varying electric field. Instead of loops of magnetic field wrapping around a physical current, the loops of magnetic field can wrap around a region of time-varying electric field!

So if a time varying electric field can generate a magnetic field just like a physical current, Maxwell decided to call this current as well—just a different kind of current. He called it the displacement current. He saw that as long as we include displacement current, current flow is everywhere continuous. This total current always flows in complete loops.

It's a pretty amazing concept—that we can have a "current" that flows with perfect ease through completely and perfectly empty space.

2.8.3 Electromagnetic Waves and Maxwell

Once Maxwell had invented the displacement current concept, the mathematics immediately led to electromagnetic waves. We can see this without ever writing a single equation. In the year Maxwell was born, Faraday had discovered that a time-varying magnetic field can generate voltage. This allowed engineers to start building electric generators and motors. Maxwell put this into his theory, as time-varying magnetic field can create an electric field.

Maxwell's displacement current says the reverse is also true. A time-varying electric field can create a magnetic field. So now it is easy! If we can set up a time-varying electric field in combination with a time-varying magnetic field so that each one exactly recreates the other as they fly through space, then we have an electromagnetic wave. Maxwell calculated the speed of the wave and found it was essentially equal to the mechanically measured speed of light. So, he concluded what Faraday had suspected all along—that light is an electromagnetic wave.

We know today that there are many more kinds of electromagnetic waves than just light. In this book, we deal with radio waves. But, of course, there are others.

As a result, the various problems addressed in this book can all be solved using the blessing of the electromagnetic theory that Maxwell bestowed upon us.

2.9 Transmission Lines Versus Antennas

Sometimes the dividing line is not clear. Even so, it is important to understand.

2.9.1 What Is an Antenna?

With the popularity of cellular telephones and other wireless equipment, the antenna has become familiar to everyone. We can understand that the electric and magnetic fields extend from the cell phone held to our ears out into free space guided by Maxwell's equations. Antennas can be thought of as converting the energy in a circuit into radiating electromagnetic energy. In a cell phone, this is desired.

At the same time, the mission of the transmission line is to carry electromagnetic energy from one place to another without attenuation. To do this, we must make sure that radiation is minimized.

A fundamental antenna is the half-wavelength-long dipole (Chapter 8). Equivalent to the half-wave dipole is the quarter-wave monopole. The quarter-wave monopole is identical to a half-wave dipole with a ground plane cutting through the center. The ground plane acts like a mirror, making the quarter-wave monopole think that the other half of the dipole is actually there. As for the required dimensions, at 1 GHz, a quarter wavelength is 7.5 cm.

At any rate, at higher operating frequencies, transmission lines can easily be one half or one quarter wavelength long. But is it really a transmission line?

2.9.2 A Transmission Line or an Antenna?

When can a transmission line actually become an antenna? One factor is the frequency. The higher the frequency, the larger the displacement current and the greater the chance of radiation. However, in order to detect such situations, it is necessary to understand why some part of the line might become an antenna. We introduce this topic here, and more details are provided in Chapter 5.

2.9.3 Discovery of Electromagnetic Waves

With the introduction of Maxwell's displacement current, the existence of electromagnetic waves was predicted. However Maxwell died at age 48 in 1879 without ever having seen experimental confirmation of his theoretical electromagnetic waves. It was not until 1888 that Maxwell's theory was experimentally verified by a German physicist Heinrich Hertz (1857–1894).

After Hertz's discovery, transmission line theory advanced rapidly. It explained how current flows on a Lecher wire, the electric field, and the magnetic field (i.e., electromagnetic wave propagates along the wire as shown in Figures 1.3 and 1.4 in Chapter 1). Oliver Heaviside explained this in 1893.

It is possible to say that energy is conveyed by the voltage and current on a transmission line. But it is also possible to say that it is propagated as the energy of the electric and magnetic field around two parallel wires in free space. We have (at least) two ways to look at the same thing. They both come up with the same answer. So, is the energy in the fields or in the current? The answer is that it does not matter.

Earlier, we explored situations where electrical energy can radiate into free space. Calling this "electricity with bad manners" might be viewed as odd or perhaps humorous by the two great scientists Maxwell and Hertz. They might be more pleased with "electricity is everywhere."

2.10 Skin Depth

The higher the frequency, the more current concentrates on the surface of a conductor. This is the skin effect. The related term skin depth indicates how deep the current penetrates into a conductor.

To see what is happening, imagine an electromagnetic wave incident on a conducting plane. If the conductor is perfect, the entire wave is immediately reflected. However, in real life, the conductor has some small amount of resistance. When the electromagnetic wave hits the metal, the electric field penetrates a short distance into the conductor. The depth at which the current amplitude falls to $1/e$ (where $e = 2.718\ldots$) of its value at the surface is the skin depth, δ.

$$\delta = \sqrt{\frac{2}{\omega\mu\sigma}}$$

Here, ω is the angular frequency (= $2\pi f$), μ is the magnetic permeability (approximately $4\pi \times 10^{-7}$ H/m in vacuum), and σ is the conductivity of the conductor.

For example, copper has conductivity $\sigma = 5.8 \times 10^7$ S/m and the previous expression is calculated as

$$\delta = \frac{0.066}{\sqrt{f}}$$

Here, f is in Hertz and the skin depth is in meters.

From this expression, we can see that current is more concentrated at the surface at higher frequency. Further it varies with the inverse of the square root of frequency. At four times the frequency, the skin depth is cut in half.

For example, at a frequency of 1 MHz (= 1,000,000 Hz), the skin depth is 0.066 mm. Concentrating all the current this close to the surface increases the resistive loss of the line. It increases the series inductance of the line as well. Finally, this also tells us that there is very little to be gained by making the line metal thicker than 0.2 or 0.3 mm. Loss will not decrease further because all the current flows near the surface.

2.11 Confirming Results of This Chapter by Simulation

Now, we show how to simulate the problem of the right-angle MSL bend described earlier in this chapter.

2.11.1 Draw a Right-Angle Bend

Figure 2.27 is an MSL model with a 90-degree bend on the same substrate we used in Chapter 1. The only difference is that it is drawn with two rectangles instead of one.

After displaying the S_{11} result, select File > Add File(s)... in the graphing program to add the result of the straight line MSL (msl.son) simulated in Chapter 1. Now, we can compare them, as shown in Figure 2.28.

The MSL right-angle bend has a large reflection. This is because the electromagnetic field is twisted out of shape at the bend and this launches reflected waves. Figure 2.29 is the current distribution on the right angle bend. In order to see distribution details in the corner, we set a finer cell of 0.02 mm. The resulting analysis requires more memory, but it is still within free Sonnet Lite restriction of 16 MB.

As expected, strong current flows on both edges of the line, and when going around the bend most of the current flows on the inside corner. This path

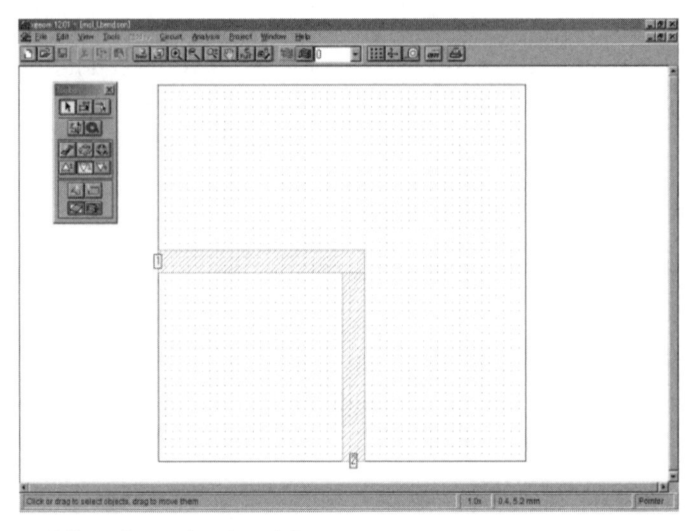

Figure 2.27 MSL 90-degree bend model.

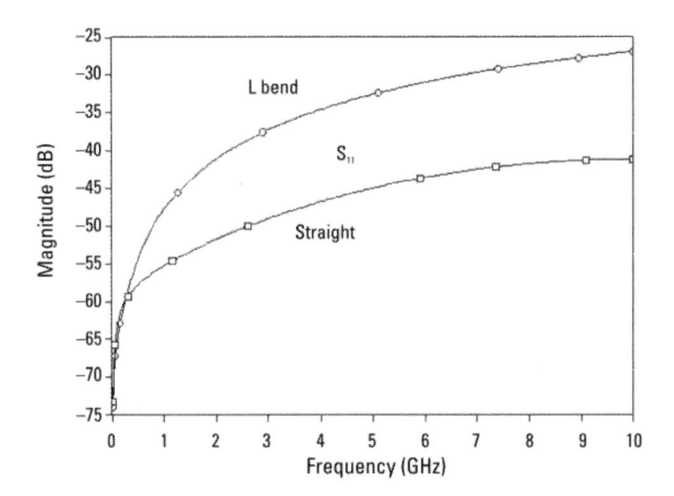

Figure 2.28 S_{11} of a straight line and an MSL right-angle bend.

is the shortest possible path through the bend, and it means that the time delay going through the bend is reduced.

2.11.2 The Effect of Cutting Corners

The current distribution on an MSL bend whose outside corner is cut is shown in Figure 2.30. This is called a mitered bend. As expected, we have strong current on both edges of the line and on the inside corner at the bend.

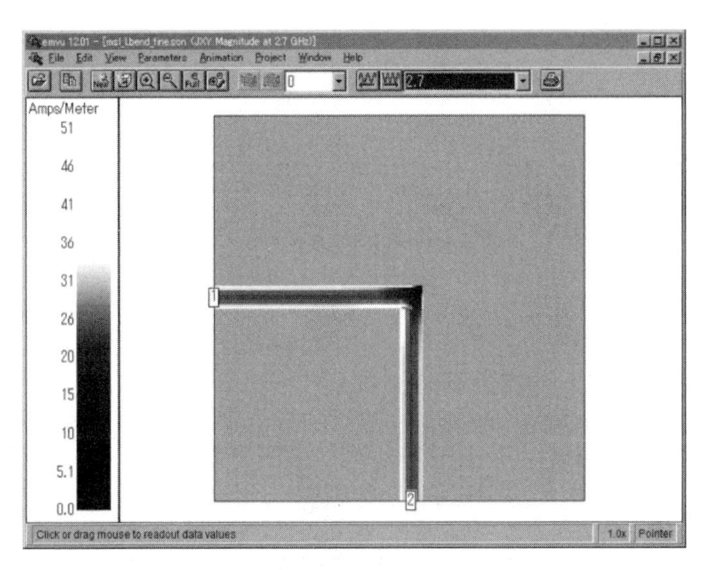

Figure 2.29 Surface current distribution on the right-angle bend.

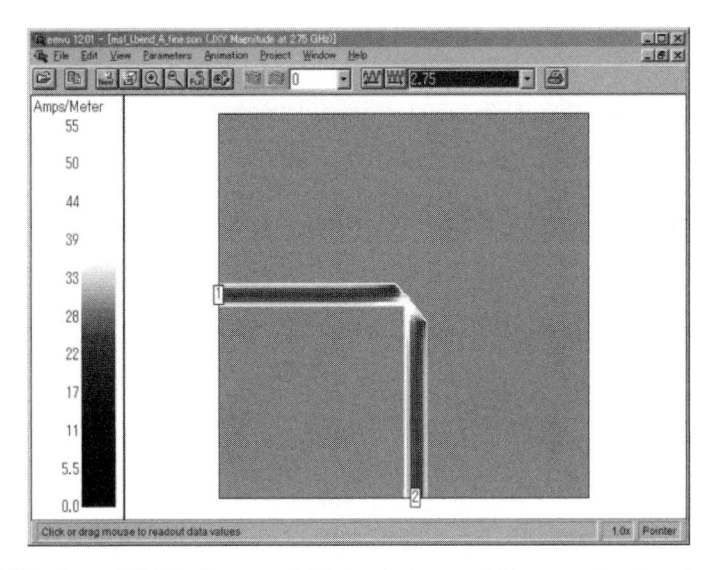

Figure 2.30 Current distribution on an MSL bend whose outside corner is mitered.

S_{11} of the MSL bend with and without mitering is compared in Figure 2.31. The mitered bend has substantially less reflection.

Both inside and outside corners of the bend are modified in Figure 2.32. Figure 2.33 shows the comparison of the phase angle of S_{21}. We see that we can achieve a larger phase delay.

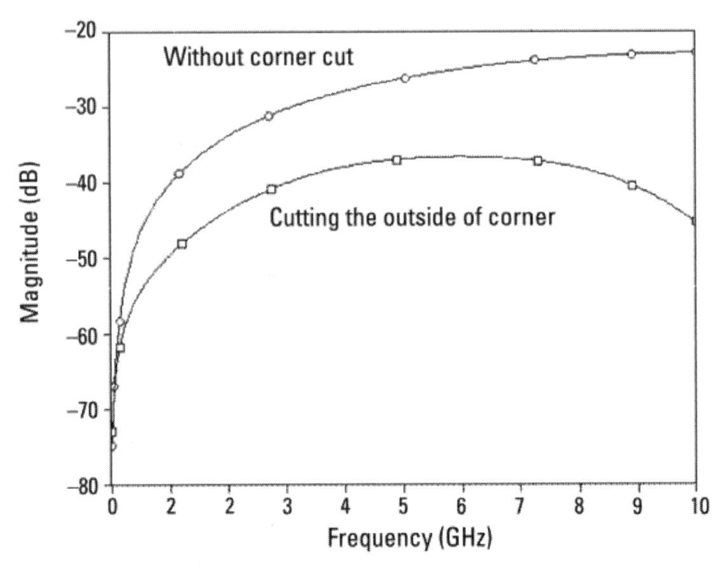

Figure 2.31 S_{11} of MSL without the corner cut and that of MSL whose outside corner is mitered.

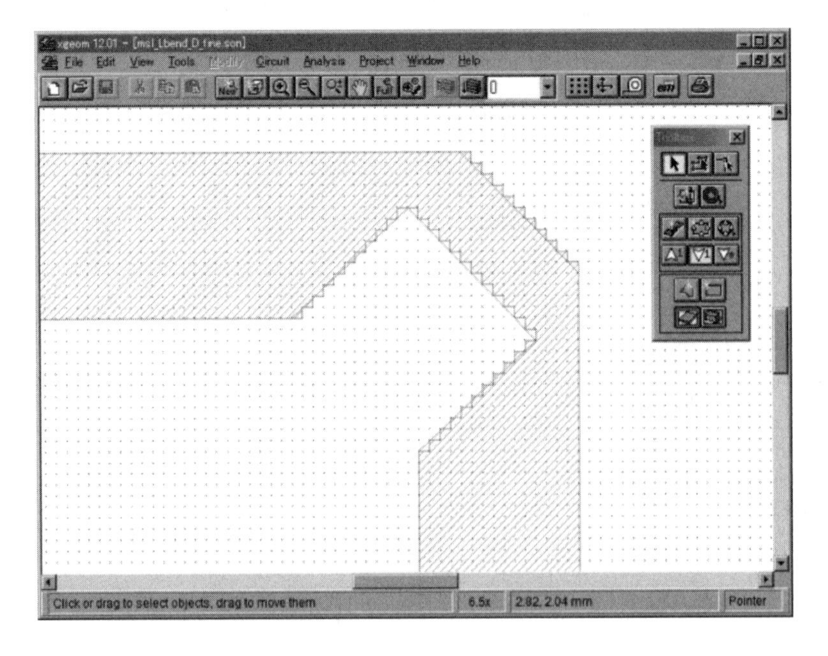

Figure 2.32 Model of a bend with both inside and outside corners modified.

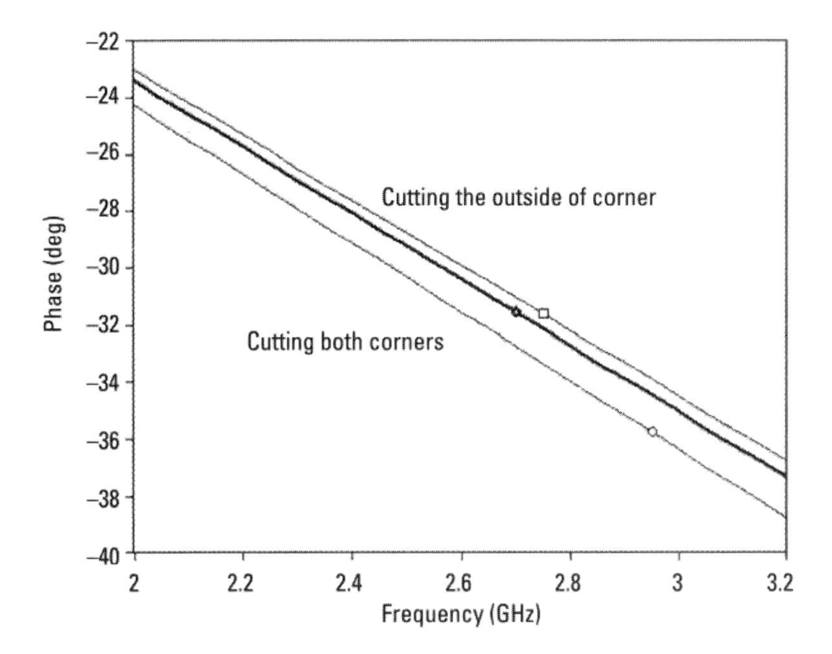

Figure 2.33 Comparison of the phase angle of S_{21} for different miters.

2.12 Summary

1. Electromagnetic energy tends to be radiated into free space from right angle bends and other discontinuities.
2. In general, the higher the frequency, the greater the displacement current and the higher the radiation.
3. At frequencies where the reflection coefficient is large, standing waves will be generated.
4. Common mode currents can induce radiation.

3

What Happens at High Frequency?

3.1 Scattering Parameters

The electromagnetic field in the vicinity of a circuit board was investigated in detail in Chapter 2. We could intuitively see that electromagnetic energy is easily radiated into space from a discontinuity like a right angle bend. In this chapter, we learn about high-frequency scattering parameters, or S-parameters, in detail. We will get a feel for how to evaluate and use the numerical values by investigating specific examples.

The electromagnetic field simulator uses ports as the signal sources or observation points to get S-parameters. In a two-port circuit, S_{11} is the reflection coefficient on the input side and S_{21} is the transmission coefficient. So let's see what we can find out about S-parameters.

3.1.1 Definition of S-Parameters

S-parameters are obtained by measuring the transmission and the reflection characteristics of a network. S-parameters are not evaluated in terms of the voltage and current at a port. Rather, they are evaluated in terms of the amplitude of traveling waves. The amplitude of traveling waves is usually specified in terms of voltage, but that is the peak (or RMS) voltage of the forward traveling wave (the incident wave) or the reverse traveling wave (the reflected wave), separately.

In order to separate the forward and reverse traveling waves, we need a length of transmission line connected to the port. Then, for example, if we see a standing wave, we know that we have two traveling waves each going in opposite directions. Then, with measurements of the standing wave, we can determine the amplitude and phase of both waves, and thus determine the S-parameters.

Since we measure the incident and reflected waves on a transmission line, we need to know the characteristic impedance of the transmission line. This is called the normalizing impedance. In almost all cases, S-parameters are reported normalized to 50Ω. Thus, during measurement, all ports must be connected with a 50Ω transmission line or be terminated in a 50Ω resistor.

In Figure 3.1, a_1 and a_2 are incident waves and b_1 and b_2 are reflected waves. For a two-port, each S-parameter is related to the incident and reflected wave amplitudes by the following expressions:

$$b_1 = S_{11}\, a_1 + S_{12}\, a_2$$

$$b_2 = S_{21}\, a_1 + S_{22}\, a_2$$

When more ports exist, like N ports in Figure 3.2, a_n denotes an incidence wave and b_n denotes a reflected wave.

The definition of two-port S-parameters is as follows:

$$S_{11} = \left.\frac{b_1}{a_1}\right|_{a2=0} \qquad S_{21} = \left.\frac{b_2}{a_1}\right|_{a2=0} \qquad S_{12} = \left.\frac{b_1}{a_1}\right|_{a1=0} \qquad S_{22} = \left.\frac{b_2}{a_2}\right|_{a1=0}$$

According to Figure 3.1, S_{11} is the voltage ratio of the reflected wave at port 1 to the incidence wave at port 1, so this represents the reflection coefficient. In addition, S_{21} is the voltage ratio of the transmitted wave coming out of port 2 to the incidence wave at port 1, so this represents the transmission coefficient. In both cases, port 2 is terminated in the normalizing impedance, so that a_2 (wave incident at port 2) is zero.

Similarly, S_{22} is the reflection coefficient at port 2, and S_{12} is the transmission coefficient in the reverse direction. Most passive circuits are reciprocal

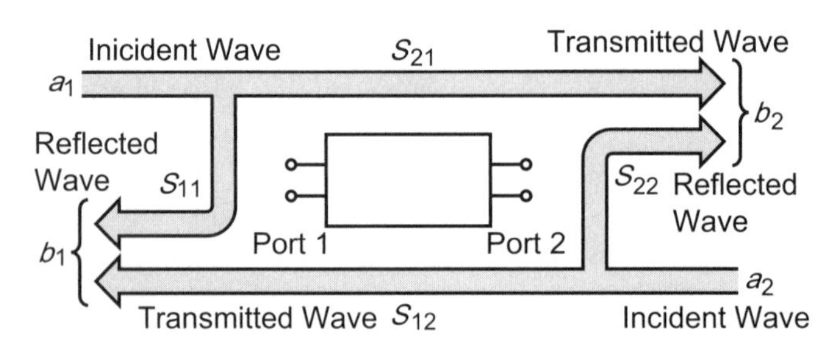

Figure 3.1 Two-port S-parameters.

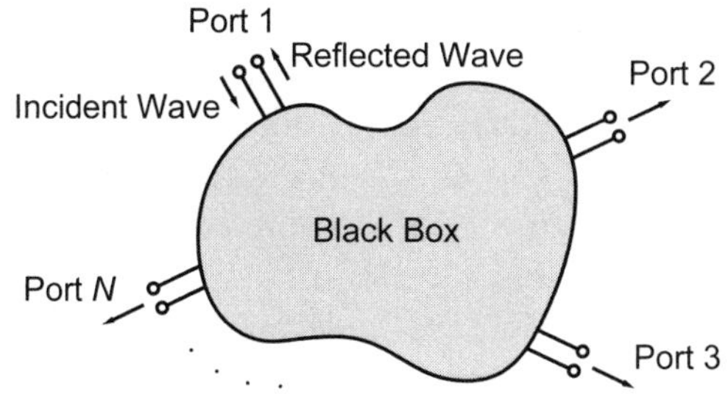

Figure 3.2 *N*-port S-parameters.

because the forward transmission coefficient is the same as the reverse transmission coefficient.

S-parameters are based on voltage. Power is proportional to voltage squared. Thus, for passive two-port devices, we have the following:

$$|S_{11}|^2 + |S_{21}|^2 \leq 1$$

In other words, the reflected power plus the transmitted power must be less than or equal to the incident power. If a circuit is perfectly lossless, then this is met in equality. This equation is valid for any normalizing impedance, as long as all the ports are normalized to the same impedance.

3.2 Let's Use the Network Analyzer

The network analyzer, shown in Figure 3.3, is an instrument that can measure S-parameters with high accuracy. Because impedance (and thus, S-parameters) is displayed using complex numbers with real and imaginary parts, or equivalently magnitude and angle, it is frequently called a vector network analyzer (VNA).

The basic component of a VNA, shown in Figure 3.4, is the reflectometer. This is also called a directional coupler. There are two of them, one on either side. The reflectometer measures the reflection coefficient. The ratio of the incidence wave and the reflected wave are determined by the directional coupler. The reflection coefficient is calculated as the ratio of the reflected wave to the incidence wave.

The frequency synthesizer is the signal source. To measure S_{11} (reflection coefficient) and S_{21} (transmission coefficient), the synthesizer is connected to the directional coupler 1 by the switch. At this time, the directional coupler 2 is terminated by the matched load, and a_2=0, as is required for an accurate measurement.

Port 1 is the left side of the device under test (DUT) and port 2 is on the right side. Incident and reflected wave ratios are first measured with the switch in the position shown. The measured wave amplitudes are used in the first two equations to determine S_{11} and S_{21}. Then the switch is moved into the other position, and wave amplitudes are measured to determine S_{12} and S_{22} using the second two of the four previous equations.

The measurement of the amplitude ratio such as b_1/a_1 and the phase difference between them (remember, this is a vector network analyzer) is performed using digital signal processing after the signals have been converted from high frequency to low frequency using the heterodyne principle. This involves mixing the RF signal (which was applied to the DUT) with a local oscillator (LO). The output of the mixer is at a frequency that is the difference between the RF and LO frequencies. This lower frequency signal is more easily and accurately processed.

3.2.1 The Importance of Calibration

In the real world, we do not have ideal components. As for the reflectometer, the complex ratio of measured wave amplitudes is determined by the characteristics of the directional coupler, the switch, and the transmission lines (sometimes quite long) that connect the directional coupler to the DUT. If these characteristics are not precisely determined, the resulting S-parameters are not accurate.

In order to determine the measurement system characteristics, several known standards are first measured. For example, in the OSL calibration technique, we measure an open circuit, a short circuit, and a load (50Ω resistor). We know what the correct answers are for these calibration standards. We also know what our uncalibrated VNA measured. From the difference, we can determine the characteristics of the VNA measurement equipment. Now, when

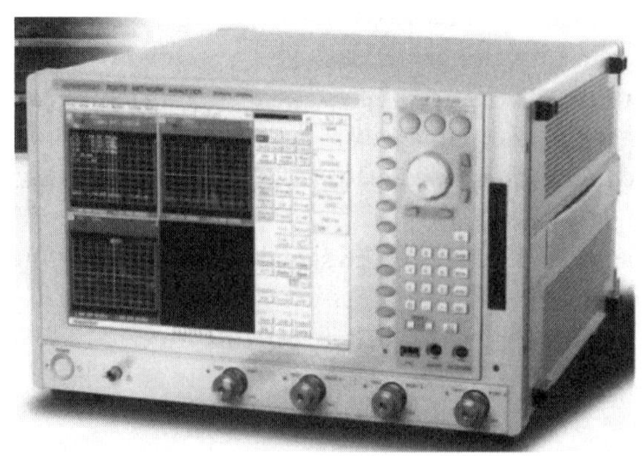

Figure 3.3 Example of a VNA.

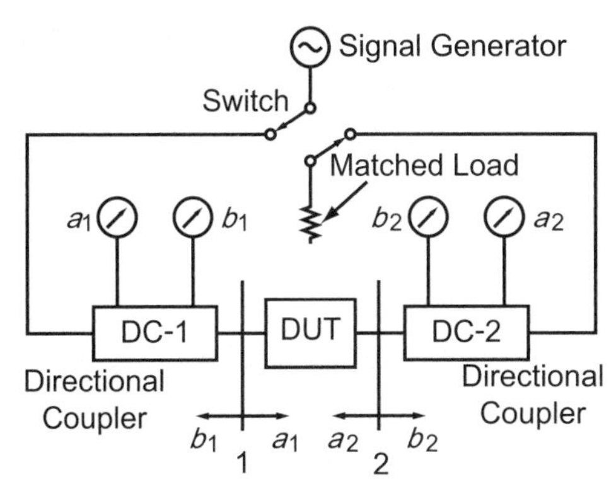

Figure 3.4 Basic components of a VNA.

we measure the DUT, we can correct the actual measurements for the VNA characteristics to give us calibrated measurements of the DUT.

There are many different calibration techniques, all with various advantages and disadvantages. The manual for a VNA will describe what calibration techniques can be used and how to do them. Read and understand the manual carefully.

3.2.2 De-Embedding and Calibration

After calibration, the DUT is connected to the VNA and measured. High-accuracy measurements are now easily obtained. This calibration is also called de-embedding. The de-embedding removes the VNA characteristics, and the effect of the transmission line connecting the VNA to the DUT.

Electromagnetic (EM) simulators also have measurement characteristics that must be removed. In addition, EM simulators also have a transmission line that connects the port to the DUT. In Chapter 2, our DUT was a right angle bend. To remove the influence of the port and of the port connecting line from the DUT, many electromagnetic field simulators provide de-embedding.

Figure 3.5 is a line discontinuity we use to demonstrate de-embedding at the end of this chapter. The patch in the center is de-embedded to the end of both arrows. Removing the effect of a length of transmission line is called shifting the reference plane.

3.3 S-Parameters of Four Bent Coupled Lines

Figure 3.6 is a model of four L-shaped bent coupled lines. Sonnet Lite allows up to four ports. In Sonnet Lite, you can model two bent coupled lines. In this example, ports 1–4 are input ports, and ports 5–8 are output ports. Though the

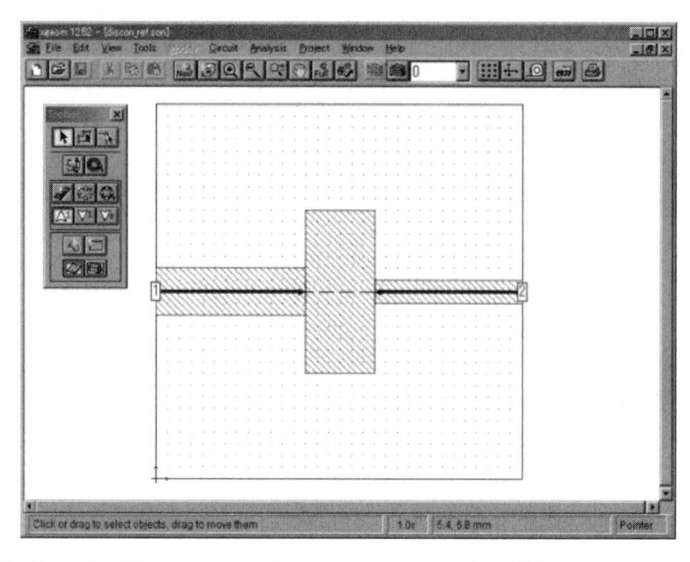

Figure 3.5 De-embedding to move reference planes up to the DUT.

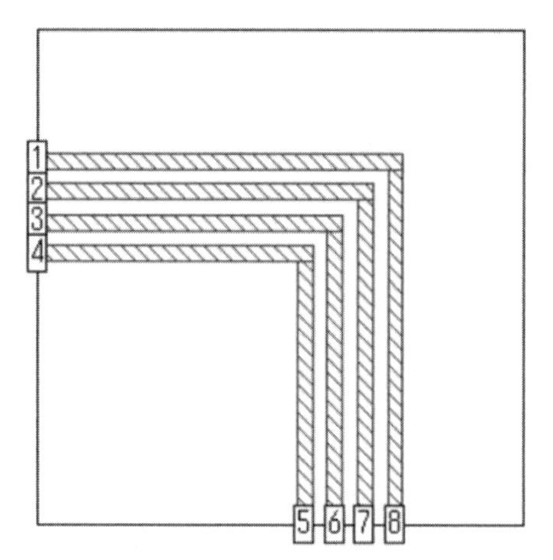

Figure 3.6 Model of four L-shaped bent coupled lines.

cell size (set in Circuit > Box) is 1.0 mm in both the *x* and *y* directions, it is not necessary to make them equal.

The dielectric substrate is 1.0 mm thick, relative permittivity is set to 4.0, and the layer under that is the ground plane. Though the parameters of loss tangent (tanδ) and others can be set, here we analyze it without loss for the moment.

Table 3.1 shows the resulting S-parameters. S-parameters are complex numbers. They are listed here in terms of their magnitude and phase angle. First

	Magnitude	Angle
S11	0.0762	79.0
S21	0.0241	76.8
S31	0.0055	72.5
S41	0.0023	72.9
S51	0.0020	-108.8
S61	0.0050	-109.1
S71	0.0149	-107.4
S81	0.9966	-10.69

Table 3.1 S-parameters Resulting from the EM Analysis of the Four Bent Coupled Lines

of all, on the last line in the S-parameter table, the magnitude of S_{81} is 0.9966 (almost the maximum value of 1.0) and the phase angle is about –10 degrees. S_{81} is a transmission coefficient where the input signal on port 1 is transmitted to port 8. If the amplitude of the wave incident on port 1 is 1.0V, a wave with an amplitude of 0.9966V comes out in port 8. As port 8 is connected directly to port 1 by the circuit, this is working as a very nice transmission line.

Next, let's look at the phase of S_{81}. Table 3.1 holds the S-parameters for 100 MHz. The phase of –10 degrees means that there is a delay of 10 degrees caused by the time it takes the 100-MHz signal to arrive at port 8. Because one cycle of 100 MHz needs $1/(100 \times 10^6)$ seconds = 10 ns (nanosecond), the time delay is 10 ns × (10 degrees/360 degrees) = 0.28 ns.

If you want to see a plot of delay, select Equation > Add Equation Curve and select Group Delay in the graphing program.

When we look at S_{71}, if an incident wave of 1.0V is applied to port 1, a 0.015V amplitude wave exits port 7. In other words, the electromagnetic coupling between two the lines of ports 1–8 and ports 2–7 is about 1.5 percent.

Because S_{71} refers to the signal that comes out of port 7, this is forward crosstalk. At this point, "forward" means that the signal that comes out of port 7 is traveling in the same direction as the signal on the ports 1–8 line.

Moreover, if a signal of 1.0V is applied to port 1, crosstalk is also seen in the immediately adjacent port, port 2. This is called backward crosstalk. "Backward" means that the signal coming out of port 2 is in the opposite direction of the signal going into port 1.

3.3.1 How to Evaluate S-Parameters

S-parameters are calculated assuming that all ports are terminated by a user-specified impedance. Nearly always, this normalizing impedance is 50Ω. If the normalizing impedance is not stated, it is nearly always safe to assume it is 50Ω. In general, S-parameters can be evaluated normalized to any impedance value.

The crosstalk we evaluated here uses 50Ω S-parameters. Thus, the numerical values we obtained correspond to the crosstalk that is seen when all the lines are terminated in 50Ω. To calculate crosstalk when the lines are terminated in some other impedance, the S-parameters should be normalized to the same impedance that actually terminates the circuit when it is in operation.

Fortunately, once you have 50Ω S-parameters, it is possible to mathematically transform the S-parameters to any other normalizing impedance. In fact, we can even have different normalizing impedances for different ports. For more information, read the chapter about ports in the *Sonnet User's Manual* (under Help > Manuals).

3.4 More Complicated Circuit Examples

Let's start getting a little closer to real-life circuits.

3.4.1 PCB Interconnect Example

As an example of a more complicated circuit, we consider the byte reverse circuit of Figure 3.7, file br32.son. The size of the substrate is 128 mm × 128 mm, and there is a layer 0.1 mm thick under the lines, while the ground plane is 1 mm below that. A full copy of Sonnet is needed to analyze this circuit.

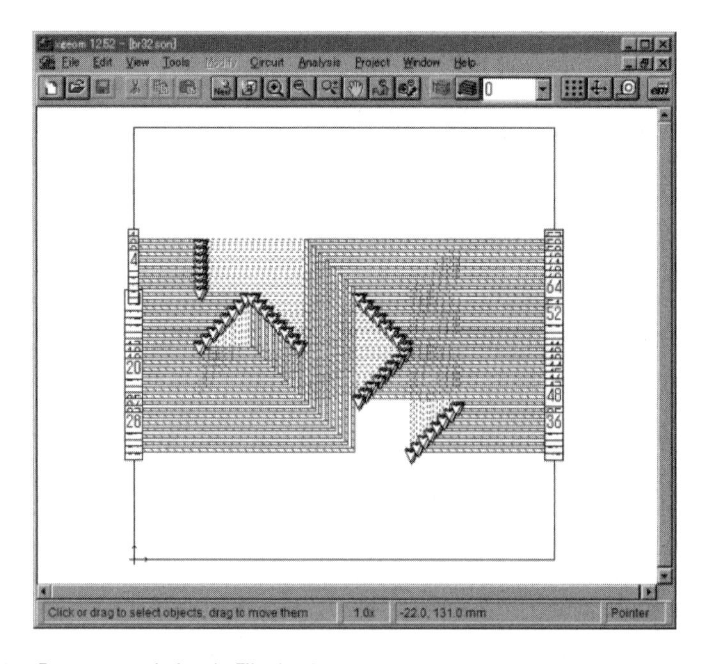

Figure 3.7 Byte reversal circuit. File: br32.son.

This circuit has 32 lines with input being ports 1 to 32 on the left side in Figure 3.7. The lines go down to the lower layer by vias at each location marked with small downward triangles. Figure 3.8 shows the lines on the lower layer. When we trace each line all the way through, we can confirm that each byte (8 lines) is reversed in order on the output.

3.4.2 Searching for Problems by Viewing the Current Distribution

Figure 3.9 is the current distribution on the lines at 15 MHz. When the signal is incident on port 1 on the left, port 33 on the right edge is displayed in red with strong current (1 A/m or more). However, the lines for ports 25–57 still display a small current (about 0.4 A/m) even though there is no input connected to those lines. It is easy to imagine crosstalk being induced from the signal line in the lower layer shown in Figure 3.8.

Normally, we think of crosstalk to nearby lines as being strongest. However, in a multilayer substrate, lines can cross over each other on different layers, causing unexpected electromagnetic coupling. One way to avoid this problem is to put a ground plane in between to shield crossing lines from each other.

Because this kind of problem is strongly dependent on frequency, it is difficult to make general statements. Examining each individual circuit with an electromagnetic field simulator is important.

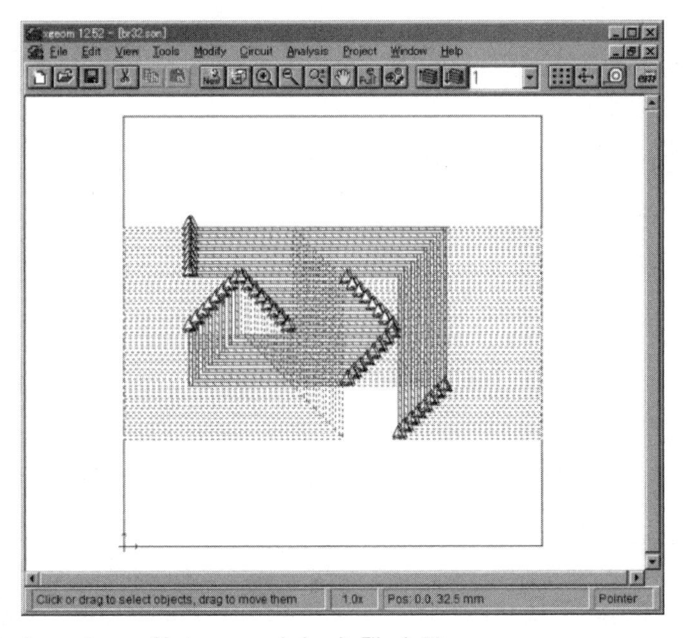

Figure 3.8 Lower layer of byte reversal circuit. File: br32.son.

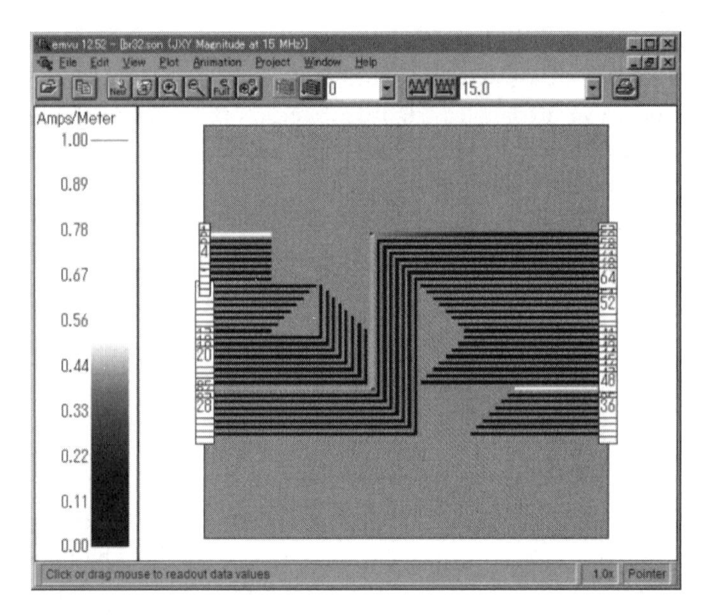

Figure 3.9 Byte reversal circuit. Surface current distribution on the lines is displayed.

3.4.3 Interpretation of S-Parameters

The current distribution gives us qualitative information about crosstalk. The S-parameters, shown in Figure 3.10, give us specific quantitative results. Let's check out the crosstalk to the port 25–57 line. This line overlays the port 1 line on the right half of the network. The forward crosstalk is S57_1, and the backward crosstalk is S25_1. The backward crosstalk to the line horizontally adjacent to the port 1 line is S_{21}; it is –35 dB at 15 MHz. As this value is 8 dB less than the –27 dB that is the backward crosstalk to port 25 in Figure 3.9, we see that vertical adjacency can be more important than horizontal adjacency.

3.5 Ground Bounce and Ground Loops

When a strong current flows in the ground plane, voltage can be impressed across the ground. In other words, different parts of the "ground" are at different voltages! This problem would not exist if the ground level were a lossless perfect conductor. However, when there is some resistance, the return current that flows in the ground plane generates voltage by Ohm's law. This voltage is called ground bounce.

To introduce this situation, we investigate a simple printed circuit board circuit shown in a cross section in Figure 3.11. Input on the PCB line goes to a via that then goes to ground. The return current flows in the ground plane

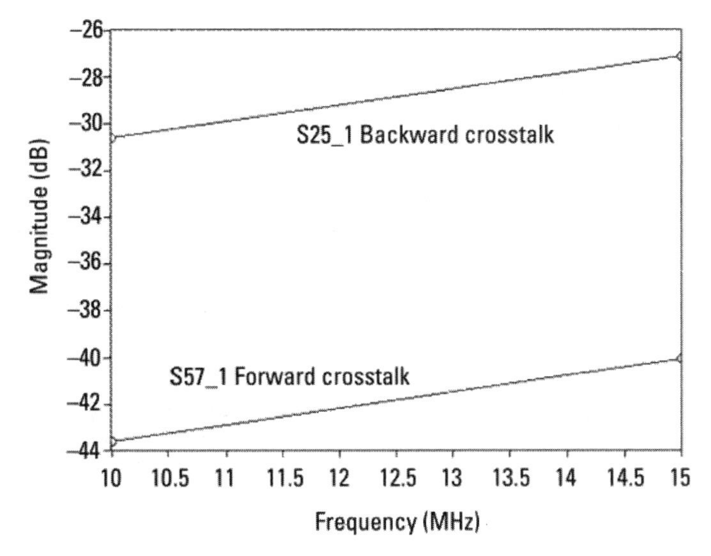

Figure 3.10 Crosstalk of the line from the port 1 line to the port 25–57 line.

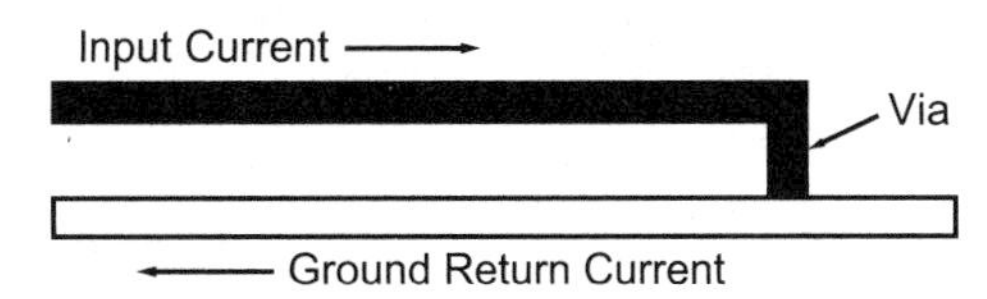

Figure 3.11 Cross section (viewed from the side) of a printed circuit board circuit used to introduce ground bounce.

under the PCB line. This current induces a voltage gradient in the ground whereever it flows.

Voltage is always measured between two points. Thus, if ground bounce is a problem, we must have two connections to the ground plane. The ground bounce voltage is the voltage between those two points.

We normally think of ground as being 0V by definition. Of course, when there is ground bounce, it is no longer 0V. The ground bounce voltage acts as an input to other circuits. Current that flows through such ground potential differences results in a problem called a ground loop. This ground loop may cause unexpected coupling between circuits, and when it is too strong, it results in design failure.

3.5.1 Example of Ground Bounce and Its Analysis

An example circuit we use to explore ground bounce is shown in Figure 3.12. Viewing the circuit from above, the input current flows from left to right, and it goes to ground through a via on the right end. The return current flows on the ground plane and returns to the input port.

The short, center circuit is parallel to the return current path, and both ends are connected to "ground" by a via. Then, the voltage of port 2 is the voltage between the two ground-attach points. This allows us to measure the ground bounce voltage.

When analyzed with all the metals as lossless perfect conductors, S_{21} is 0.013 at 100 MHz, as shown in Figure 3.13. Because there is no loss in the ground plane, this value is not due to the ground bounce. Rather, it is due crosstalk coupling between the lines. Next, the loss in the ground plane was set to a high value (for illustration) of 1 Ω/□ (Ohms per square) and simulated. (The file is bounce2.son.)

With loss, S_{21} increases 0.027, as seen in Figure 3.12. This increased voltage is due to ground bounce. Because the voltage input for this simulation is 1V, we conclude that the 0.014V that we attribute to ground bounce means we have 1.4 percent ground bounce.

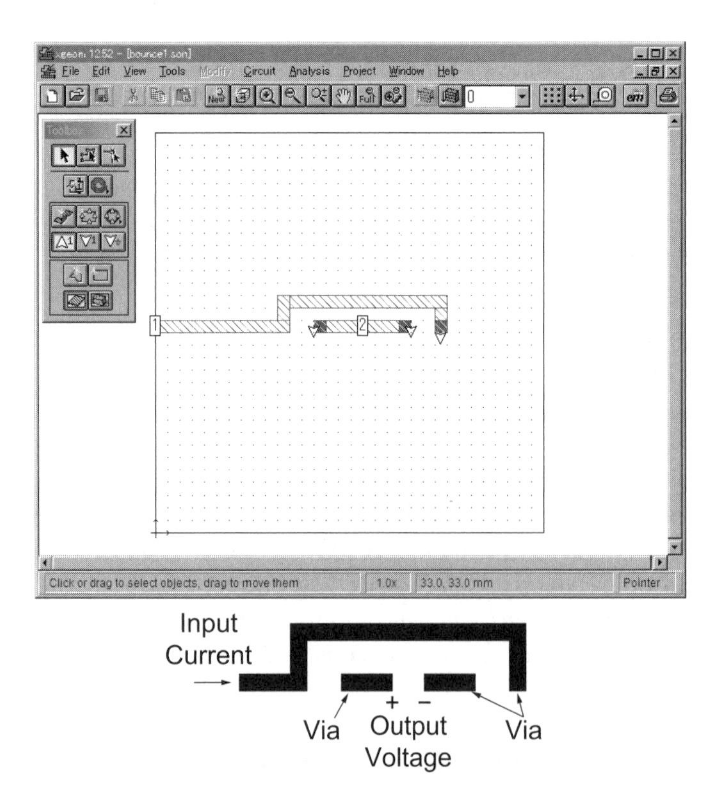

Figure 3.12 A simple PCB circuit model (viewed from above) used to investigate ground bounce. File: bounce1.son.

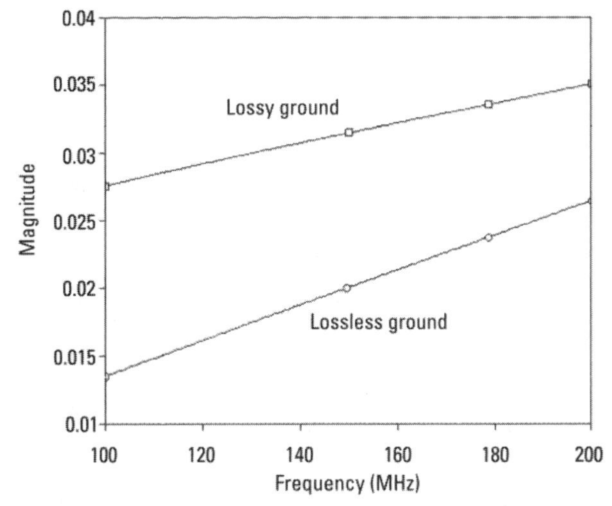

Figure 3.13 Comparing S_{21}. File: bounce1.son (lossless ground) and bounce2.son (lossy ground).

Figure 3.14 is the ground plane return current with the surface resistance set to 1 Ω/□ (File: bounce2.son). At this frequency, most of the return current is under the input line almost like we have a mirror right under the line. The return current under the short, ground bounce detecting line is small but

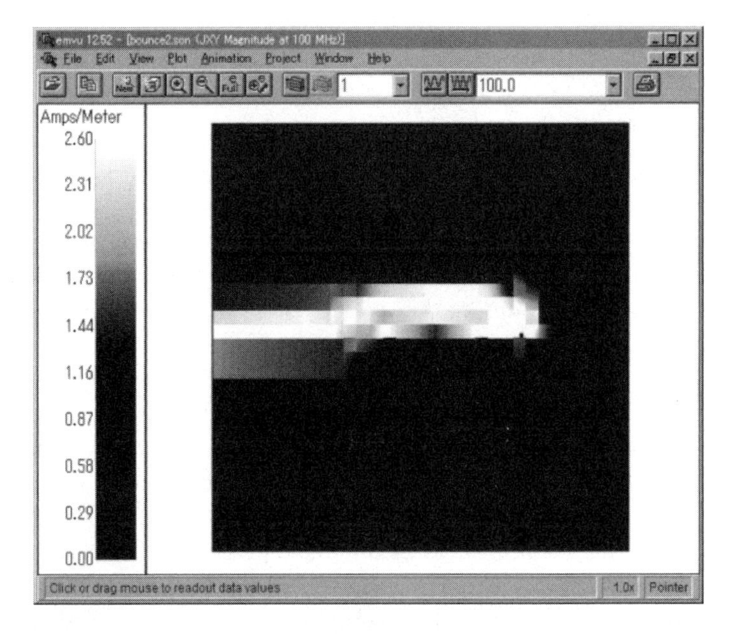

Figure 3.14 Return current on the ground. File: bounce2.son.

significant. Keep in mind at high frequencies, and with more complex circuits, the input circuit's ground return current might not be as nicely behaved.

3.6 Some Definitions of Characteristic Impedance

We keep talking about this concept of characteristic impedance, usually denoted as Z_0. It is important because if we know the value of Z_0, we need to measure only one of the following quantities for a traveling wave on a transmission line—line voltage, line current, or power flow—and we can determine all the others.

3.6.1 Characteristic Impedance of the Waveguide Tube

As described in Chapter 1, the MSL is a quasi-TEM transmission line. To the degree that the TEM approximation is valid, the current on the line can be obtained by simply integrating the current density over the cross section of the line. For example, if the current density everywhere in the metal is 1,000 A/m^2 and the line is 1 mm × 1 mm in cross section, the total current is 1 mA.

The voltage is the potential difference between point B and point A. Note that we must always have two points specified in order to determine voltage. We can determine voltage in an EM simulator by "adding up" the value of the electric field over a path between the line and the ground. This process corresponds to the following expression:

$$V_{BA} = -\int_{A}^{B} E \cdot ds$$

For example, if we have a TEM transmission line with an electric field of 50 V/m over a distance of 1 mm, the line voltage is 50 mV. If this is the same TEM line we just looked at for current, the characteristic impedance is 50 mV/1 mA = 50Ω (assuming the voltage and current are measured with no standing wave present).

Unfortunately, if this method is applied to the waveguide tube as it is, how do we choose the voltage integration path? And what is the current?

For waveguide tubes, because the definition of voltage is not easy to see, we go to a more fundamental definition, as shown in Figure 3.15. The characteristic impedance of any electromagnetic wave is defined as the ratio of the electric field to the magnetic field in the plane perpendicular to the direction of travel of the electromagnetic radiation. Inside a waveguide tube, it happens that this ratio is the same everywhere. This is called the wave impedance or the characteristic impedance.

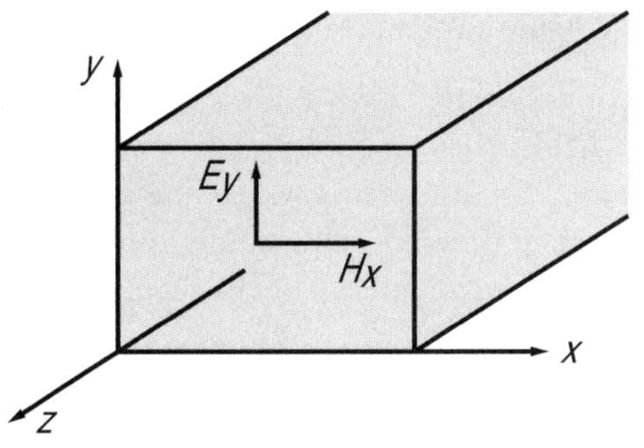

Figure 3.15 Characteristic impedance of the TE_{10} mode of a rectangular waveguide tube is determined by the ratio of electric to magnetic field.

The characteristic impedance of the TE_{10} mode of a rectangular waveguide tube (Chapter 6) is obtained by the following expression. We have the wavelength, λ, in the expression; thus, it cannot be a TEM mode because there is a frequency dependency.

$$Z_0 = \frac{120\pi}{\sqrt{1-\left(\frac{\lambda}{2a}\right)^2}}$$

Here, a is the dimension of the longer side, and λ is the wavelength.

In an empty waveguide tube, the characteristic impedance η_{TE} and η_{TM} of the TE mode and the TM modes are given by the following expressions:

$$\eta_{TE} = \frac{120\pi}{\sqrt{1-\left(\frac{f_c}{f}\right)^2}}$$

$$\eta_{TM} = 120\pi\sqrt{1-\left(\frac{f_c}{f}\right)^2}$$

Here, f_c is the cutoff frequency. The rectangular waveguide cannot propagate any power below the cutoff frequency. Basically, the wavelength of lower frequencies is just too big to fit in the tube. Notice that η_{TE} is always greater than 120π, and η_{TM} is always less. It is also no accident that $120\pi = 377\Omega$, which is the ratio of the electric to magnetic field for a plane wave in free space (the kind that our cell phones receive). A little bit of trivia that is important to remember if you start doing high-accuracy work: the 120π is a convenient approximation—it is not exact. The exact value is $\mu_0 c = 4\pi \times 10^{-7} \times 299{,}792{,}458.0 = 376.730\ldots$, while $120\pi = 376.991\ldots$.

3.6.2 So How Do We Apply a Voltage to a Waveguide Tube?

When a battery is connected to a two-conductor transmission line such as two parallel wires or an MSL, then DC current flows in the load. On the other hand, when we connect a battery to a waveguide tube, as in Figure 3.16, all we get is a large current in the side walls, simply shorting the battery out. This current flows around the circumference of the tube. It does not flow along the length of the tube.

As mentioned earlier, waveguide tubes do not propagate signals when the wavelength is too large. As we can see, DC (with infinitely long wavelength) simply will not propagate in a waveguide tube. To insert a propagating signal into a waveguide tube, we first select a frequency with a sufficiently short wavelength. Then we place what amounts to an antenna inside the waveguide. This antenna is a waveguide probe. There are many ways to do this, which we do not detail here.

3.6.3 A Useful Method Due to Heaviside

Oliver Heaviside (1850–1925) was an eccentric self-taught British mathematical genius who placed Maxwell's equations in their modern form. He is famous as the inventor of the telegrapher's equation, which determines how voltage and current propagate along a transmission line. According to Heaviside, a short (i.e., a tiny fraction of a wavelength long) section of transmission line with a propagating TEM wave, like the Lecher wire of Chapter 1, can be modeled with a lumped circuit, as shown in Figure 3.17.

To model a long transmission line, simply connect many of these models together, one after another. For an exact model, Heaviside showed how to take the length of each segment down to zero and simultaneously take the number of segments up to infinity. Such a line is defined based on per unit length (p. u. l.) values. For example, specifying p. u. l. capacitance and inductance completely specifies a lossless TEM line.

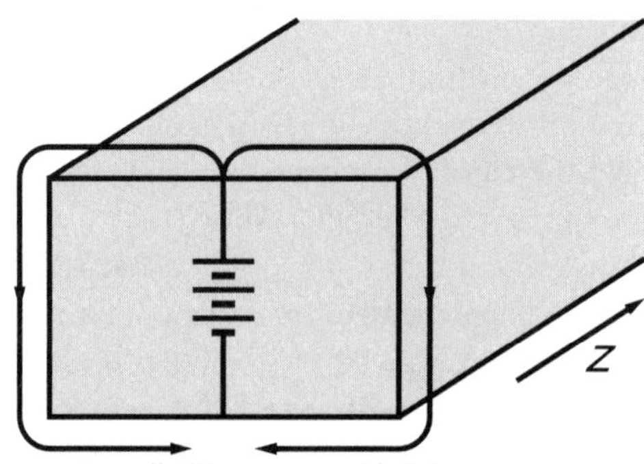

Figure 3.16 Direct current applied to a waveguide tube.

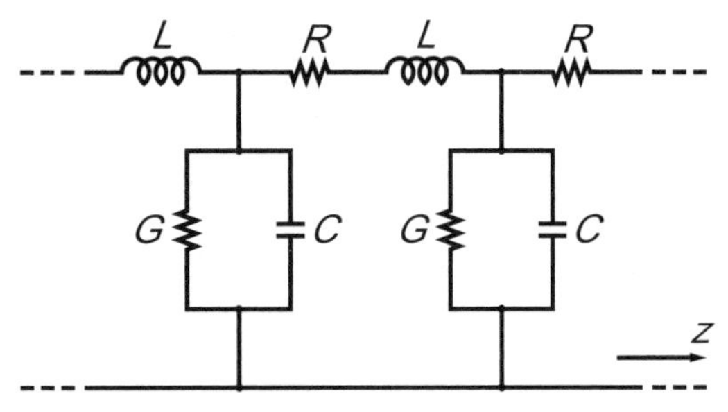

Figure 3.17 A short section of a Lecher wire (or any transmission line) can be modeled as a lumped circuit.

Figure 3.18 shows two parallel lines of constant cross section extended to infinity in the *z* direction. When the series impedance distributed along the lines in Figure 3.17 is *Z* per unit length and the parallel admittance is *Y* per unit length, Ohm's law is applied to the infinitesimal section d*z*, using $I(z)$ that flows on the line, voltage $V(z)$ between lines, the small change in current $dI(z)$, and voltage $dV(z)$ along the length of infinitesimal section d*z*.

We will discuss the ideas of Heaviside using the per unit length impedance and the admittance of a transmission line.

$$Z = R + j\omega L$$

$$Y = G + j\omega C$$

Here, *R*, *L*, *G*, and *C* are the resistance, inductance, conductance, and capacitance per unit length, and ω is an angular frequency (= $2\pi f$). Using these simple concepts, it becomes much easier to understand the complicated inter-

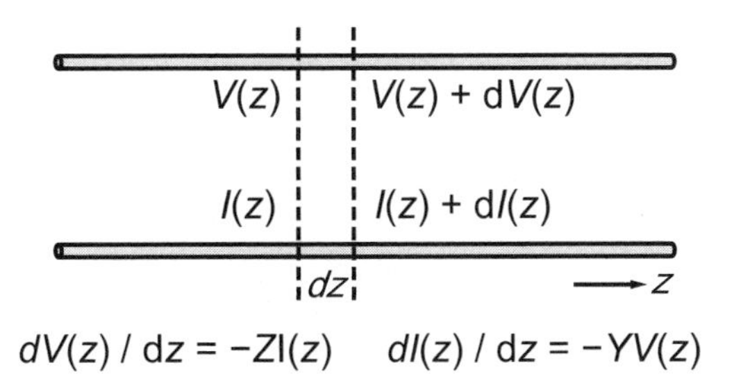

Figure 3.18 Two parallel lines of the constant sectional area extended infinitely to the *z* direction.

actions between electric and magnetic fields using Heaviside's circuit theory model.

British physicist William Thomson (Lord Kelvin), a good friend of Maxwell, is well known today, as his name is now used for the unit of absolute temperature, K (Kelvin). He also devised a model of the trans-Atlantic cable around the same time that Heaviside devised his model. However, Lord Kelvin's model was simpler than Heaviside's. There was no inductance, only capacitance. This resulted in what is known as the diffusion equation instead of Heaviside's wave equation. Kelvin thought that electrical pulses "oozed" into the transatlantic cable. Kelvin's approximation worked well for the high capacitance undersea cables, but failed for in-the-air land-based telegraph lines.

For the TEM transmission line, the characteristic impedance is shown by the following expression using the per unit length lumped model in Figure 3.19:

$$Z_0 = \sqrt{\frac{R + j\omega L}{G + j\omega C}}$$

Note that whenever there is loss, Z_0 is complex. It is common to ignore the imaginary part, something that should not be done if high accuracy is needed. When there is no loss in the line, R and G are zero, and the previous expression is written as follows:

$$Z_0 = \sqrt{\frac{j\omega L}{j\omega C}} = \sqrt{\frac{L}{C}}$$

3.6.4 Matching Source to Load

When thinking about the electric power P supplied by a power supply with internal resistance R_i to the load R_L, as seen in Figure 3.19, P is written as follows:

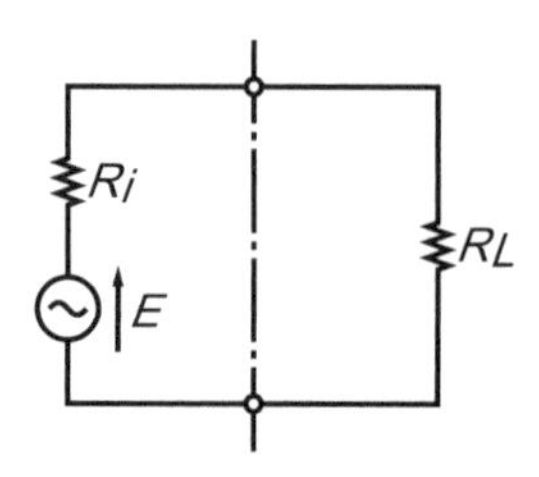

Figure 3.19 Investigating the power supplied to the load.

$$P = \left(\frac{V}{R_i + R_L} \right)^2 R_L$$

Given that R_i and V are fixed, P is maximized when $R_i = R_L$. (This is for a pure resistive load. For reactive loads, the relation is a little more complicated.) For microwave purposes, the load is usually some distance from the source, and they are connected by a transmission line. In order to maximize transfer of power to the transmission line, we must select the characteristic impedance of the line to be R_i. In order to maximize transfer of power from the line to the load, we likewise need to make the load impedance R_i. In this case, the load is said to be matched.

3.6.5 Why Is 50Ω the Standard?

Figure 3.20 is a simulation result showing the electric and the magnetic vector fields in a coaxial line. The ratio of the electric field **E** to the magnetic field **H** represents Z_0. For coaxial cable, this ratio is the same everywhere. Of course, only one traveling wave (no standing wave) is assumed.

In addition, C and L for each unit length of a coaxial cable that uses air dielectric with the inside conductor diameter a, and outside conductor diameter b are as follows:

$$C_0 = \frac{2\pi\varepsilon_0}{\ln\left(\frac{b}{a}\right)}$$

$$L_0 = \frac{\mu_0}{2\pi}\ln\left(\frac{b}{a}\right)$$

For any lossless transmission line, the characteristic impedance is as follows:

$$Z_0 = \sqrt{\frac{j\omega L}{j\omega C}} = \sqrt{\frac{L}{C}}$$

Then, Z_0, obtained by substituting the expressions for L and C into this expression, is as follows:

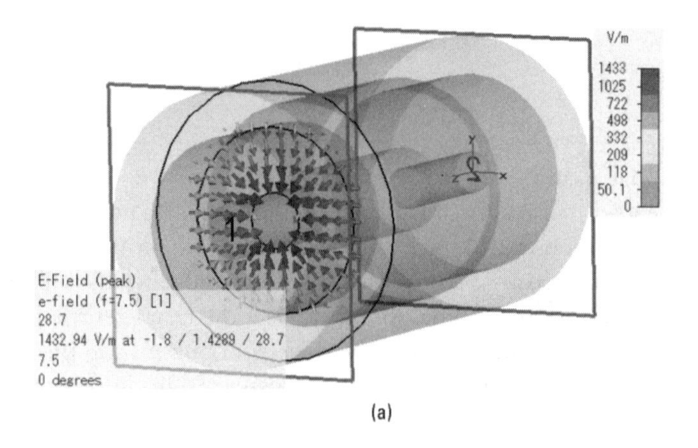

(a)

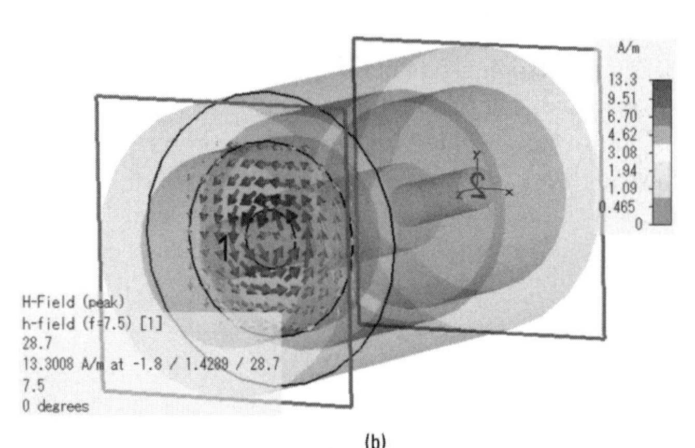

(b)

Figure 3.20 (a, b) Simulation result of the electric and magnetic vector fields in a coaxial line. Note that the far (upper right) square indicates the port 2 plane, which is labeled as the *x*-*y* plane. The center conductor is along the *z*-axis. The port 1 plane is nearer to us, and we are viewing the port 2 plane from behind, thus the label appears backwards. There is a reduction in the width of the center conductor that is attached to port 2.

$$Z_0 = \frac{1}{2\pi}\sqrt{\frac{\mu_0}{\varepsilon_0}}\ln\left(\frac{b}{a}\right)$$

In coaxial line, it is known that there is a characteristic impedance that minimizes the transmission loss. Physically, if Z_0 is too high, the center conductor becomes very thin and its resistance becomes too high. If Z_0 is too low, then the current on the line becomes too high and even a tiny amount of resistance becomes very lossy. If we solve mathematically for minimum loss, the optimum value is about 75Ω for air-filled coaxial line. The optimum becomes about 50Ω when using a dielectric of relative permittivity about 3. Many reasons have been

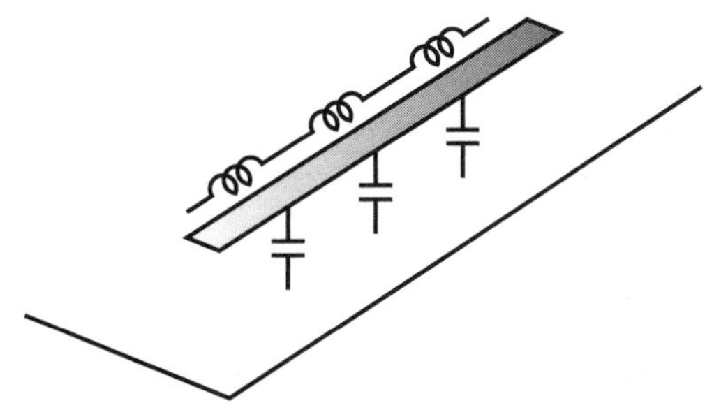

Figure 3.21 Distributed per unit length lumped model of an MSL.

proposed for why 50Ω is the most common standard, and there is no longer any way to historically verify why this choice was made. The reason cited earlier is one of the most convincing.

The value of *b*/*a* for a 50Ω coaxial line is about 2.3.

3.6.6 Characteristic Impedance of a Microstrip Line

The characteristic impedance of an MSL is determined by the square root of the ratio of *L* to *C*. The capacitance between the line and the ground plane and the series inductance are as shown in Figure 3.21. Increasing the width of the line, for example, increases the capacitance and decreases the inductance. Both changes decrease the characteristic impedance. Thus, we can control the characteristic impedance of an MSL by adjusting its dimensions and thus match the input impedance of a load.

The precise characteristic impedance of a microstrip line structure can be determined with an electromagnetic field simulator. The following approximate expression is sometimes used for a good first guess.

$$Z_0 = 30\ln\left[1+\frac{4h}{W_0}\left\{\frac{8h}{W_0}+\sqrt{\left(\frac{8h}{W_0}\right)^2+\pi^2}\right\}\right]$$

This is just one of many expressions, and this one assumes air for a substrate. Each expression can have different levels of accuracy and different levels of complexity. There are many free programs online to calculate microstrip characteristic impedance, and these should be used when you need some numbers. In any case, because the distribution of the electric and the magnetic fields around the line changes when the thickness of the substrate and the line width are changed, we know that the characteristic impedance changes, too.

The electromagnetic field in free space and in the dielectric substrate travel at different velocities. This means that microstrip, which has field present in two different dielectrics, cannot propagate a TEM wave. The overall net velocity of the wave and the characteristic impedance are functions of frequency. This means that any formula that models this frequency dependence is both complicated and approximate. Dispersion means that peculiar results can sometimes be seen. Designing circuits for low frequency was easy by comparison. We must now be prepared for a strange new world.

3.7 Confirming Results of This Chapter by Simulation

First, we explore de-embedding the transmission line discontinuity described earlier in this chapter.

3.7.1 Modeling the Discontinuity

Figure 3.22 is a discontinuity that can be viewed as a cross junction with two open-circuited stubs (one connected on either side). It can also be viewed as a cascade of two step-in-line-width (usually called steps for short) discontinuities. The substrate size is 6.4 mm × 6.4 mm, as shown in Figure 3.23, and the cell size is 0.2 mm for both the *x* and *y* directions.

Because Symmetry is checked at lower right, a dotted horizontal line appears at the center of the substrate. This feature can be used for symmetrical geometries. A virtual wall (it is also called magnetic wall) divides the box along the dotted line. In this case, only the upper half model is calculated, so the memory for the simulation can be cut to one quarter and solving the matrix is

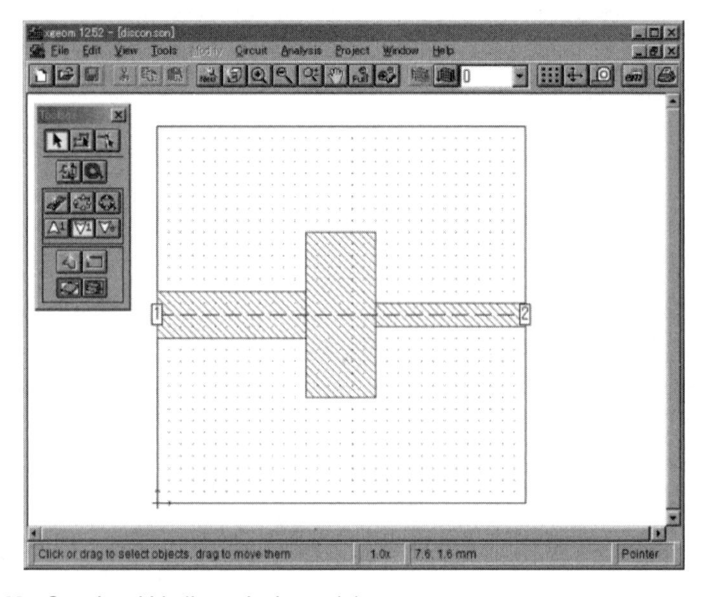

Figure 3.22 Step-in-width discontinuity model.

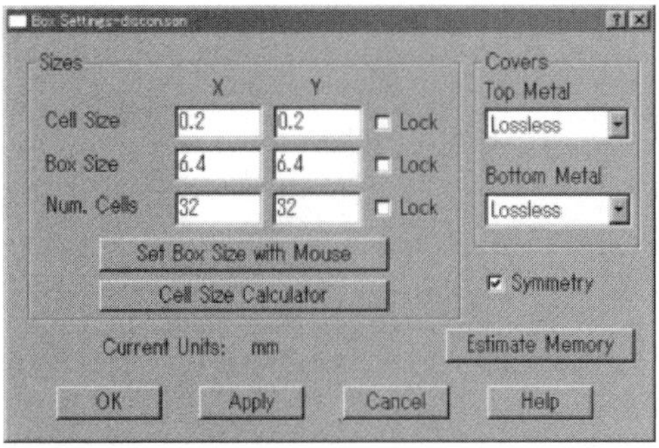

Figure 3.23 Dimensions of the Sonnet Box.

eight times faster. With the Symmetry option enabled, the bottom half is assumed to be exactly the same as the upper half.

Next, the thickness of the dielectric substrate is set to 0.4 mm and the relative permittivity is set to 5.0, as shown in Figure 3.24. Above the level (Level 0) where the line was drawn, we have an air layer 100 mm thick, and relative permittivity = 1.0. This is plenty to keep the top cover from influencing the circuit. Usually ten times the substrate thickness is more than enough.

3.7.2 Simulation Result Including Port Connecting Transmission Lines

Select Analysis > Setup..., as in Figure 3.25, and input the range of 1 GHz to 5 GHz, for example. Analysis Control is left at the default of Adaptive Sweep (ABS). Check the box at the top left, Compute Current Density, so that we can display the current density after the analysis is complete.

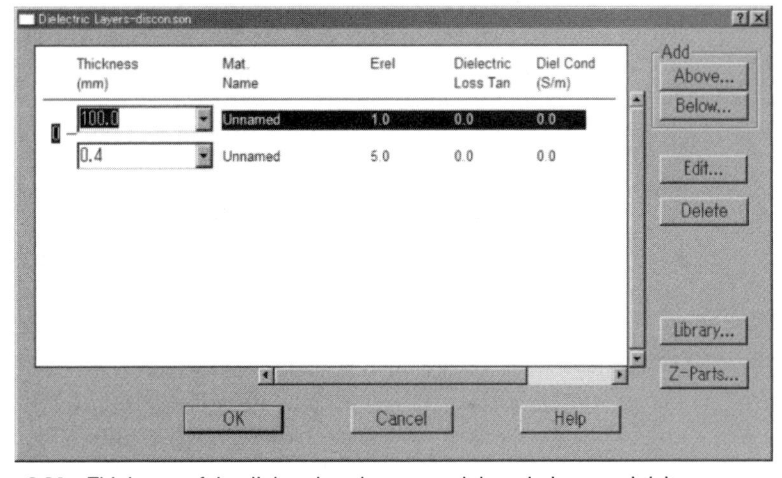

Figure 3.24 Thickness of the dielectric substrate and the relative permittivity.

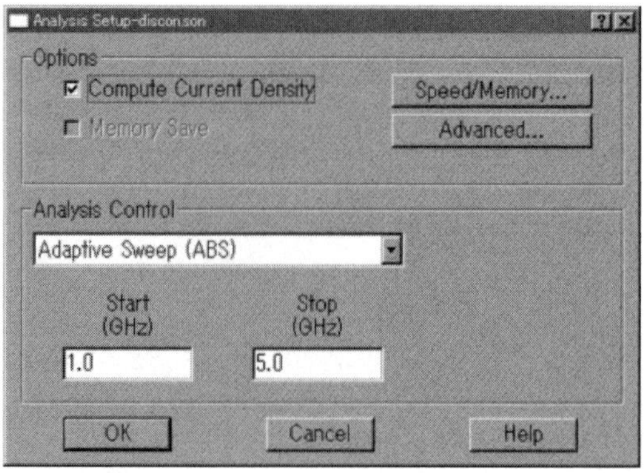

Figure 3.25 Specify a range of frequencies for analysis.

Select Project > Analyze to start the simulation. When it is complete, select Project > View Response > Add to Graph to display the resulting S-parameters.

The default plot shows only S_{11} (reflection coefficient). Double click DB[S11] on the top left of the graph window and move DB[S21] to the right into the Selected window, and click the OK button. Now, S_{21} (transmission coefficient to port 2) is displayed too, as shown in Figure 3.26. The S_{11} shows that at high frequency, more of the signal is reflected from port 1 as it approaches 0 dB.

3.7.3 Surface Current Distribution

Select Project > View Current, or click the View Current button, and the current distribution is displayed, as in Figure 3.27. The color bar on the left shows that red is high current and blue is low current. In the figure, we can clearly see the edge singularity (the high current at the edge). The output line is entirely

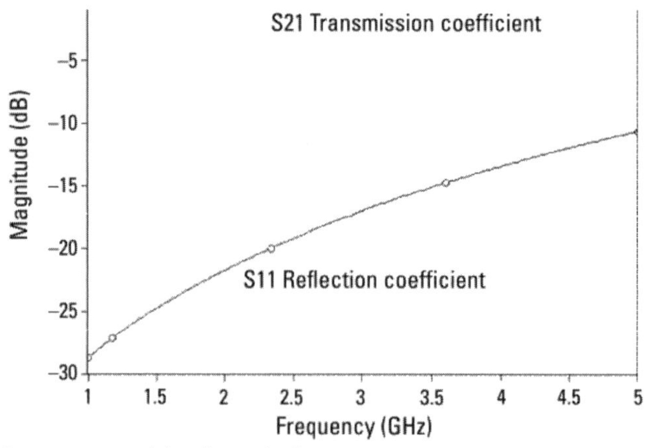

Figure 3.26 S-parameters of the discontinuity.

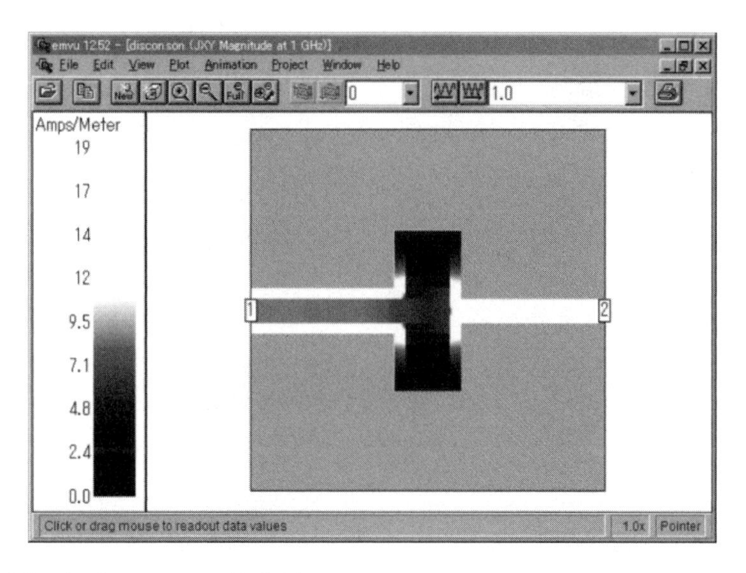

Figure 3.27 Surface current distribution.

red (on your computer screen) because we are using a large cell size. If smaller cells are specified (Circuit > Box), the accuracy of the analysis, and the current distribution, is improved.

3.7.4 Current Distribution Animation

When we have analyzed several frequencies, we can animate the current distribution as a function of frequency. With the current distribution displayed, click Animation > Settings…, as in Figure 3.28. You can now select Frequency or Time animation. If we chose time animation, we can see how the current distribution changes through one cycle of a sine wave at a single frequency. Select Frequency animation so we can see how the current distribution changes with frequency.

Next, click Animate View in the Animation pull-down menu, and the small dialog box for Animation Controls is displayed, as in Figure 3.29. Use the controls to display the animation from 1 GHz to 5 GHz. All analyzed frequencies will be displayed.

3.7.5 Display Subsections

EM analysis divides the metal of our circuit into small subsections. Larger subsections analyze faster, but give a less accurate answer. Smaller subsections analyze slower, but give a more accurate answer. You can set the minimum subsection size in Circuit > Box. The minimum subsection size is called the cell size. If you want higher accuracy, or a more detailed and accurate current distribution,

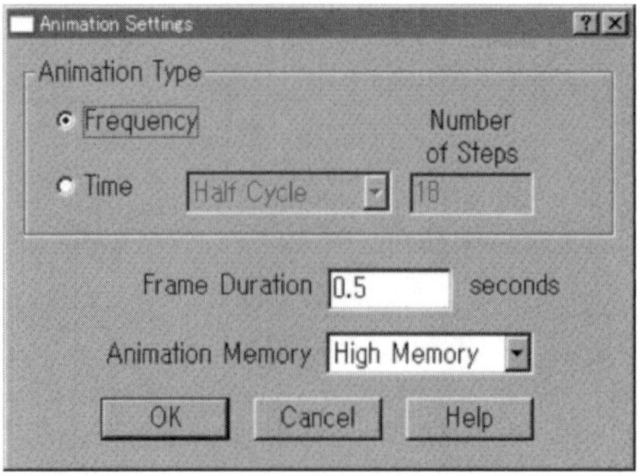

Figure 3.28 Animation type menu.

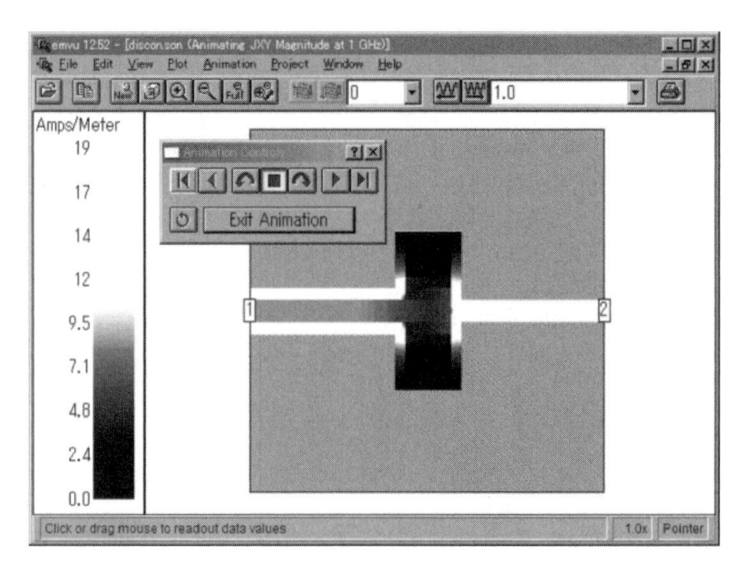

Figure 3.29 Animation controls.

set a smaller cell size. To see the actual subsections, select View > Subsections, as in Figure 3.30. Since we have Symmetry turned on for this circuit, subsections below the center line are not displayed, and the circuit below the center line is assumed to be an exact mirror image of the circuit above the center line.

3.7.6 Simulation with Removal of Port Connecting Transmission Lines

Our analysis so far includes the transmission lines that connect the ports (on the edge of the substrate) to the discontinuity. Notice the arrows extending from the ports in Figure 3.31. This indicates that we have moved the reference planes for each port up to the start of the discontinuity. The EM analysis will

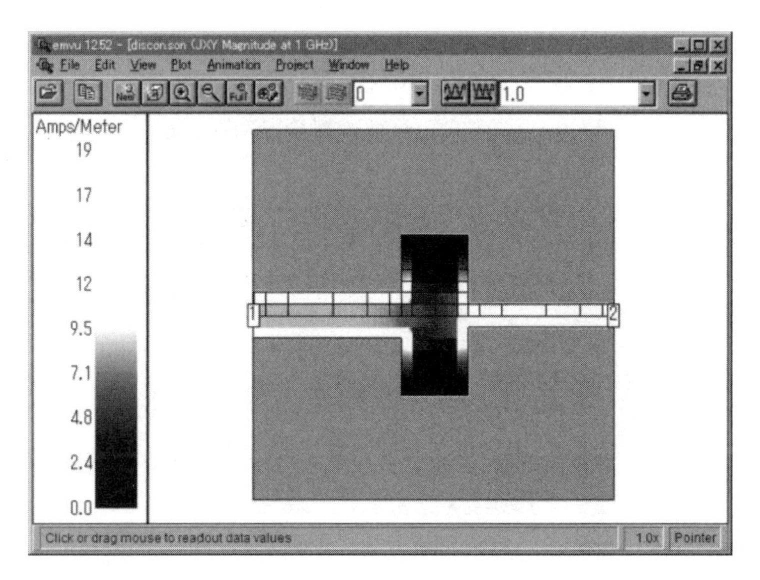

Figure 3.30 Subsection display.

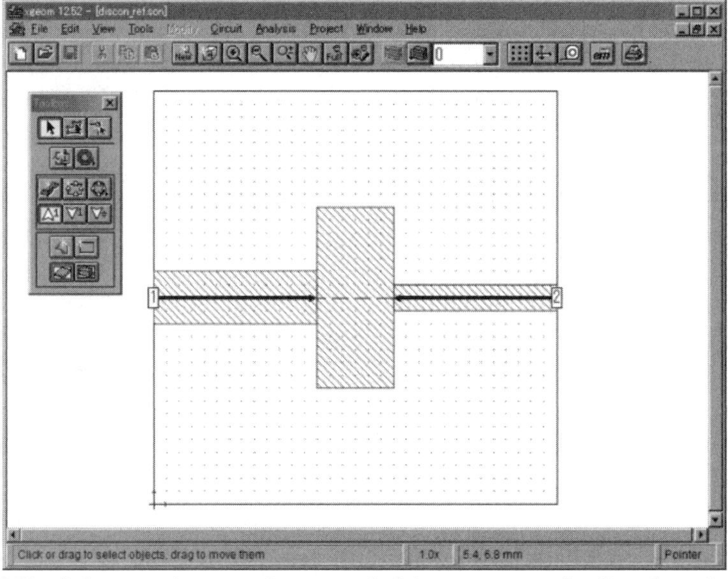

Figure 3.31 Reference planes set for removal of the port connecting lines from the result. File: discon_ref.son.

still analyze the entire circuit, including the port connecting transmission lines. However, with the reference planes as indicated, the effect of the port connecting lines will be removed, leaving just the S-parameters of the discontinuity all by itself.

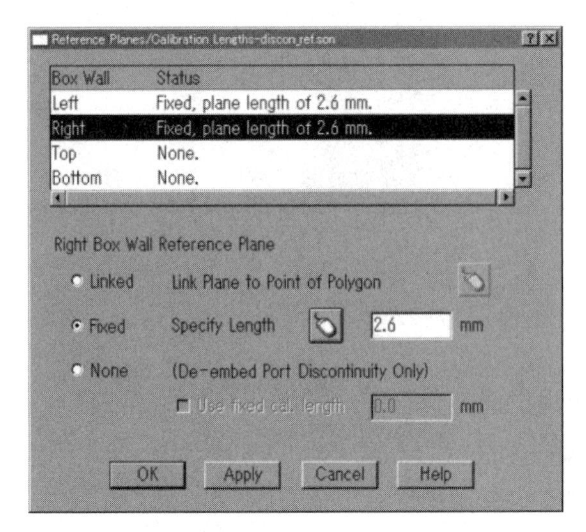

Figure 3.32 Set the reference plane. Check the Fixed radio button.

This is just like the calibration of a network analyzer described earlier. For the network analyzer, we must remove the effect of the cables that connect the network analyzer to the DUT. Mathematically, this is exactly the same problem, and, in fact, we even use almost exactly the same mathematics.

Let's set the new reference planes. With the circuit layout displayed, select Circuit > Ref. Planes/Cal. Length..., as in Figure 3.32. Check the Fixed radio button and click on the mouse icon. The cursor changes to a + mark. Then, click where you want the reference plane for port 1 to be set. The distance from the port is measured, and then the reference plane is set after selecting the OK button.

To set the reference plane for port 2, select the Right line at the top and repeat for port 2. The reference plane for all ports on each side of the box, Left, Right, Top, and Bottom (as viewed on your screen) must all be the same. This means if you have several ports on one side of the box, they must all be the same reference plane. Note that the de-embedding even removes the effect of multiple tightly coupled lines.

3.7.7 Comparing S-Parameter Results

Let's compare results for this circuit with and without the effect of the port connecting lines, as in Figure 3.33. To add results from other projects to this graph, select File > Add File(s)..., and add the previous project, discon.son. Next, using the options on the left side of the plot, add that project's results to the graph, as shown in Figure 3.33.

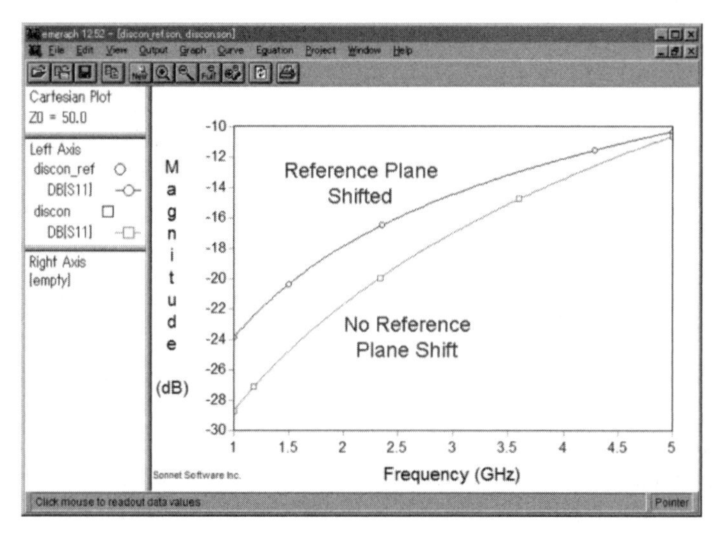

Figure 3.33 Comparing S_{11} with and without the connecting transmission lines.

3.7.8 Coupled-Line Right-Angle MSL Bend

Figure 3.34 is the same coupled-line right-angle MSL bend we looked at earlier, except that this one has only two lines instead of four. This is so we can analyze it in Sonnet Lite, which allows up to four ports. First, load the msl_Lbend.son that we did in Chapter 2 (it is also on this book's DVD). After saving (File > Save As…) under a new name, add one more line. To do this, click on the draw rectangle icon, the last icon of the next to last row of the Toolbox, or select Tools > Add Metalization > Rectangle, and input the dimensions for Width and

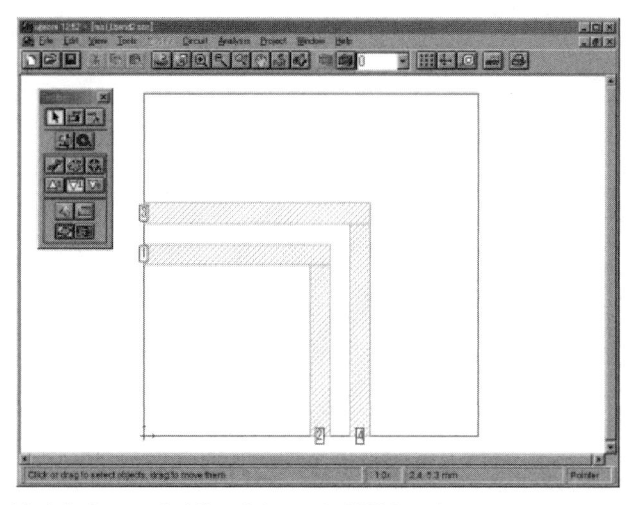

Figure 3.34 Model of a coupled-line right-angle MSL bend.

Height. The distance between lines and the line widths are both 0.3 mm. Then add ports 3 and 4 to each end of the line (use the Toolbox port tool).

The crosstalk is shown in Figure 3.35. We see that the forward crosstalk becomes larger than the backward crosstalk at around 8 GHz. The crosstalk is caused by the proximity of the lines. You can conduct numerical experiments by plotting the crosstalk as a function of the distance between the lines. Crosstalk becomes less as the lines get farther apart.

3.7.9 MSL Crosstalk Reduction

Electric field lines of an MSL flow between closely spaced transmission lines and to the underlying ground conductor, as in Figure 3.36. If there is another line in the vicinity, some of the electric field lines also connect to it. If this electric field increases, crosstalk also increases. Perhaps we can reduce it by setting a vertical wall between the lines, as shown in Figure 3.37. Electric field lines from the transmission lines are pulled into the wall surface, and coupling to the next line is reduced.

To include a via wall between the lines, we need a smaller cell size. Set the cell size to 0.1 mm for both *x* and *y* in Circuit > Box.

In Sonnet, the vertical wall is put in place by drawing a via. First, with the layout displayed, press the down arrow key or CTRL-D to move to the GND level. Next, in the Tool Box, make sure the Up One Level button is selected (first icon, fourth row), as in Figure 3.38. Then select the Draw Via Rect button (middle icon, third row). Draw two via rectangles on the GND level, as indicated in Figure 3.37.

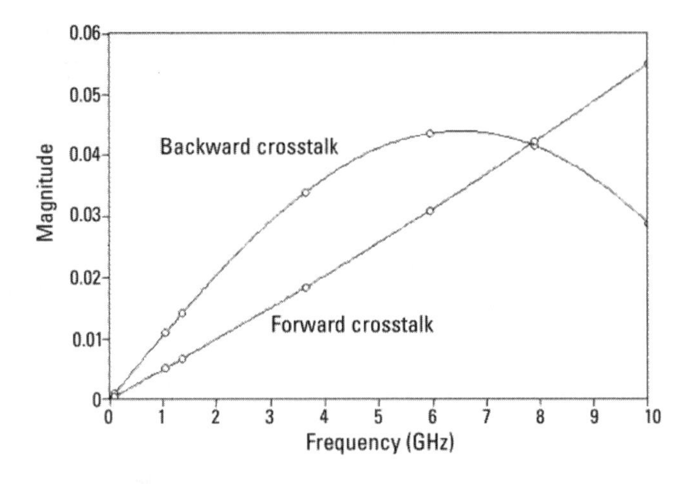

Figure 3.35 Crosstalk of a coupled-line right-angle MSL bend.

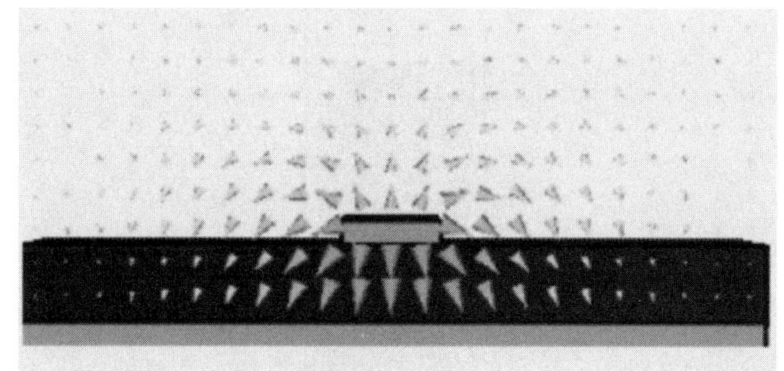

Figure 3.36 Electric vector field around an MSL.

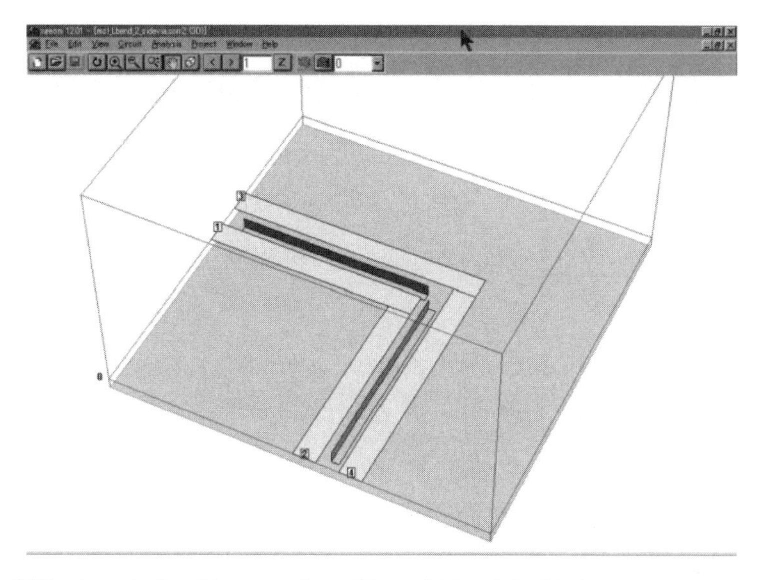

Figure 3.37 A vertical wall between lines. File: msl_Lbend_2_sidevia.son.

Figure 3.39 shows S_{31}, backward crosstalk. We have reduced it by up to 7 dB by adding the via wall. Depending on the frequency, improvement is also realized by allocating individual vias at equal intervals. You can try different configurations and investigate the effect.

3.7.10 Simulation of Crosstalk for More Than Two Lines

Sonnet Lite is restricted to no more than four ports. However, it is still possible to evaluate the crosstalk between more than two lines. Just put in all the lines you want to evaluate. Then add up to four ports for the S-parameters that you want. Finally, place 50Ω resistors in all the other port locations, as in Figure 3.40.

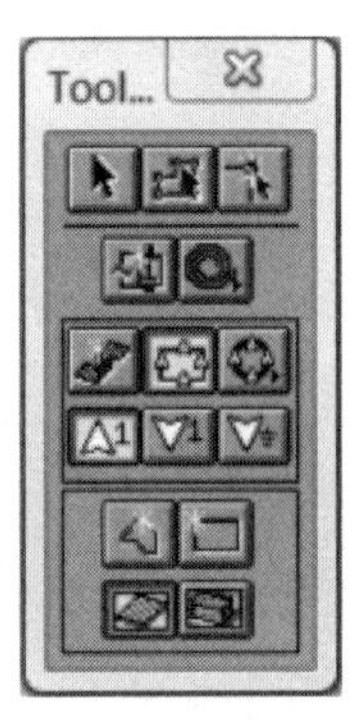

Figure 3.38 Up One Level button.

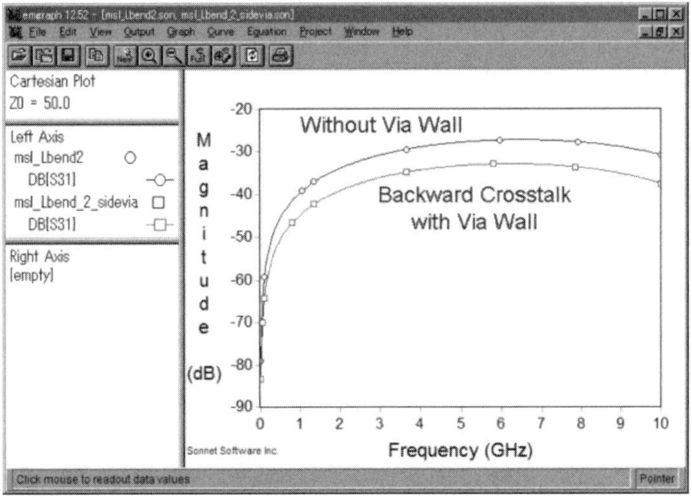

Figure 3.39 Backward crosstalk with and without a via wall. File: msl_Lbend_2_sidevia2.son and msl_Lbend2.son.

To add resistors, select Circuit > Metal Type..., Figure 3.41. Set the metal type to resistor and set Rdc to 50 Ω/□. The units mean that a square patch of this metal is a 50Ω resistor.

Now, when you double click on a polygon, you can specify the resistor metal. Add squares to the unused port locations (Figure 3.42). You can place squares there by simply using the razor knife tool (Edit > Divide Polygons) to cut square sections out of the lines. Then double click on the sections and change the metal type to the resistor type we just specified.

Be careful to make the resistor metal areas square. For example, if we were to make the resistor regions twice as long, we have two 50Ω resistors in series and the total resistance is now 100Ω.

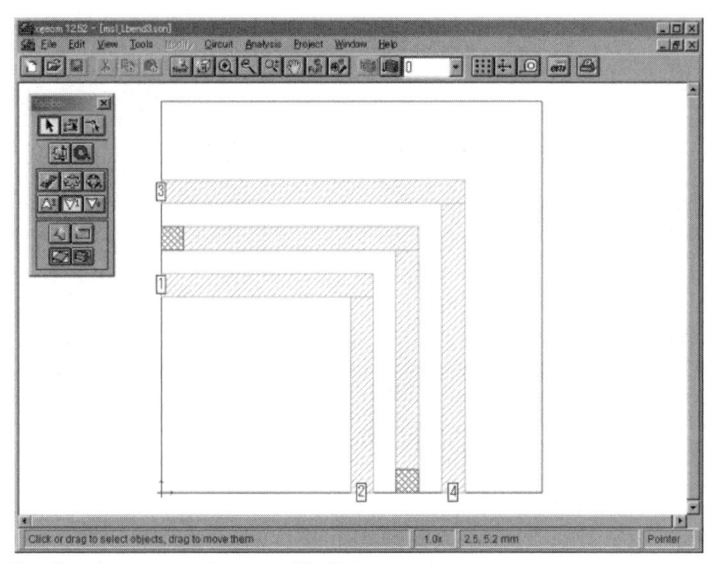

Figure 3.40 Terminate unused ports with 50 Ω/sq resistance.

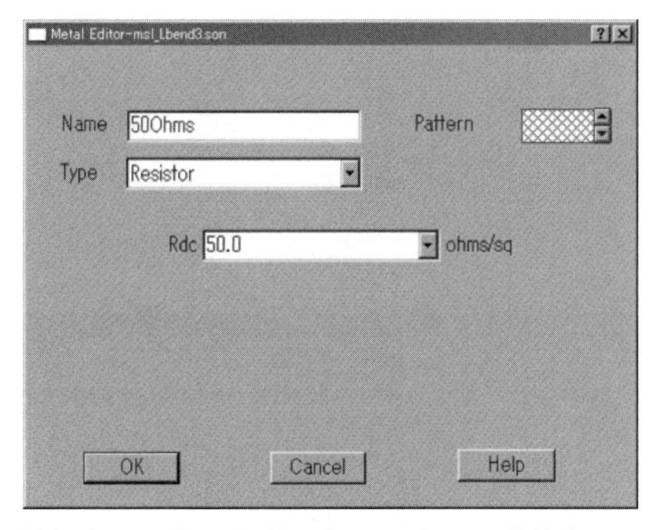

Figure 3.41 Dialog box to make a 50 Ω/□ resistor metal type available for use.

Figure 3.43 shows two graphs of S_{41}. The result from Sonnet Professional (using ports) and from using this terminating resistor model are almost identical.

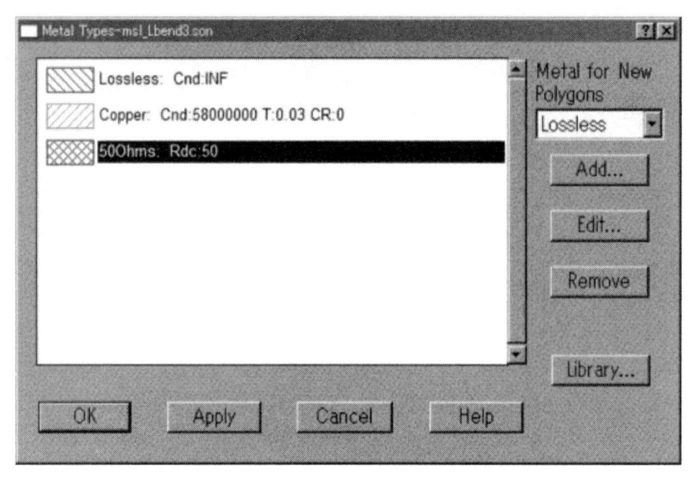

Figure 3.42 The new metal type is added.

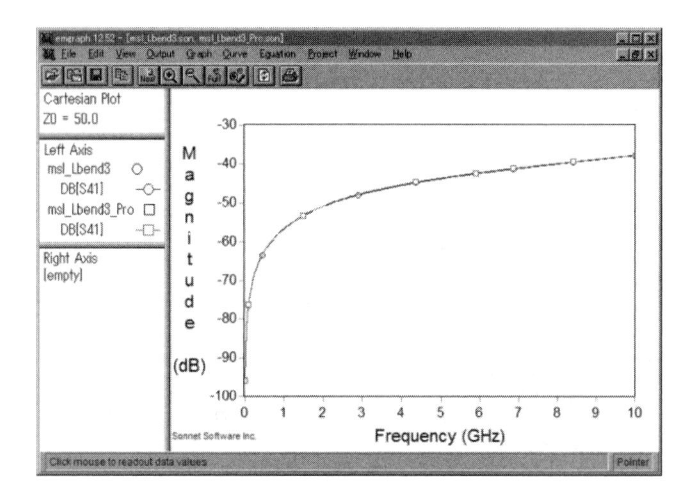

Figure 3.43 S_{41} crosstalk result using Sonnet Professional and using the terminating resistor model are almost identical.

3.8 Summary

1. The engineer needs quantitative, accurate, numerical values in order to make decisions.
2. Delay is calculated from the phase angle of the S-parameters.

3. If there is a requirement on the allowed delay, it might be necessary to change the length of a line for the correct phase angle of the S-parameter.
4. By examining the magnitude of an S-parameter, we can determine if it exceeds the crosstalk value that can be allowed. If it does, it is necessary to modify the design, perhaps by separating the offending lines.
5. The characteristic impedance of a transmission line is important for efficiently transmitting electromagnetic energy to the load.

4

What Is Different About High-Frequency Circuits?

4.1 Electromagnetics and Circuit Theory

In Chapter 3, we learned how characteristic impedance and S-parameters are useful in quantitatively evaluating a high-frequency circuit. Circuit simulators like Simulation Program with Integrated Circuit Emphasis (SPICE) are widely used to evaluate additional quantitative aspects of circuits. When we use our favorite flavor of SPICE in combination with electromagnetic field simulators, we can achieve especially deep understanding.

Most electromagnetic field simulators can generate lumped SPICE models automatically. In this chapter we will learn how to utilize SPICE based on some practical high-frequency circuits.

4.1.1 Microstrip Lines Are Distributed Circuits

Many books are published about PSpice and SPICE, and these tools have become widespread. SPICE solves Kirchhoff's law, which states that if you add up all the current going into a node and subtract all the current going out, the total must be exactly zero. The basic components that SPICE-based tools handle are resistors (R), inductors (L), and capacitors (C). If we are to analyze high-frequency distributed circuits (i.e., transmission lines), it is useful to represent these circuits using lumped R, L, and C components.

Frequently, we will want to use the same component repeatedly, in which case we specify the RLC model and label it as a single unit called a subcircuit.

A transmission line on a substrate can be taken as a lumped equivalent circuit, as shown in Figure 4.1. It can be convenient to represent this as a subcircuit. Then we use the subcircuit to connect to other circuit elements. Each LC section can represent only a short (with respect to wavelength) section of the transmission line. So, keep in mind that long transmission lines will require many sections. We will discuss this later.

As for an analog circuit of low operating frequency, even a simple SPICE lumped parameter model can give a good answer. However, when the frequency becomes high, a resistor no longer acts like a resistor because it also includes inductance and capacitance. In addition, the electromagnetic coupling to the ground plane and the DC power distribution plane can make the entire circuit model complicated.

4.1.2 Treating a Circuit on a Printed Circuit Board as a Transmission Line

Transmission lines should be modeled as distributed circuits. The lumped model is appropriate as long as each LC section is short compared to wavelength. A transmission line induces delay and the signal reflects from the far end when it is not terminated properly. In addition, electromagnetic coupling also occurs between transmission lines, which gives us crosstalk.

Long transmission line models become complicated. At a sufficiently high frequency, transmission lines gradually become antennas, further complicating models. For digital circuits, frequencies at several times the clock frequency (depending on how fast the rise and fall times are) can become important. If all of these effects are not properly included in a SPICE model, then there could be unexpected problems for which SPICE gives no clue.

4.2 Design of Microwave Circuits

For over half a century, microwave circuit designers have been designing complicated highly integrated circuits from several hundred megahertz to several

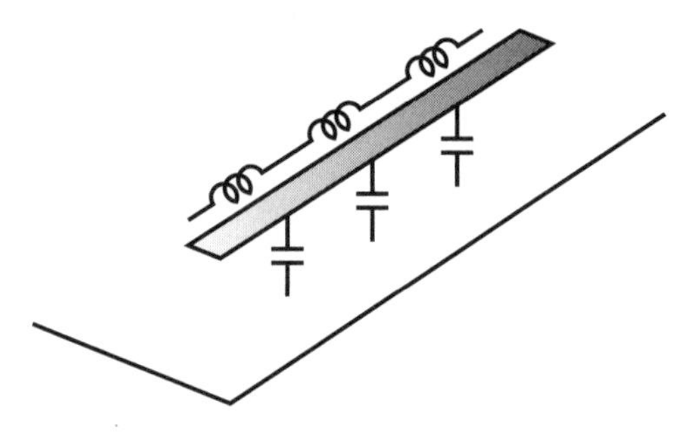

Figure 4.1 A microstrip line can be described by a lumped equivalent circuit.

dozen gigahertz for communication systems and radars. They have often solved problems by trial and error in creating better SPICE models. This is expensive and can take a lot of time.

For the last quarter century, microwave circuit designers have been using electromagnetic field simulators. Today, they first analyze nearly all designs with electromagnetic field simulators before fabricating the actual microwave circuits. They can now find the source of any problems before they fabricate, rather than after.

In addition to finding problems, they can now design their entire circuit much more compactly by bringing circuit components and transmission lines closer together. The effect of compacting the circuit is taken into account by repeating simulations and appropriately adjusting circuit dimensions. If a problem is found when running electromagnetic field simulations, designers can correct it before production.

We can benefit from everything that microwave designers have learned about how to make good use of electromagnetic field simulators. Taking advantage of all this know-how, it is easy to see that these simulators are most useful.

4.3 Bent Coupled Lines Again

Now we return to the model of four coupled lines with a right-angle bend, described in Chapter 3. Figure 4.2 is a Sonnet model of these lines. This circuit has eight ports and the input ports are numbered from 1 to 4. Output ports are numbered from 5 to 8.

Because the number of ports exceeds the Sonnet Lite limit of four, if we are going to use Sonnet Lite, we need to select the four ports for which we want

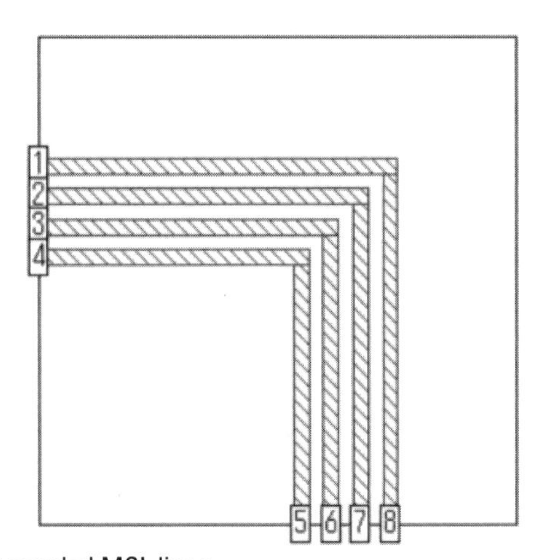

Figure 4.2 Four bent coupled MSL lines.

S-parameters, and terminate the remaining ports with 50Ω resistive squares, as shown in Chapter 3. This step is not required if you can use the professional version of Sonnet, which is what we do here.

Cell size in this example is 1.0 mm in both the *x* (horizontal, as displayed on the computer screen) direction and *y* (vertical, as displayed) direction. The dielectric is 1.0 mm thick, the relative permittivity is 4.0, and the lower layer is a ground plane. Though we could set parameters like dielectric loss tangent (tanδ), here we model the circuit as lossless.

In order to get a SPICE model using one of Sonnet's modeling options, we must analyze at two frequencies. In this case, we select 100 MHz and 110 MHz. To set the output file name, select Analysis > Output Files...and click the PI Model... button.

Figure 4.3 shows the SPICE subcircuit generated for this circuit, and we selected PSpice output format. As shown in Figure 4.2, even for a simple circuit like this, it is not easy to build a SPICE model by hand and include all the couplings between lines.

```
*  Spice Data
* Limits: C>0.01pF L<100.0nH R<1000.0Ohms K>0.01
    *  Analysis frequencies: 100.0, 110.0 MHz
.subckt bend4a_0 1 2 3 4 5 6 7 8 GND
C_C1 1 GND 1.571353pf
C_C2 1 2 0.102519pf
C_C3 1 7 0.049192pf
C_C4 2 GND 1.30278pf
C_C5 2 3 0.092879pf
C_C6 2 6 0.044453pf
C_C7 2 8 0.049192pf
C_C8 3 GND 1.178695pf
C_C9 3 4 0.084424pf
C_C10 3 5 0.039983pf
C_C11 3 7 0.044456pf
C_C12 4 GND 1.151346pf
C_C13 4 6 0.03998pf
C_C14 5 GND 1.151345pf
C_C15 5 6 0.084422pf
C_C16 6 GND 1.178719pf
C_C17 6 7 0.092878pf
C_C18 7 GND 1.302779pf
C_C19 7 8 0.102519pf
C_C20 8 GND 1.571352pf
L_L1 1 8 20.18877nh
L_L2 2 7 18.46781nh
L_L3 3 6 16.73613nh
L_L4 4 5 14.99663nh
Kn_K1 L_L1 L_L2 0.157053
Kn_K2 L_L1 L_L3 0.04465
Kn_K3 L_L1 L_L4 0.018848
Kn_K4 L_L2 L_L3 0.156923
Kn_K5 L_L2 L_L4 0.044569
Kn_K6 L_L3 L_L4 0.156418
.ends bend4a_0
```

Figure 4.3 SPICE subcircuit of four coupled bent MSL lines.

For a lumped model to be valid, the circuit must be small compared to the wavelength of the highest simulation frequency. Setting an upper limit of one-tenth of a wavelength is reasonable for most applications. Sonnet can extract a SPICE PI-model automatically, as we did here. This is good for simple circuits but fails for more complicated circuits. A general "Broadband Spice Extractor" is available in the professional version. It is appropriate for circuits of any complexity.

The PI-model synthesis method was invented by James Rautio and has been available in Sonnet for some time.

4.3.1 Analysis Using SPICE

Figure 4.3 shows the PSpice subcircuit for our coupled line bend. The node numbers are the same as the port numbers of Figure 4.2, and ground is node 0. For example, C_C1 is a 1.57 pF capacitor between port 1 and ground. L_L1 is a 20.19 nH inductor from port 1 to port 8. In addition, Kn_K1 represents the mutual inductance L_L1 and L_L2.

Mutual inductance is when current through one inductor generates voltage across another inductor. This is in contrast to self-inductance, where current in an inductor generates a voltage across itself. Mutual inductance is current operated. Mutual capacitance (capacitors connected from one line to another) are voltage operated. Mutual inductance and capacitance cause crosstalk.

When comparing this SPICE model to Figure 4.2, we can see that the longer line has the larger inductance. It also has greater capacitance between lines as well.

The 0.1 pF capacitance between ports 1 and 2, C_C2, is extremely small. This would be important only at a very high frequency and only when there is a large voltage difference between the lines. Looking at all the mutual inductances that are listed, we can see that the inductive coupling between closer lines is larger.

4.4 A High-Speed Digital Circuit

Figure 4.4 shows the byte reversal circuit we first saw in Chapter 3. This circuit has 32 input ports and 32 output ports, 64 ports in total. First, 8 bits (from upper left, ports 1 to 8) shift down to the next level (Level 1) through vias (Figure 4.5). After going under all the other lines, they come out on the right side as the last 8 bits.

Note that there is no ground plane separating the two levels of circuitry. This increases crosstalk substantially. This is not good for an actual circuit. However, this circuit works very well for demonstrating the ability to model crosstalk, simply because there is so much of it.

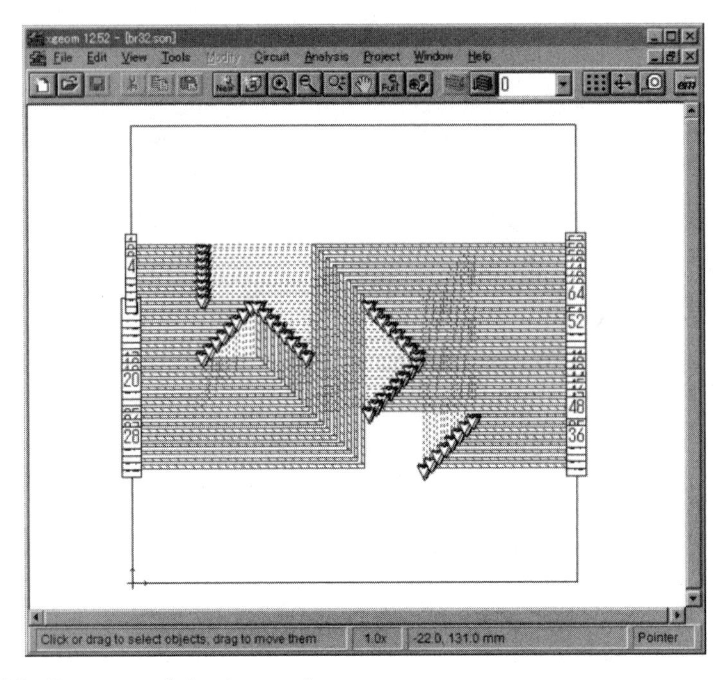

Figure 4.4 Byte reversal circuit, upper layer.

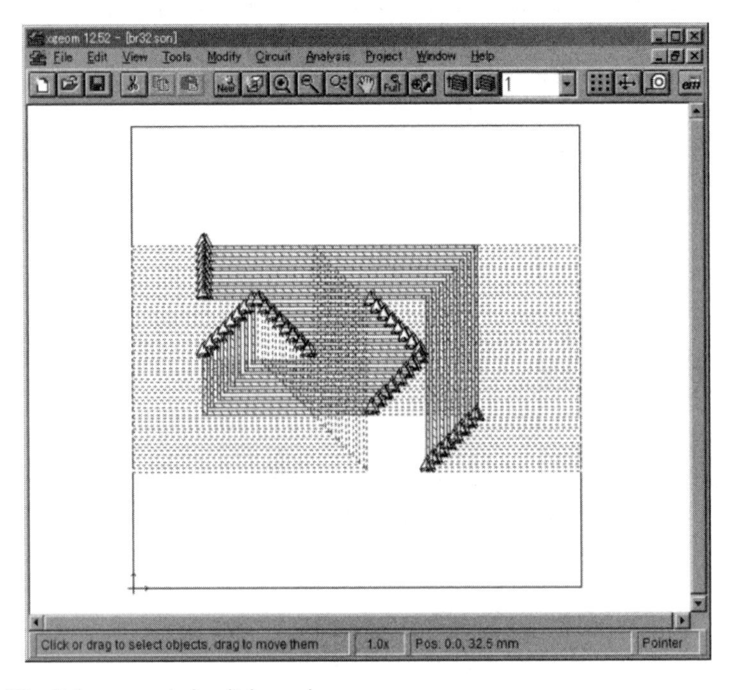

Figure 4.5 Byte reversal circuit, lower layer.

This model is analyzed at two frequencies, 10 MHz and 15 MHz, to get the SPICE subcircuit (it is included in the example files on this book's DVD). Unfortunately, this cannot be viewed by Sonnet Lite because the circuit has more than four ports.

Figure 4.6 shows a portion of the subcircuit derived automatically after simulating this model. The complete model has more than 1,600 lines of element description. Manual generation of this model is simply not possible.

Nodes 1 to 64 correspond to the port numbers of Figure 4.4 and Figure 4.5. When checking the mutual inductance (represented as coupling coefficient *k*) of this subcircuit, we can identify the lines of large crosstalk when we see a large coupling coefficient between lines.

All model capacitances are included in the model if they are greater than the value of Cmin, which was specified when we set the PI Model output file. The default is 0.01 pF, which is extremely small. To obtain a simpler model, we can set the value of Cmin larger. The tiny capacitors are then left out of the model. If they are not needed for whatever level of accuracy is important to you, you can take advantage of a simpler model.

4.4.1 Guidelines for Frequency Selection

Sonnet Lite can synthesize lumped models and output them in the widely used OrCAD PSpice format. As described earlier, Sonnet Lite must first simulate

```
.subckt br32_0 1 2 3 4 ... 63 64 GND
C_C1 1 GND 14.4632pf
C_C2 1 2 1.064645pf
C_C3 1 9 0.044716pf
.
.
C_C1426 64 GND 9.099054pf
L_L1 1 33 77.25012nh
L_L2 2 34 77.30183nh
L_L3 3 35 77.35662nh
.
.
L_L32 32 64 79.6777nh
Kn_K1 L_L1 L_L2 0.166483
Kn_K2 L_L1 L_L3 0.050502
Kn_K3 L_L1 L_L4 0.02287
.
.
Kn_K176 L_L31 L_L32 0.177254
.ends br32_0
```

Figure 4.6 A portion of the PSpice output for the byte reversal network.

two frequencies in order to synthesize a model. The guidelines for simulation are as follows:

1. Select two frequencies that are at least 10 percent apart and are not above the highest frequency of interest.
2. When cell size is less than 0.00001 wavelength, EM simulation accuracy becomes an issue. When cell size (or the height of via) is 1 mm, for example, selecting an analysis frequency less than 1 MHz is not advisable.
3. When a simulation is finished, do a "reality check." If you get a strange result, it is possible that one of guidelines was violated.
4. A second reality check can be done by generating a second model using different analysis frequencies. The two models should be fairly similar. Two models are generated automatically if you analyzed four frequencies, for example.

4.4.2 Generation of a SPICE File

Transmission lines (e.g., two parallel wires) are distributed circuits that can be modeled as a per unit length inductance, L, in series and a per unit length capacitance, C, in shunt, Chapter 3.

A wire that is short compared to wavelength can be modeled as a simple lumped L and C. Here we introduce a method that approximates a distributed circuit by combining lumped elements to generate a SPICE model.

Figure 4.7 is a microstrip line modeled using Sonnet Lite. It is a length of line whose far end is shorted to ground with a via. The substrate is 0.25 mm thick and the relative permittivity is 4.9. We assume lossless metal and dielectric. The line width is 1 mm and the line length is 5 mm. An equivalent circuit at 3 GHz, for example, is a parallel LC, as shown in Figure 4.8.

We can see the inductance by imagining the magnetic field wrapping around the line. We can see the capacitance by imagining the electric field going from the line to ground. Thus, we have both magnetic and electric energy existing together. This must also be the case for our lumped model. This is why we have both a capacitor and an inductor.

First, we write an equation for the admittance Y of our lumped model:

$$Y = j\omega C - \frac{j}{\omega L} = j\left(\omega C - \frac{1}{\omega L}\right)$$

where, $\omega = 2\pi f$ is the angular frequency.

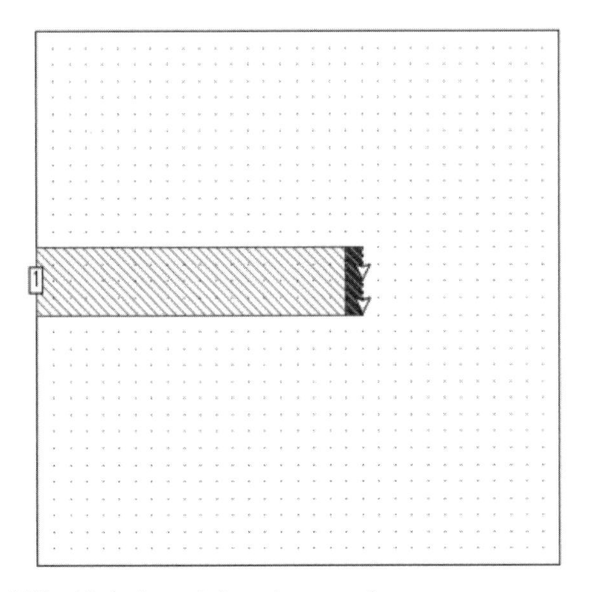

Figure 4.7 An MSL with the far end shorted to ground.

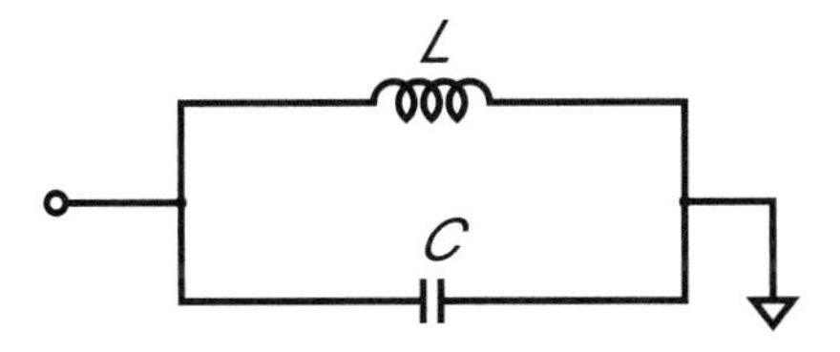

Figure 4.8 Equivalent circuit of an MSL with the far end shorted to ground.

Next, we analyze at 3 GHz using Sonnet Lite. Then, instead of outputting S-parameters, we output Y-parameters. In general, a Y-parameter is $y_0 = g_0 + jb_0$. Here, the subscript refers to the result at 3 GHz and g is called the conductance, while b is called the susceptance. Since we are using a lossless model, we can simplify to $y_0 = jb_0$.

To build a lumped model, we need a second frequency. Let's pick a frequency 10 percent higher, 3.3 GHz. We analyze at 3.3 GHz, and we get the result $y_1 = jb_1$. Simplifying our previous equation for the admittance of our parallel LC, and evaluating at each of the two frequencies, we have

$$b_0 = \omega_0 C - \frac{1}{\omega_0 L}$$

$$b_1 = \omega_1 C - \frac{1}{\omega_1 L}$$

From the Sonnet Lite analysis, we have values for b_0 and b_1. We do not yet have values for *L* and *C*. This is two simultaneous equations with two unknowns. So, we solve for the unknown *L* and *C*. The equations are not linear, so their solution takes a little work. We have done that work and the answer is

$$L = \frac{\dfrac{\omega_0}{\omega_1} - \dfrac{\omega_1}{\omega_0}}{b_0\omega_1 - b_1\omega_0}$$

$$C = \frac{b_0\omega_0 - b_1\omega_0}{\omega_0^{\,2} - \omega_1^{\,2}}$$

If we know admittances of two frequencies that are slightly different, we can obtain the equivalent circuit from the previous expression.

Here we solved with a lossless model to simplify the explanation. More generally, we can derive equations for a more complicated RLGC circuit, Figure 4.9, to obtain *R* and *G* as well. More recent research has extended this technique to a wide variety of RLGC circuits.

By using the automatic SPICE lumped model synthesis in Sonnet Lite, we obtain the parallel LC circuit of Figure 4.10, valid for frequencies around 3 GHz. Remember, for validity, the circuit must be small with respect to wavelength. In this case, the transmission line is just under 0.1 wavelength long, so this simple model should be reasonable.

Figure 4.11 is a model of two parallel lines (a differential pair) using Sonnet Lite. An internal port is used for both ports with the right-hand port shown in the detail. For this example, the two copper lines are 5 mm wide and 1,000 mm long, with 50-mm spacing. The conductors attached directly to each port are lossless.

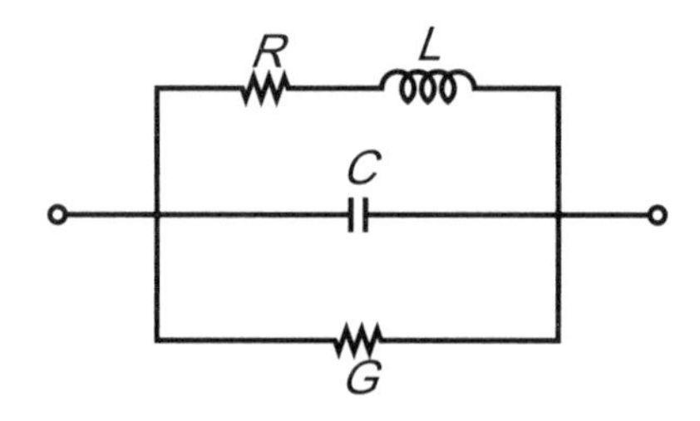

Figure 4.9 A more general RLGC circuit model.

```
* Spice Data
* Limits: C>0.01pF L<100.0nH R<1000.0Ohms K>0.01
     * Analysis frequencies: 3000.0, 3300.0 MHz
.subckt SonData 1
C1 1 0 0.381291pf
L1 1 0 0.970331nh
.ends SonData
```

Figure 4.10 Equivalent circuit netlist and values of each element.

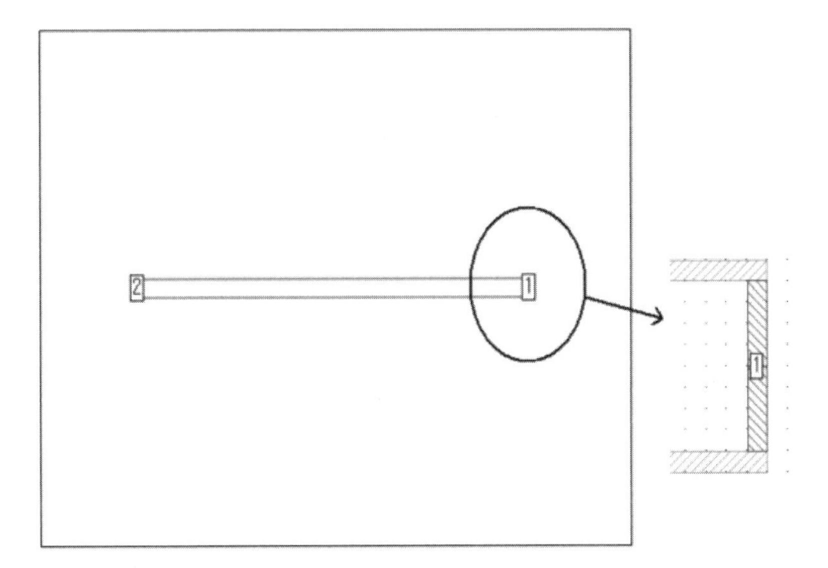

Figure 4.11 Model of two parallel lines in Sonnet Lite.

The subcircuit around 10 MHz generated with this model is shown in Figure 4.12. This is the PI Model synthesized equivalent circuit (PI Model assigned in Analysis > Output File(s)…).

4.4.3 The RLGC Matrix

We mentioned the problem earlier that lumped models are valid for circuits that are small with respect to wavelength. In this section, we introduce a solution to that problem for transmission lines.

To invoke this solution, select Analysis > Output Files… and select N-Coupled Line Model…. In this case an RLGC matrix is output.

Figure 4.13 shows an equivalent circuit based on the RLGC matrix of four coupled microstrip lines. This model shows our LC model of a transmission line with coupling between them modeled by interconnecting inductors and capacitors. When we use an RLGC matrix in SPICE, it actually takes the limit

```
* Spice Data
* Limits: C>0.1pF L<10000.0nH R<1000.0Ohms K>0.01
    * Analysis frequencies: 10.0, 11.0 MHz
.subckt SonData 1 2
C1 1 0 3.605256pf
C2 2 0 3.613033pf
L1 1 3 1501.366nh
RL1 3 2 0.247014
.ends SonData
```

Figure 4.12 Equivalent circuit netlist and values of each element.

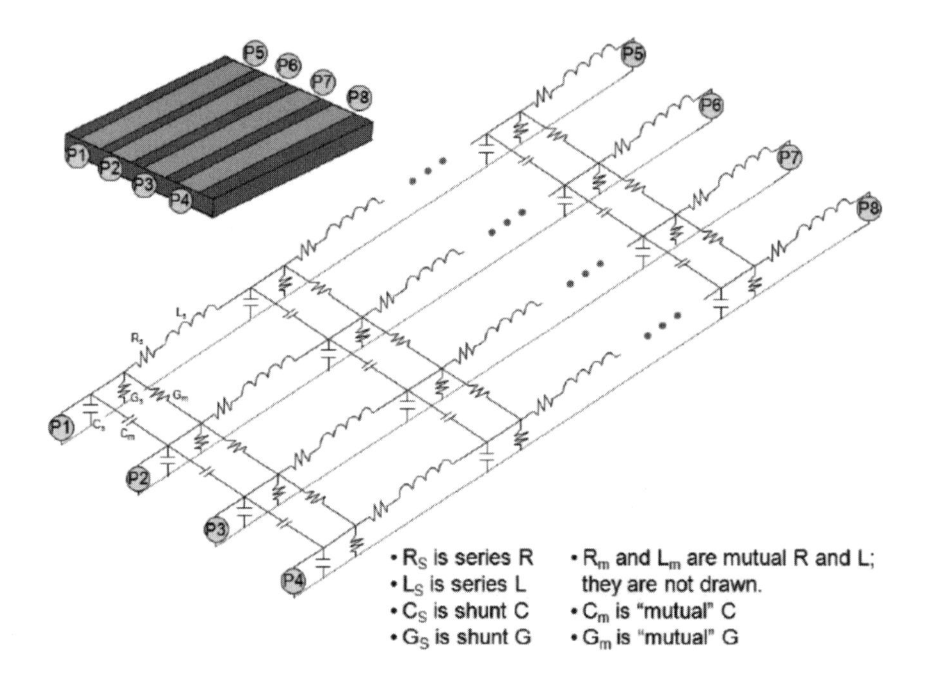

Figure 4.13 Equivalent circuit based on an RLGC matrix of four lines MSL.

of making each section infinitely small and then making an infinite number of them, just like Heaviside did (Chapter 3). With an accurate RLGC matrix, we can now just tell SPICE what the length of the line is and SPICE analyzes the line correctly.

The problem is in getting a good RLGC matrix. For most transmission lines, the values of RLGC change with frequency. If the change is not properly evaluated by the EM analysis, then issues of causality (the line gives an output before it is possible), stability (the model starts oscillating), and passivity (the model shows gain at some frequency) arise. In these situations, the absolute highest EM analysis accuracy is required, and this is where Sonnet is very strong.

4.5 Simulation of Filters

We mentioned that microwave engineers have learned a lot about EM analysis since the ealy 1980s. A lot of that knowledge has been gained analyzing filters.

4.5.1 Analysis of a Bandpass Filter

Now we examine a bandpass filter with six resonators in Figure 4.14. For proper operation, every resonator length and every gap between resonators must be set exactly correctly. To load this filter into Sonnet Lite, select Help > Examples... > Filters. When clicking the COPY EXAMPLE button, you can save a copy of the bpfilter folder and you will have all the files associated with this example. This folder is also on this book's accompanying DVD.

If you try to analyze this filter as it is, it will exceed the memory limit of Sonnet Lite. So we use a method to divide the filter into several pieces. After simulating them separately, we connect them all back together to obtain the result for the entire filter.

4.5.2 Dividing a Circuit

For this example, we will repeatedly divide the circuit into many pieces. Since each piece is EM analyzed separately, electromagnetic coupling between the pieces is not included. So, it is important to divide the circuit in a way that does not remove important coupling.

This filter is symmetric in that the right and left halves are identical. So the first division is to cut the circuit in half, with the left half as shown in Figure 4.15. After analyzing this half in Sonnet Lite, we connect it twice in a Sonnet Netlist Project. We then obtain the result for the entire filter.

However, even half of the filter is too much for the memory limit of Sonnet Lite. For that reason, we divide the model of Figure 4.15 into four pieces and connect them together after simulating them separately.

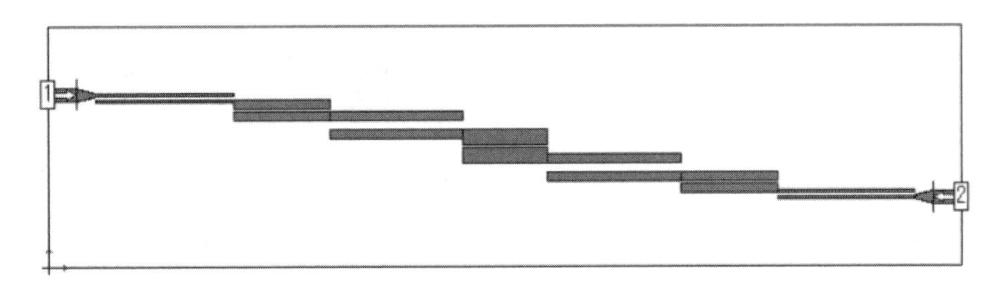

Figure 4.14 A bandpass filter example.

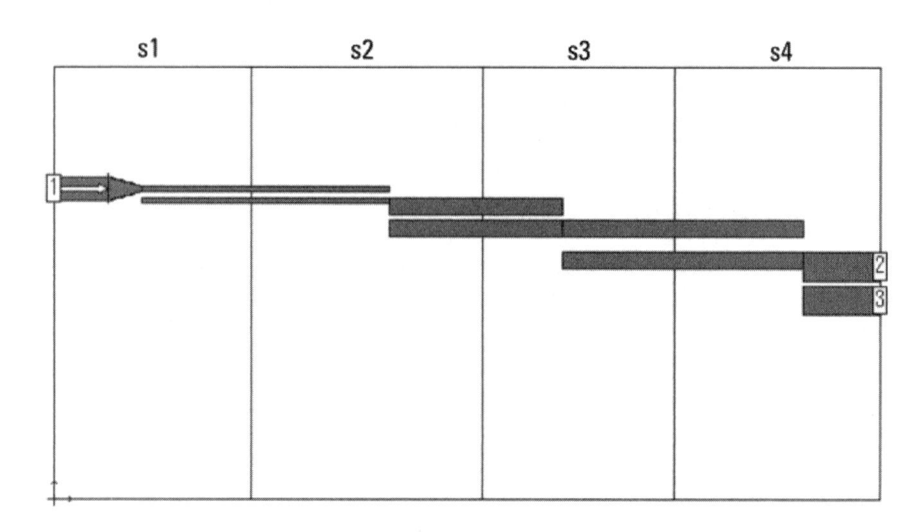

Figure 4.15 The left half of the filter (shown here) is identical to the right half.

By the way, this circuit is symmetric about the *y*-axis (vertical axis, as displayed). It is not symmetric about the *x*-axis (horizontal axis, as displayed). In addition, the ports are not in the center, on the horizontal symmetry axis. For this reason, we cannot check Symmetry in Circuit > Box.

Division of the circuit should be done where the electromagnetic coupling is not an important factor. For example, do not divide a circuit along the gap that separates two resonators, as the resonators will no longer couple! Highly coupled circuitry that cannot be separated should be kept unsplit inside each section. By doing this, we can obtain high-accuracy electromagnetic results much faster.

4.5.3 Creation of Geometry File

The process of dividing a circuit up so many times, and then making a netlist to connect it all back together again, sounds very tedious. Fortunately, most of this procedure has been automated. For details on how to actually do the division automatically (after you have manually specified where to divide it), see Help > Manuals. Select the User's Guide and read the two chapters on "Circuit Subdivision." The following description shows the highlights of the process. For example, we divide the half filter into four sections, as in Figure 4.15, by selecting Tools > Add Subdivider.

After specifying the dividing lines, we have to divide the circuit into separate EM analysis files and put ports on each section so that we can connect them back together again with a netlist. Figures 4.16(a–d) are the sections of the half filter of Figure 4.15. These files are in the bpfilter folder that we saved

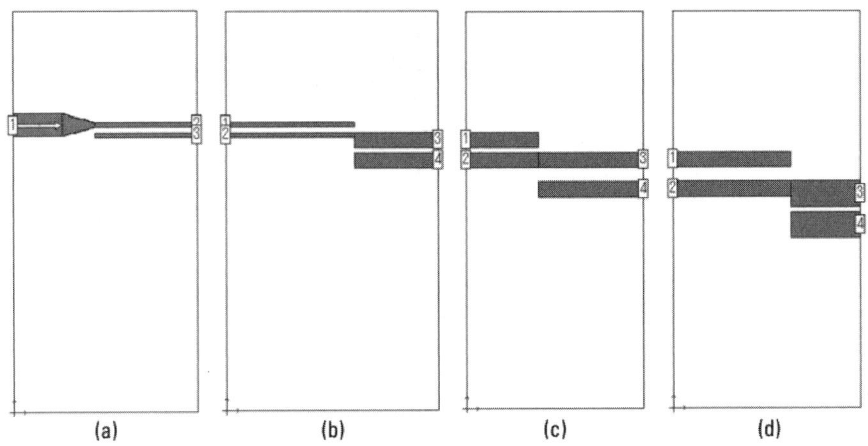

Figure 4.16 (a) Divided circuit. File: halfnet_s1.son. (b) Divided circuit. File: halfnet_s2.son. (c) Divided circuit. File: halfnet_s3.son. (d) Divided circuit. File: halfnet_s4.son.

at the beginning of this example. As described in detail in the Sonnet manual, this process is entirely automated. The user simply selects Tools > Subdivide Circuit....

Normally, the cell size for an entire EM analysis project must be the same everywhere. However, after we have divided the project into separate sections, as we have done here, it is no problem to manually specify different cell sizes for each section, as long as the physical shape and size of the circuit are maintained. If there are sections of the project with very fine geometry, you can select a small cell size. Other sections, with large structures, can be analyzed with a larger cell size. For this filter, we still keep the same cell size everywhere.

4.5.4 Simulating the Divided Circuit

In this example, we simulate each section from 6 GHz to 12 GHz with adaptive Band Sweep (ABS). After simulating each file, we create a netlist project that connects the results together, giving us the response of the entire filter.

Figure 4.17 is the automatically created netlist project to create the half filter of Figure 4.15 from the separate EM analyses of the four sections. We next want to connect two of these together to obtain the result for the entire filter. We must manually add to the netlist of Figure 4.15 to do this. We shall now learn how to do this.

The modified netlist, shown in Figure 4.18, is executed from the first line in the following order:

1. First four lines: import the data from four filter sections (this was automatically created) and connect them together.

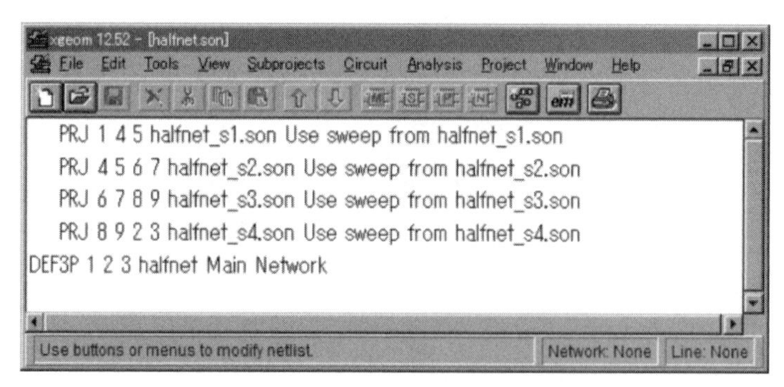

Figure 4.17 Netlist project to connect the four sections into one half of the symmetric filter. File: halfnet.son.

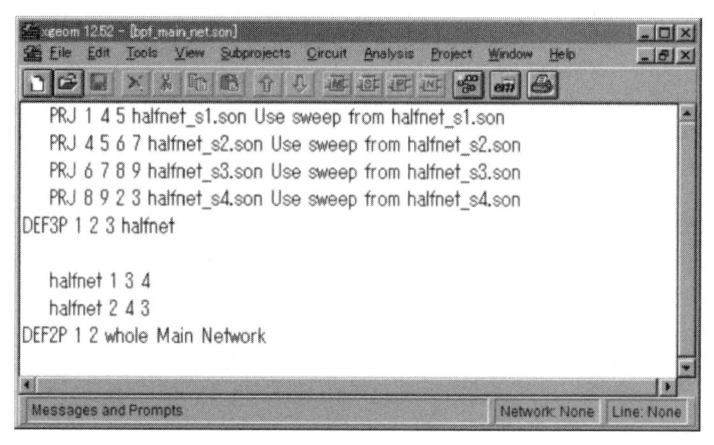

Figure 4.18 Netlist project to connect all the sections back together, giving the response of the entire filter. File: bpf_main_net.son.

2. Fifth line: define a three-port network, halfnet, which represents the half filter.
3. Last three lines: define the result, which consists of two halfnets connected together to form the response of the entire filter.
4. Output from 6.0 GHz to 12.0 GHz with 0.05 GHz step using Linear Frequency Sweep. This is set in Analysis > Setup.

4.5.5 Key Point on Where to Set Dividing Lines

The netlist project connects S-parameter data that has already been calculated, and this is very fast. As in this example, we can convert a large problem into several smaller problems that can be done using Sonnet Lite. Even if a problem can fit in available memory, this "divide and conquer" method can save simulation time and reduce memory, as long as you do not mind a little extra setup time. However, deciding where to divide a circuit requires some care, as shown in Figures 4.19 and 4.20. Do not put dividing lines across regions of important coupling.

4.5.6 Evaluation of the Result of Simulation

Figure 4.21 is the result of the bandpass filter simulation, showing the plots of S_{11} (reflection coefficient) and S_{21} (transmission coefficient). From the result of S_{21}, the input signal is transferred to port 2 output in the range of 8–10 GHz. This band is called the pass band.

Now we examine the structure of bandpass filter. The idea of this bandpass filter is that each resonator (there are six of them) is about one half wavelength long at the center frequency of the filter. When the input is at a frequency that is near resonance, they all resonate and couple energy from one to the next. When the frequency is far enough below or above resonance, they do not resonate and no energy makes it through the filter.

Figure 4.22 is the current distribution of the filter at 6 GHz. It shows strong current flowing only on the input port line, upper left. However, the input signal completely reflects and returns to port 1. Next, Figure 4.23 is the result at 9 GHz. Now the input port line is resonant, and it creates very strong fields that couple to the adjacent resonator. That resonator is also resonant, and it couples to the next one, and so on. We see that all that resonators are electromagnetically coupled and strong current flows on to the output port, lower right.

When we closely examine the region near the input port in Figure 4.22, we see that it electromagnetically couples to the nearby portion of the first resonator and strong current flows on the edge. But this is well below the resonant

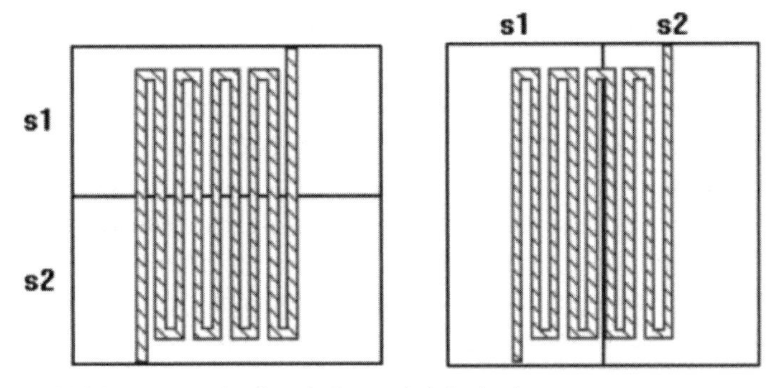

Figure 4.19 Dividing a meander line. Left: good; right: bad.

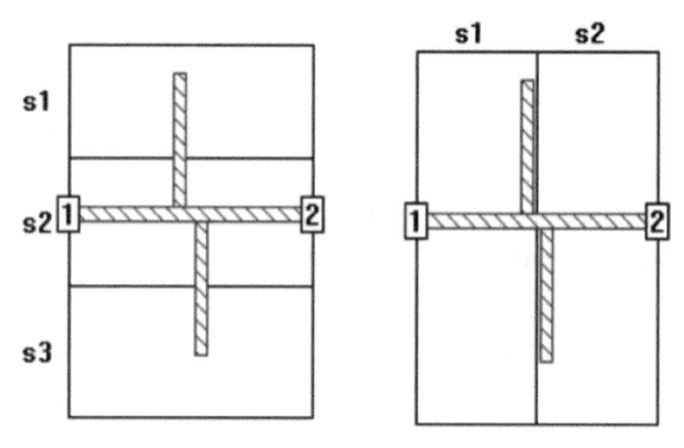

Figure 4.20 Dividing a stub line. Left: good; right: bad.

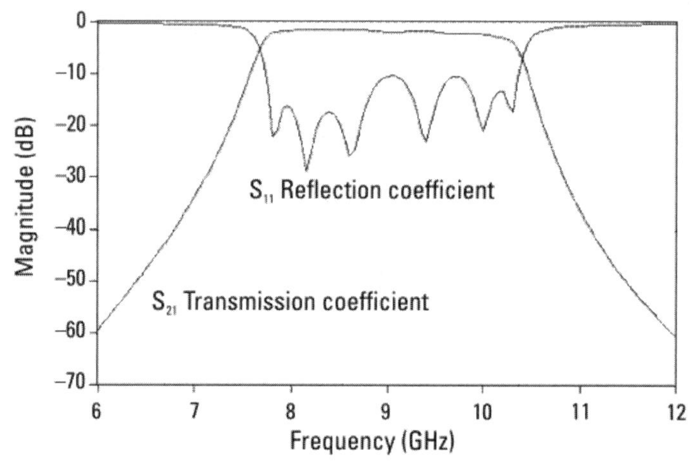

Figure 4.21 S_{11} and S_{21} result of the bandpass filter simulation.

frequency of the first resonator, so it does not electromagnetically couple to the next resonator.

As mentioned previously, the resonators resonate when they are about one half wavelength long. The line lengths are 6.1–6.4 mm. However, one-half wavelength of 9 GHz is 17 mm in free space. The half-wave resonant length is reduced because the relative permittivity of the alumina substrate used in this filter is 9.9. The dielectric constant of the substrate shortens the wavelength.

At the resonant frequency, strong electromagnetic energy accumulates around the resonating object and its fields extend over a large area. Thus, it electromagnetically couples to any nearby lines, and the signal transfers just like a bucket brigade to the output side. The length of each resonator is slightly different. In addition, the gaps between resonators are also slightly different.

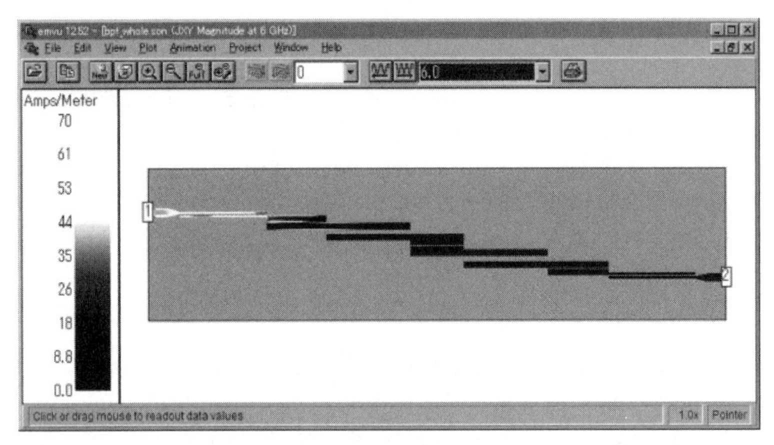

Figure 4.22 Current distribution of the bandpass filter at 6 GHz.

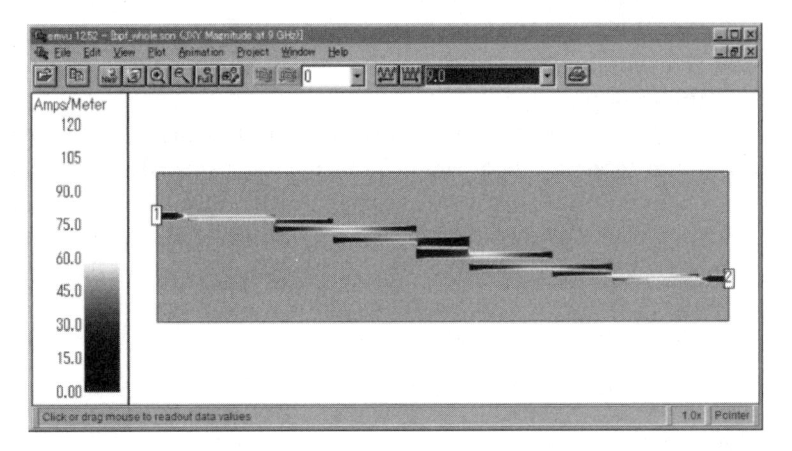

Figure 4.23 Current distribution of the bandpass filter at 9 GHz.

These are all carefully adjusted during design to obtain the desired bandwidth and center frequency, among other things.

4.6 Simulation of a T-Type Attenuator

In a netlist project, it is possible to insert lumped elements, *R* (resistor), *L* (coil), and *C* (capacitor) into an EM analysis. To show how to use lumped constant elements, we simulate a T-type attenuator. Here we insert three resistors in a netlist project as lumped elements. Different from the technique described next, this task can also be accomplished using the formal Sonnet "Component" concept and the Sonnet "Co-calibrated Port" concept, as described in the Sonnet manual. These two approaches should be used when the highest accuracy is required.

Unfortunately, the two methods described next require more than four ports, so Sonnet Lite does not permit simulation. If you do not have access to Sonnet Professional, you can substitute resistive metal for two of the resistors and use ports for the third resistor. For the second method we discuss, you can use ports for up to two of the resistors.

4.6.1 Construction of a Circuit

Figure 4.24 is the layout of a circuit that includes lumped resistors. After simulating a project file consisting of the surrounding transmission lines, we insert three resistors. The netlist file is used to calculate the two-port S-parameters of the entire circuit.

To load this circuit into Sonnet, select Help > Examples… > Circuit Theory in Sonnet. When clicking the COPY EXAMPLE button, you can select where you want the folder, named "att," to be placed. This folder contains all the files in this example.

Figure 4.25 is a model to simulate the circuit of Figure 4.24. It has three pairs of ports to insert lumped parameter elements (resistors). Ports 3 and 4, ports 5 and 6, and Ports 7 and 8 are all represented with small white triangles. These are auto-grounded ports that automatically make a port whose voltage is between the line and ground. When lumped elements are inserted in a netlist project, they are connected between the specified auto-grounded ports.

4.6.2 Creation of a Netlist File

Figure 4.26 is the netlist project used in this example, which performs its tasks as follows:

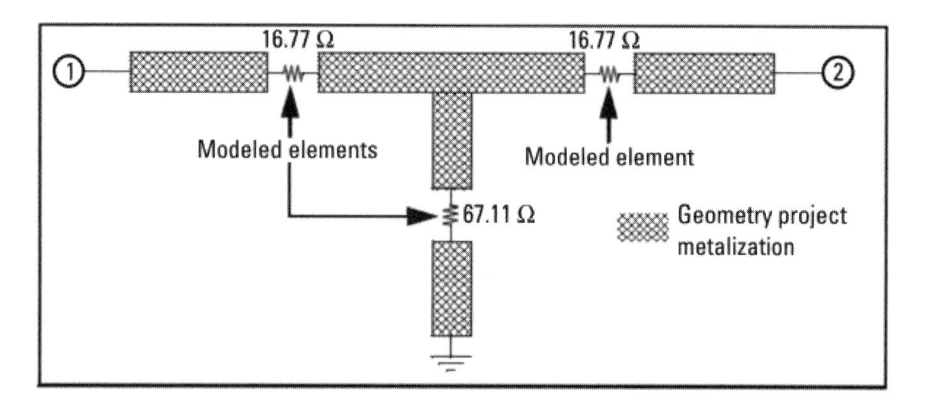

Figure 4.24 T-type attenuator with embedded resistors.

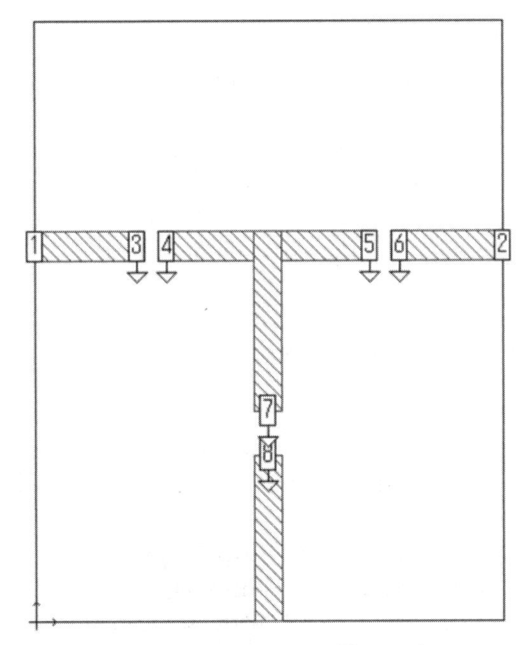

Figure 4.25 Model to simulate the T-type attenuator. File: att_lgeo.son.

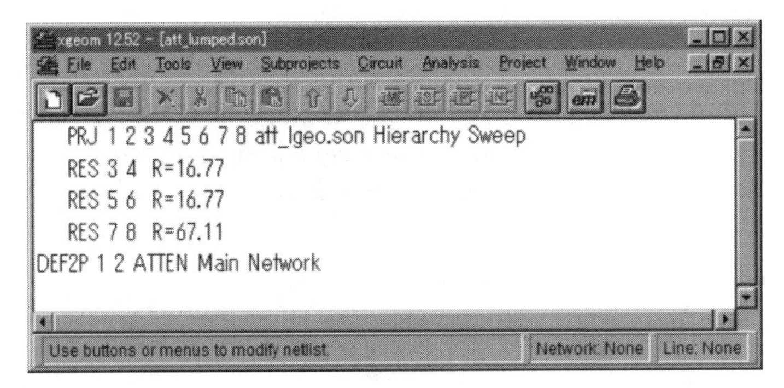

Figure 4.26 Netlist project used for the attenuator in this example.

1. Run the electromagnetic field simulation for att_lgeo.son (if the simulation is already done, go to the next step).
2. Insert a 16.77Ω resistance between nodes 3 and 4 (assign the resistance value directly in the netlist project).
3. Insert a 16.77Ω resistance between nodes 5 and 6.
4. Insert a 67.11Ω resistance between nodes 7 and 8.
5. Analyze and output the S-parameters of the attenuator.

4.6.3 Method Using Internal Ports

Figure 4.27 shows internal ports at the position of the resistors. Figure 4.28 is the netlist for this method. This case is a little more sophisticated. The three resistances are described not as fixed values but as variables like Z3, Z4, and Z5. By selecting Circuit > Add Variable... we can define variables. This is useful when using the same value at several locations, or if we want to automatically sweep a value over a range and then plot all the results.

4.7 Simulation of Meta-Materials

What are meta-materials?

Recently, research and development of meta-materials is drawing a lot of interest. Practical applications in transmission lines and antennas are being developed.

Meta-materials are media that artificially produce special physical phenomenon. These are generated by aligning a regular internal structure of metals, dielectrics, and magnetic materials with an incident electromagnetic field. These are referred to as artificial media because they do not exist in the natural world.

4.7.1 What Is a Left-Handed System?

Figure 4.29 shows a thumb, index finger, and middle finger of a right hand. They represent the electric field *E*, the magnetic field *H*, the group velocity (velocity of energy transfer) v_g, and the phase velocity (velocity of the move-

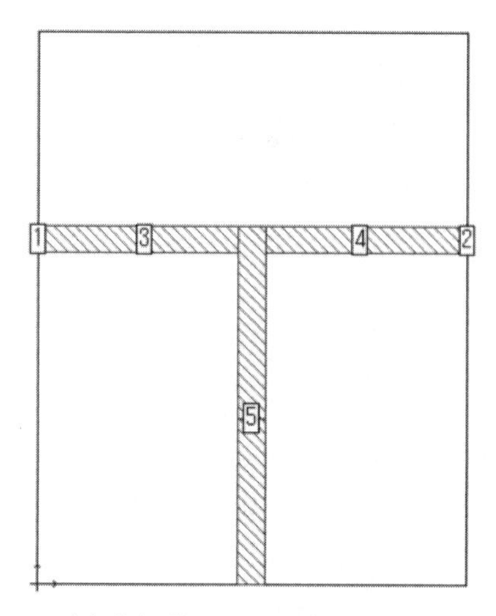

Figure 4.27 Internal port model of the T-type attenuator. File: att_lgeo2.son.

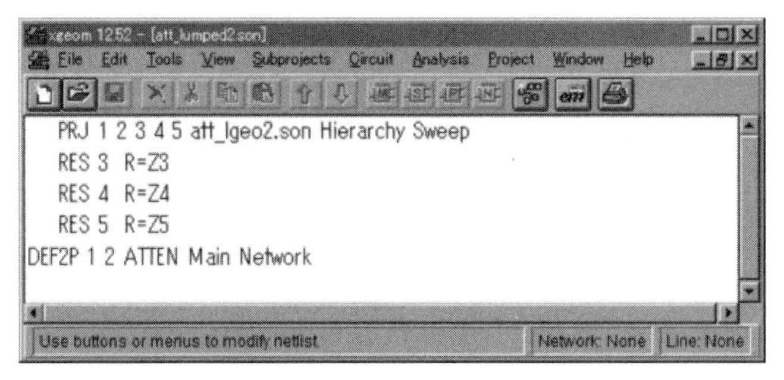

Figure 4.28 Netlist project for the internal port model of the T-type attenuator. File: att_lumped2.son.

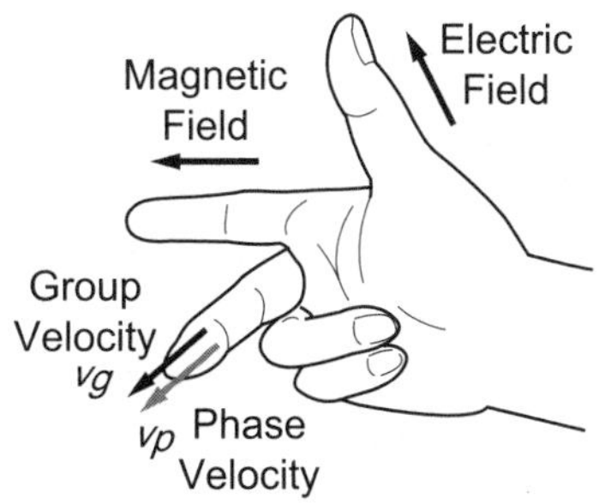

Figure 4.29 Right-handed system.

ment of wave peak or valley) v_p of an electromagnetic wave propagating through a medium. The three fingers are orthogonal to each other. This is similar to Fleming's right-hand rule, which tells us which direction current flows in a wire when it is moved through a magnetic field.

An electromagnetic wave traveling along two parallel lines has the relations in Figure 4.29 as examined in Chapter 1. It is called a right-handed system. Electromagnetic waves traveling in the natural world are all right-handed waves.

In contrast, Figure 4.30 shows a left-handed system, which indicates that the phase velocity and the group velocity are in opposite directions. This also implies a negative index of refraction. It is this singular phenomenon that appears with electromagnetic waves traveling in an artificial medium that is the goal of meta-materials.

4.7.2 Realization of Meta-Materials

Figure 4.31 shows a meta-material that consists of split rings combined with a ring-shaped metal and a thin line. If it is illuminated by microwave energy at a particular frequency, a negative refractive index is seen.

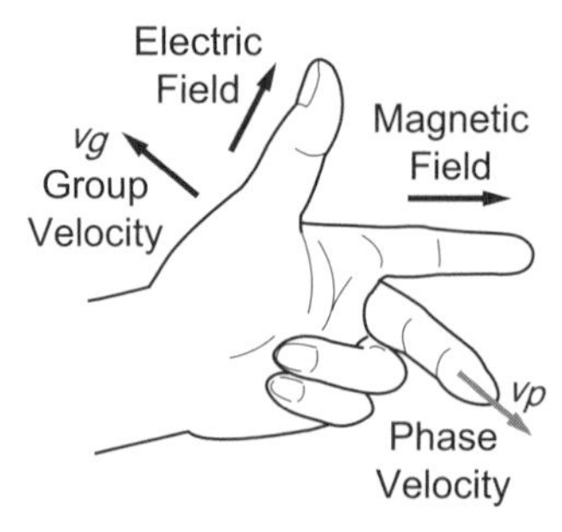

Figure 4.30 Left-handed system.

Figure 4.31 Meta-material that consists of split rings with ring-shaped metal and a thin line (from http://www.sonnetsoftware.com/).

When an electromagnetic wave is present, current flows in metallic split rings, and they resonate, provided the correct frequency is incident. The thin metallic lines also resonate, and a special band might appear where the equivalent magnetic permeability and permittivity of the meta-material both appear to be negative, again provided the right frequency is used to excite the material.

Because a meta-material can have a negative refractive index when electromagnetic waves pass through, they can be focused to a point, even though the meta-material is completely flat.

4.7.3 Fields in a Left-Handed System

The circuit shown in Figure 4.32 looks like the lumped model of a transmission line (Section 3.5 in Chapter 3) that Heaviside devised. However, the C (now in series) and L (now in shunt) have swapped positions. Series C indicates negative magnetic permeability, and shunt L indicates negative permittivity.

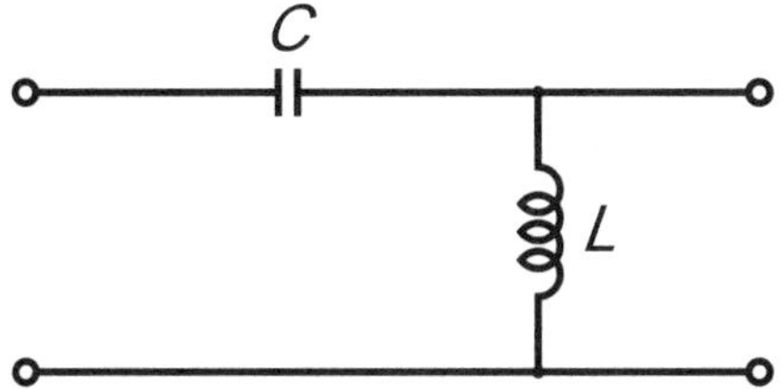

Figure 4.32 Equivalent circuit of a transmission line in a left-handed system.

When trying to create an actual left-handed line, it is difficult to realize the circuit in Figure 4.32. One idea is to add series C_0 and shunt L_0 to the conventional series L and shunt C like the left half of Figure 4.33. Figure 4.34 illustrates the structure of a left-handed system designed for microwave frequencies. It places the load capacitance C_0 and the shorted stub L_0 on an MSL. And Figure 4.35 shows an example, modeled using Sonnet, with a C_0 of 2 pF. The length of the stub is 5 mm; the line width is 1 mm; and L, the repeated dimension, is 6 mm.

Figure 4.36 is the result of simulating the model of Figure 4.35 in Sonnet. The substrate size is 87.5 mm × 32 mm, the relative permittivity of the substrate is 2.6, and tanδ is 0.01. Because the research paper does not specify

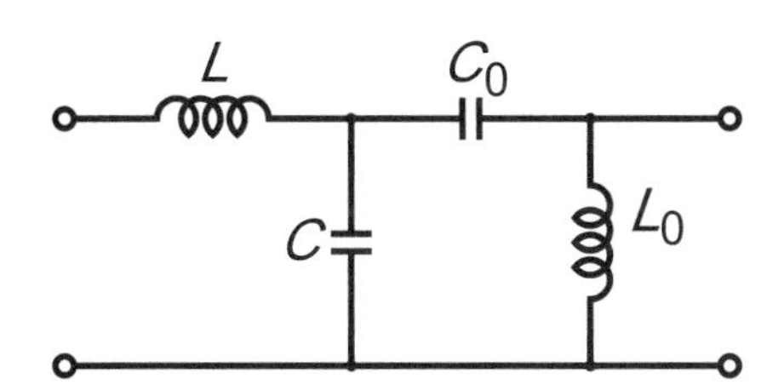

Figure 4.33 Realization of a left-handed system.

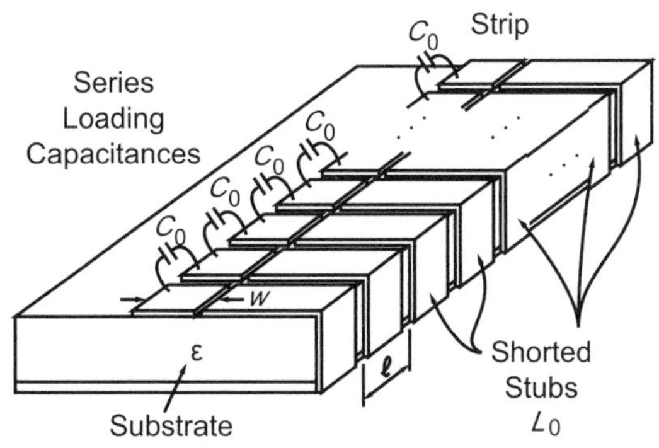

Figure 4.34 A left-handed transmission line designed for the microwave band.

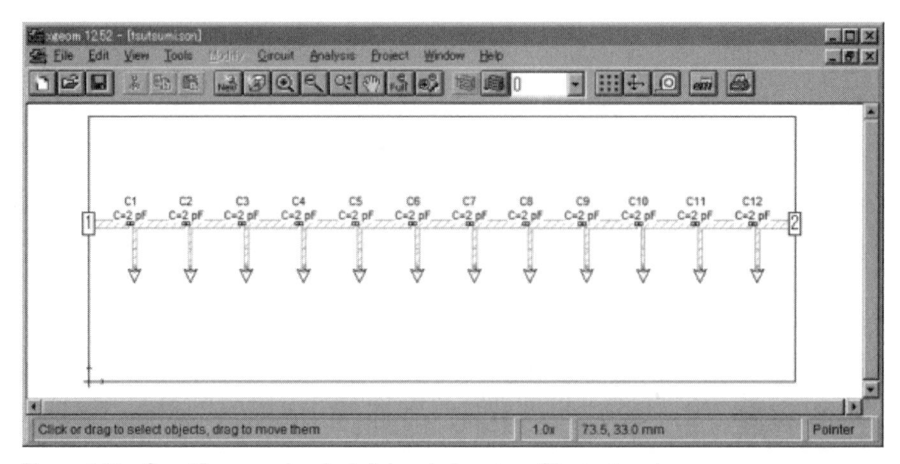

Figure 4.35 Specific example of a left-handed system. File: tsutsumi.son.

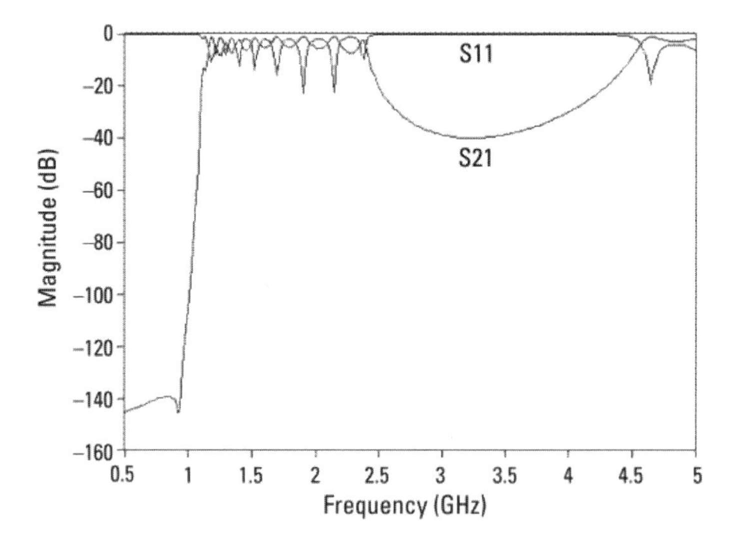

Figure 4.36 Result of simulation by Sonnet.

the substrate thickness, we set it to 400 μm and the result is almost the same as what was published (Tsutsumi Makoto, "Negative Refractive Index Transmission Media and Its Applications to Microwave Circuits," *IEICE Transactions*, June 2005, in Japanese).

We can see that the left-handed meta-material (LHM) nature emerges in the range 1–2.5 GHz. Also, the right-handed meta-material (RHM) nature emerges in the region above 4.5 GHz. The band between them is cut off, and minimal transmission occurs.

One of the characteristics of this left-handed system is the wavelength shortening effect of the meta-material, which is larger than expected. Research to apply this to miniaturizing antennas is active.

Figure 4.37 is a plot showing the electric field intensity 3m away from the substrate. We see peak radiation for the LHM (left side of the plot) and of the RHM (right side of the plot). The RHM level is higher than the LHM level. This transmission line is not intended to be an antenna; however, by modifying the design, we might design miniaturized antennas.

4.8 Confirming This Chapter by Simulation

High-frequency circuits work by means of the transmission of electromagnetic energy over electric and magnetic fields. The electric field corresponds to capacitance and charges. The magnetic field corresponds to inductance and current. If we think of the physical meanings of Cs and Ls of the lumped equivalent circuit, we can understand high-frequency circuits more clearly.

4.8.1 SPICE Subcircuit of a Right-Angle MSL Bend

An MSL with a right-angle bend, shown in Figure 4.38, was simulated in the exercise at the end of Chapter 3. Now we will output a SPICE subcircuit for this model.

First select Analysis > Output Files…, as in Figure 4.39. Next click the PI Model…, as in Figure 4.40. We see that PSpice (default) is selected as default,

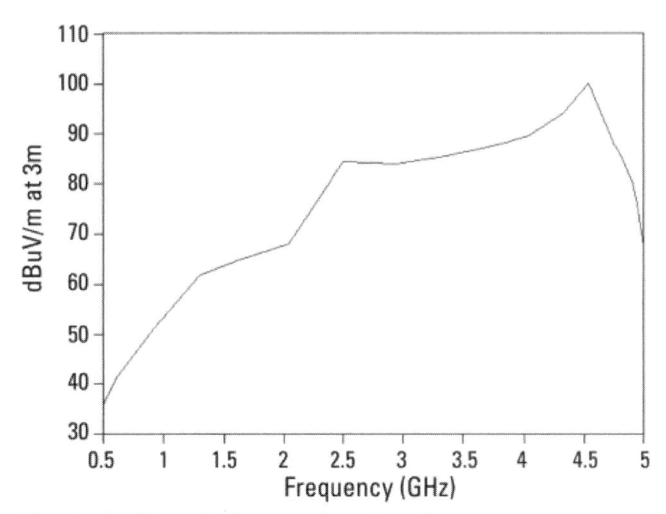

Figure 4.37 Electric field intensity 3m away from the substrate.

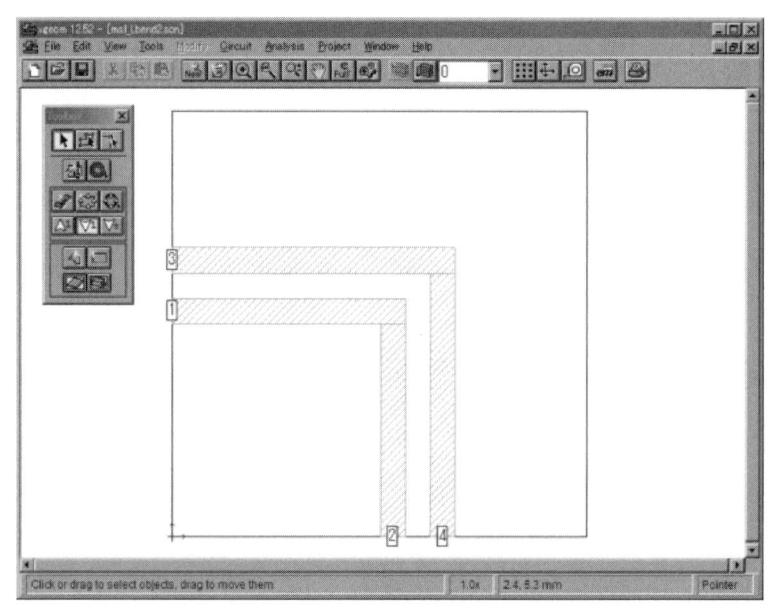

Figure 4.38 Sonnet model of an MSL with a right-angle bend. File: msl_Lbend2.son.

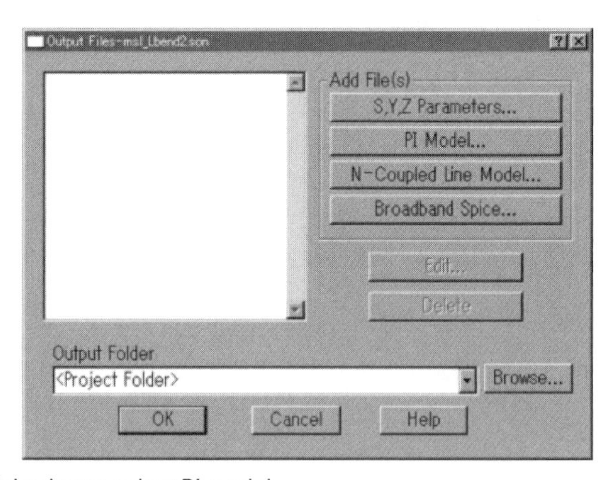

Figure 4.39 Dialog box to select PI-model.

upper left. If desired, we can also select the Spectre format of Cadence Design Systems.

Data Type set to De-embedded means the ports are fully calibrated. Even when we do not have an internal reference plane set, as in the exercise at the end of Chapter 3, De-embedded is selected to remove the electromagnetic fringing fields around the port on the edge of substrate. (Ports on the edge are called Boxwall ports in Sonnet.)

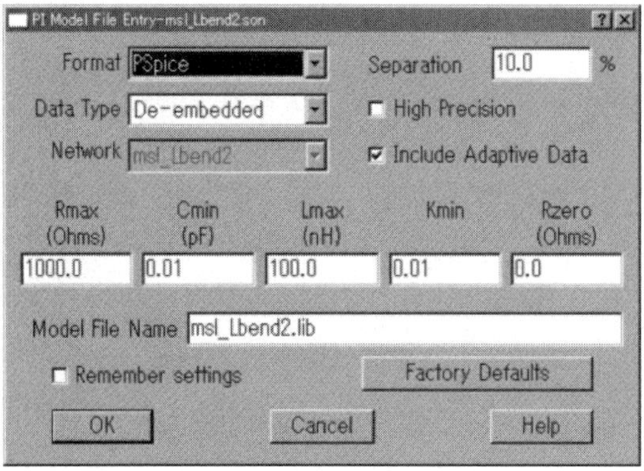

Figure 4.40 PSpice (default) is selected.

As for Rmax and Cmin in middle row, we set the range of allowed values for elements that we expect will have no effect on the equivalent circuit response. In other words, if a capacitance is too small, leave it out. Here we keep the default values as shown in Figure 4.40.

In Figure 4.41, we set the two frequencies needed to generate SPICE subcircuits. In Analysis > Setup, we can set various frequency ranges under Analysis Control. Here we select Linear Frequency Sweep and assign 1 GHz and 1.1 GHz.

Figure 4.42 is the PSpice subcircuit output. First, four Cs model the capacitance between each port and ground. This corresponds to the charges distributed between the line and the ground plane.

There are two Ls. As for L_L1, R_RL1 of 0.14Ω is connected to the 1.58 nH L in series. L_L1 corresponds to the magnetic field generated around the

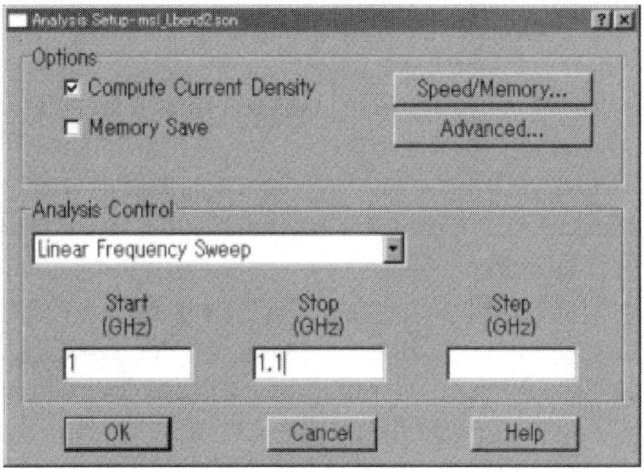

Figure 4.41 Two frequencies are set to generate SPICE subcircuits.

```
*  Spice Data
* Limits: C>0.01pF L<100.0nH R<1000.0Ohms K>0.01
    *  Analysis frequencies: 1000.0, 1100.0 MHz
.subckt msl_Lbend2_0 1 2 3 4 GND
C_C1 1 GND 0.324801pf
C_C2 2 GND 0.324801pf
C_C3 3 GND 0.402241pf
C_C4 4 GND 0.402241pf
L_L1 1 5 1.578284nh
R_RL1 5 2 0.141044
L_L2 3 6 1.940506nh
R_RL2 6 4 0.172347|
Kn_K1 L_L1 L_L2 0.06951
.ends msl_Lbend2_0
```

Figure 4.42 PSpice subcircuit output. File: msl_Lbend2.lib.

line by the current flowing on the line through the right angle bend. R_RL1 corresponds to the resistance of the copper line.

Finally, Kn_K1 is the coupling coefficient that indicates the degree of electromagnetic coupling between two lines. Two perfectly coupled inductors would have a coefficient near 1. However, as for closely located lines, the value is small, like 0.07. Small values mean less crosstalk.

4.8.2 Create a SPICE Subcircuit as a Symbol

In order to use the PSpice subcircuit, shown in Figure 4.42, in a schematic, we must create a schematic "symbol." For this example, we use OrCAD PSpice Lite Edition 9.2.

The model name msl_Lbend2_0, after .subckt in Figure 4.42, is the name generated automatically by Sonnet. It is useful to change it to something like LBEND2. We must change it in two places, in the .subckt line and in the .ends line. Create a folder in the install directory of OrCAD Lite named, for example User_Lib in C:\Program Files\OrcadLite\Capture\Library\PSpice and save it as LBEND2.lib.

Next, we create an original symbol. First, after selecting File > New > Library in the OrCAD Capture menu bar, the library manager appears. Select library1.olb as shown in Figure 4.43.

Next, select Design > New Part… in the Capture menu bar and the New Part Properties dialog box appears, as shown in Figure 4.44. Input LBEND2 in the Name section as the part's name.

Input T in the Part Reference Prefix: in Figure 4.44, which indicates a transmission line. When clicking the Attach Implement button in the right middle, the Attach Implementation dialog box is displayed, as shown in Figure 4.45. Select PSpice Model as the Implementation Type and input LBEND2 in Implementation name section.

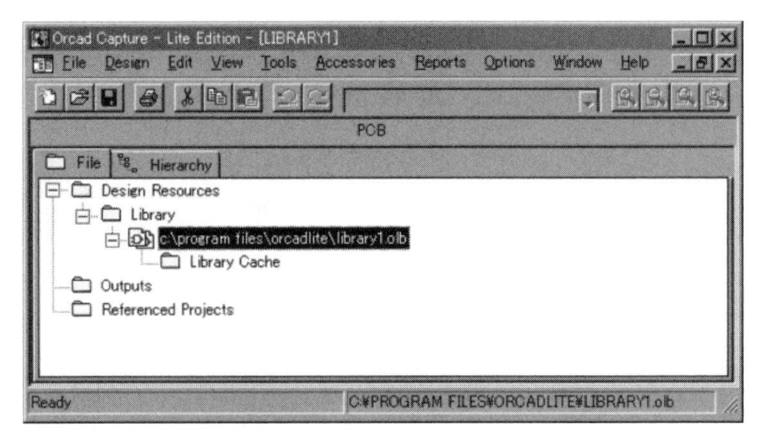

Figure 4.43 Select library1.olb.

New Part Properties
Name: LBEND2
Part Reference Prefix: T
PCB Footprint:
Create Convert View
Multiple-Part Package
Parts per Pkg: 1
Package Type
Homogeneous
Heterogeneous
Part Numbering
Alphabetic
Numeric
OK
Cancel
Part Aliases...
Attach Implementation...
Help
Pin Number Visible
C:\PROGRAM FILES\ORCADLITE\LIBRARY1.OLB

Figure 4.44 New Part Properties dialog box.

After clicking OK here, and clicking OK again in the New Part Properties dialog box, the parts editor, shown in Figure 4.46, appears.

Now we draw a simple image to represent the right angle MSL bend, as seen in Figure 4.47. Next, select Place > Pin and input the name and number of pins and then select Line as Shape: to allocate pins at five locations, as shown in Figure 4.48.

After double clicking the left mouse button on a blank spot where no parts are drawn, click the New button, shown in Figure 4.48, and the dialog box of Figure 4.49 appears.

Next, in the New Property dialog box, shown in Figure 4.49, input PSpiceTemplate in the Name section and also input X^@REFDES %In1 % Out1 %In2 %Out2 %SonnetGND @MODEL in the Value section. X indicates that this part is a subcircuit. As for the name following the %, we input the name of

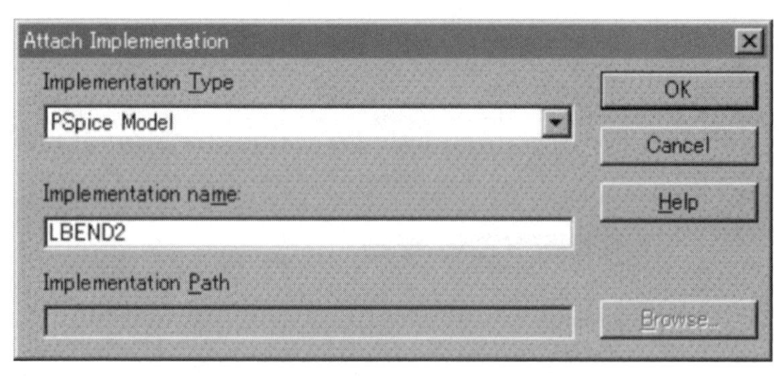

Figure 4.45 Attach Implementation dialog box.

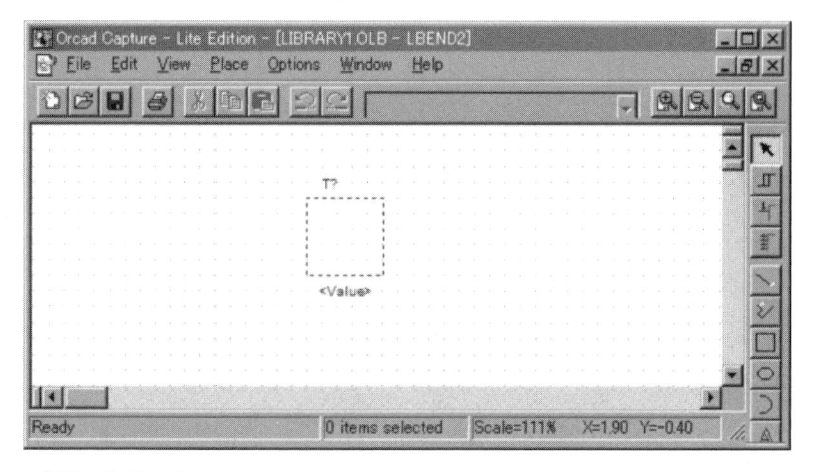

Figure 4.46 Parts editor.

a pin that was named when selecting Place > Pin. What is important here is that we should describe pins in the same order as the numbers followed by .subckt LBEND2 in Figure 4.42.

The default library name is library1.olb in the library manager screen, shown in Figure 4.43. So select this and then select File > Save As… and change the name to LBEND2.OLB. Save it in the User_Lib folder created earlier.

4.8.3 Using Parts

After inputting a project name in the dialog box displayed after selecting File > New > Project, assign a folder to save the project. When selecting Analog or Mixed-A/D in the Create a New Project Using section, click OK. The Create PSpice Project dialog box appears. Select Create a blank project.

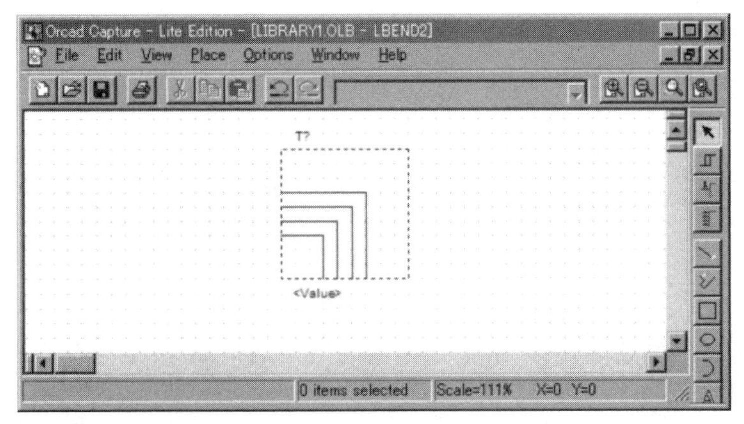

Figure 4.47 Draw the part's symbol.

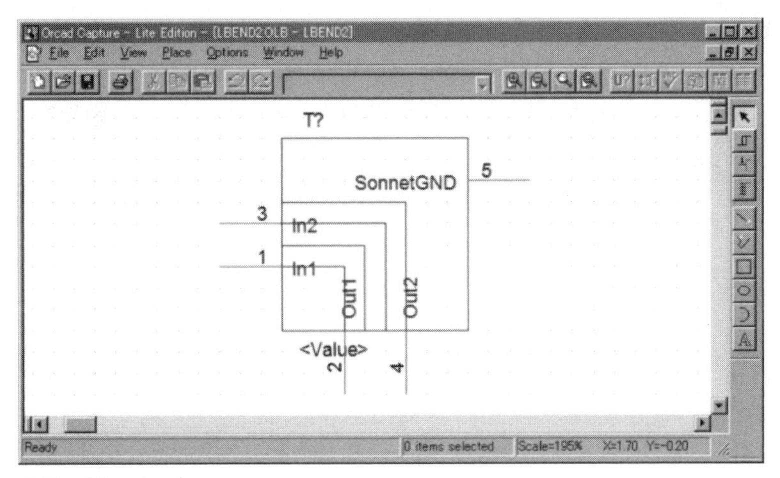

Figure 4.48 Allocate pins.

Clicking OK displays a window to draw a circuit. Figure 4.50 shows a circuit drawn in this window. This is a simple circuit to verify the behavior of a model. We input a signal on the MSL right angle bend and observe the output.

Select Place > Part… and in the Place Part dialog box, select LBEND2 from the library of registered models and then click OK. Then we can place this model in the schematic.

Select ANALOG from Libraries:, select resistor R and draw four resistors. As for the power supply, select SOURCE from Libraries: and select VRWL from the Part List:. Because this is a pulsed signal source, we set only one pulse whose rise time and fall time is 0.5n (nS), and amplitude is 5V, as shown in Fig-

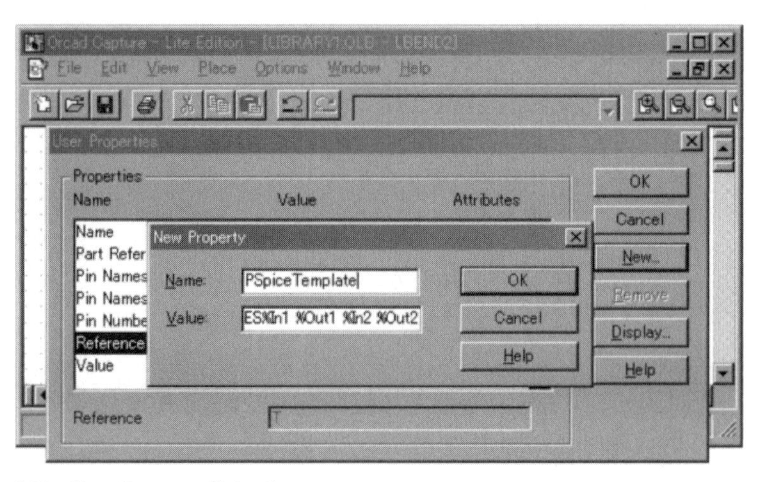

Figure 4.49 New Property dialog box.

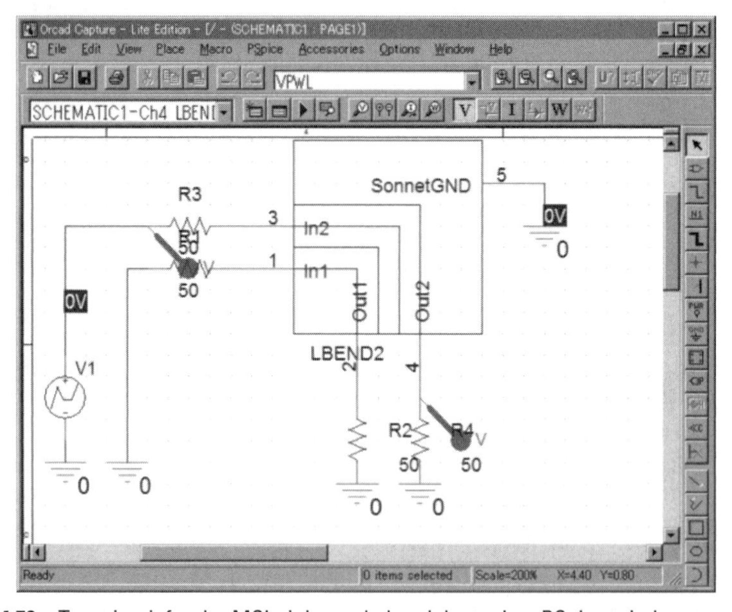

Figure 4.50 Test circuit for the MSL right angle bend drawn in a PSpice window.

ure 4.51. In addition, we set two voltage markers by selecting PSpice > Markers > Voltage Level.

4.8.4 Register the Subcircuit LBEND2.lib in the Circuit Schematic

Select PSpice > Edit Simulation Profile and then click the Library tab in the Simulation Settings dialog box (Figure 4.52).

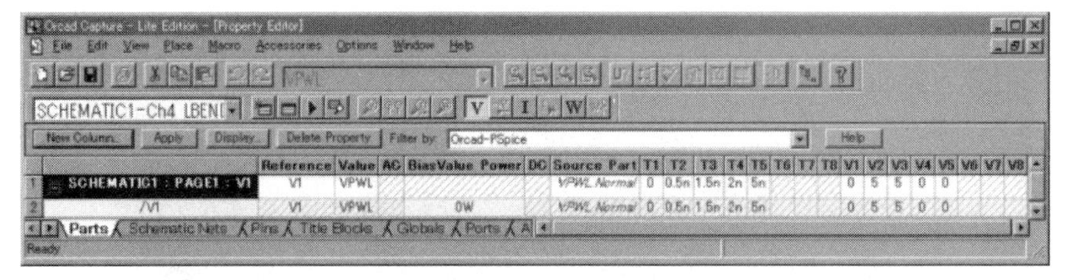

Figure 4.51 Specify a pulsed signal source.

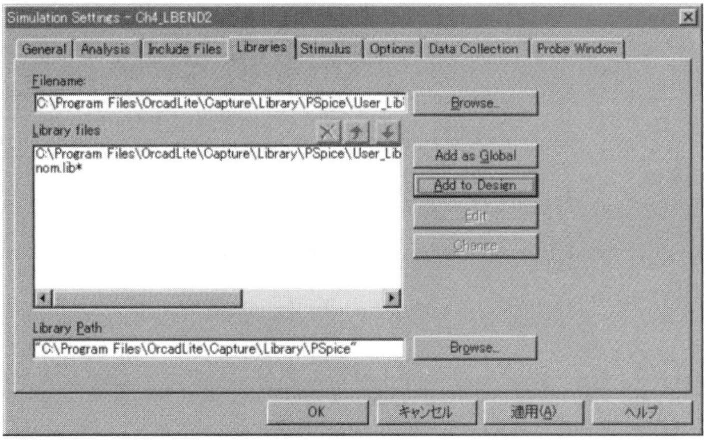

Figure 4.52 Simulation Settings dialog box.

Next, click the Browse… button and select the LBEND2.lib that was saved before in the User_Lib folder. Next, click the Add to Design button in the right middle and register this model. Now the model file LBEND2.lib is referenced from this circuit. When clicking the Add as Global button, LBEND2.lib can be referred in all circuit schematics.

Now, we will execute a transient response analysis. Select PSpice > Edit Simulation Settings, and then select Analysis type: Time Domain (Transient). Set Run to time: to 10n (nS) on the right side and Start saving data after: to 0. Finally, select PSpice > Run in the menu, and a simulation result like Figure 4.53 is displayed.

After selecting Plot > Axis Settings… to adjust the displayed range of the X and Y axes, detail of the wave shapes can be clearly seen, as shown in Figure 4.54. The wave shape of 5V amplitude is the input signal in port 3 (In2). And the wave shape of 2.5V amplitude is the wave shape of port 4 (Out2). It indicates a slight shift to the right. This is the phase delay of output signal. The corners of the waveform are also slightly rounded. This indicates attenuation of higher frequencies.

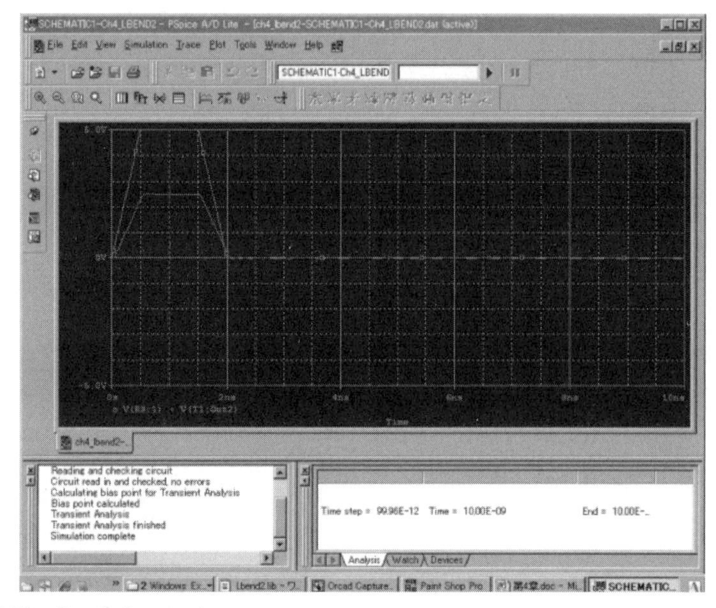

Figure 4.53 Simulation result.

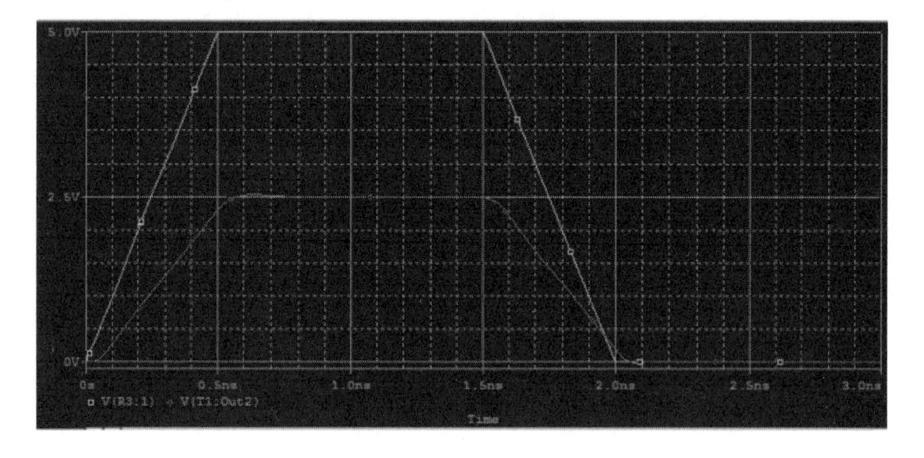

Figure 4.54 Adjusting the displayed range of the X and Y axes.

4.9 Summary

1. Some electromagnetic field simulators have a feature to generate the SPICE subcircuit automatically.

2. Even in a circuit with more than 100 ports, SPICE models can be generated, including all couplings, with the help of electromagnetic field simulators.
3. The analysis time and required memory of large circuits can be significantly reduced by dividing the circuit into sections.
4. Electromagnetic field simulators can include lumped elements in a circuit.
5. Meta-materials (artificial media) can be created by aligning orderly structures periodically.

5

High Frequencies and Undesired Radiation

5.1 Understanding Through Visualization

In Chapter 4, we began to gain a deep understanding of high-frequency circuits by considering physical meanings of the Ls and Cs in lumped models of distributed circuits. One way we did this was to view SPICE subcircuits synthesized by an electromagnetic field simulator. In this chapter, we examine visual information provided by simulation, such as the current distribution, the electromagnetic field distribution, and the radiation pattern. By examining the electromagnetic field as it transitions to space and the electromagnetic field distribution around the substrate, we will learn how to find the origin of undesired radiation.

5.1.1 Necessity of Impedance Matching

Various problems have been seen in the past with double-sided printed circuit boards as CPU clock frequencies became higher. As we learned in Chapter 1, because the wiring pattern of the double-sided printed circuit board was originally thought to be predominantly inductive, the characteristic impedance was not controlled. We also learned in Chapter 3 that if the source and load are not matched to the impedance of the line, signal reflection occurs, and a portion of the transmitted signal is sent back in the direction from which it came.

Then, use of the MSL, with a ground plane directly under the line, became popular. The characteristic impedance of MSL, shown in Figure 5.1, is determined by the width of the line W, thickness of the dielectric substrate h,

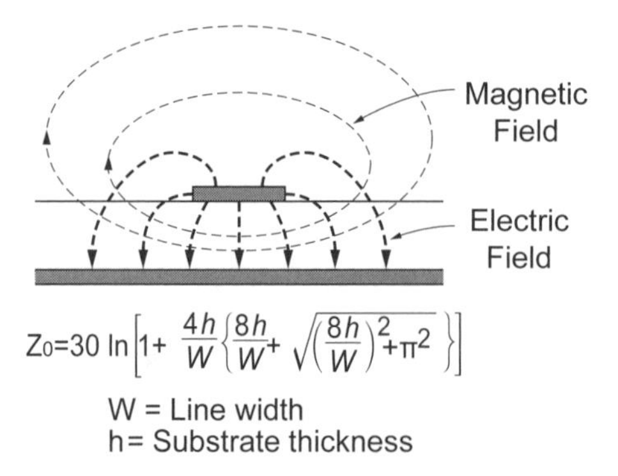

Figure 5.1 Characteristic impedance of the MSL.

and the relative permittivity ε_r of the dielectric substrate, so it is ideal for designs that match the input impedance of the load to the characteristic impedance of the line.

5.1.2 Simulation of MSLs

Figure 5.2 shows the electric field vector around a straight MSL. In this simulation, both ends of the line are terminated with the resistance of 50Ω. So if the MSL is sized so that the MSL characteristic impedance is 50Ω, the input signal is 100 percent expended in the load (ignoring transmission line loss) because there is no reflected power.

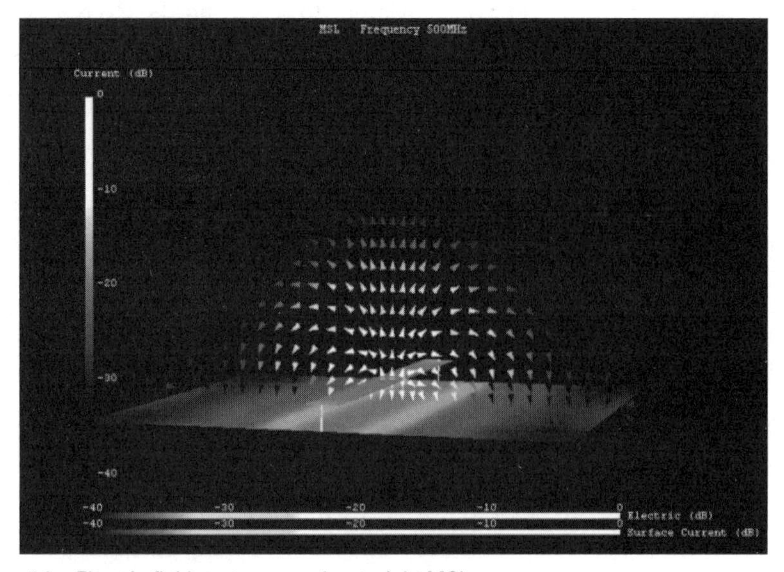

Figure 5.2 Electric field vector around a straight MSL.

Coaxial cable used in the earlier days of Ethernet local area networks (LANs) has a characteristic impedance of 50Ω and requires a termination resistance of 50Ω at both ends to keep packets of data from reflecting and bouncing around what would be a giant electrical echo chamber. If the termination resistance is removed by accident, data packets reflect, and communication to every PC on the LAN comes to a halt.

5.1.3 Current on the Ground Plane

When we look at the current on the ground plane, we see that most of it flows almost like a mirror image of the current on the MSL directly above it. The ground plane current flows in the opposite direction; thus, we can say that the ground plane current is the return current for the MSL.

As described in Chapter 2, it is simple for electromagnetic field simulators to model the current on conductors with zero thickness. We call this the surface current, and the units are amperes per meter (A/m). If we look at the current distribution on the ground plane, and we see light blue, dark blue, or even black, the current is very low. Sometimes, however, we see bright red (light gray or white in the figures in this book) flowing right along the edges. This means that current is high. High current causes radiation.

5.2 Simple MSL Model

Now we simulate a straight line MSL, seen in Figure 5.2. The substrate dimension is 30 mm × 30 mm, the line width is 1 mm, the dielectric thickness is 300 μm, and the relative permittivity is 4.8.

We selected these values arbitrarily, and the characteristic impedance is not 50Ω. For the straight line MSL, there are several formulas for characteristic impedance, including the one shown in Figure 5.1. Figure 5.3 shows calculating the characteristic impedance using Agilent AppCAD (free download, http://www.hp.woodshot.com/). The user inputs the line width, the substrate thickness, and the relative permittivity.

Sonnet Lite calculates the characteristic impedance for this MSL, shown in Figure 5.4, at 1 GHz to be 34.5Ω, shown in Figure 5.5.

The Sonnet Lite result of a wideband simulation is shown in Figure 5.6. Most electromagnetic field simulators, including Sonnet, calculate S-parameters assuming all ports are terminated in 50Ω loads. Such S-parameters are called 50Ω normalized S-parameters. As this MSL has a characteristic impedance of 34.5Ω, it seems that the reflection (S_{11}) is large throughout the band due to mismatch. As is typical in these situations, there are periodic frequencies with very little reflection, which is to say these few frequencies are matched frequencies. However, this "match" is a function of how long the line is, and

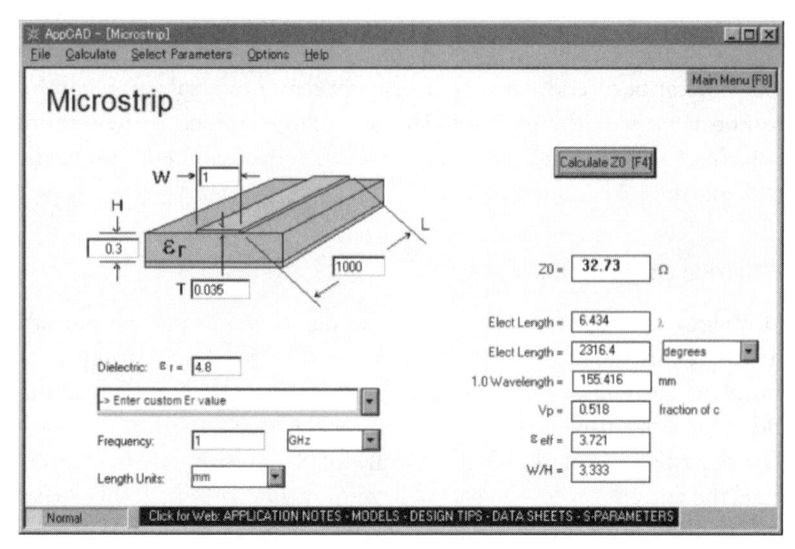

Figure 5.3 Calculating the characteristic impedance using Agilent AppCAD.

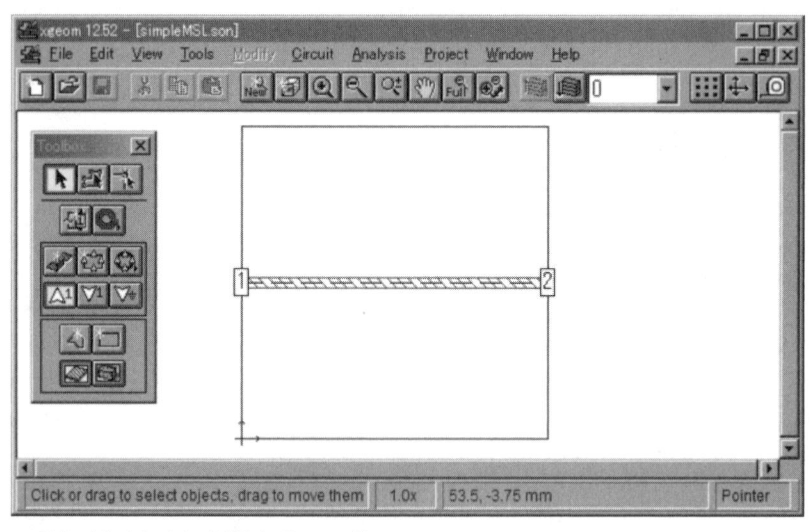

Figure 5.4 Model of the MSL in Sonnet Lite.

unless the line length was selected to realize this kind of match, it can be viewed as an "accidental" match.

From the viewpoint of a microwave designer, these are well-known frequencies where the line length corresponds to an integral multiple of half wavelengths. This line is 30 mm long. If this length is one half wavelength, then one wavelength should be 60 mm. The speed of the electromagnetic wave in free

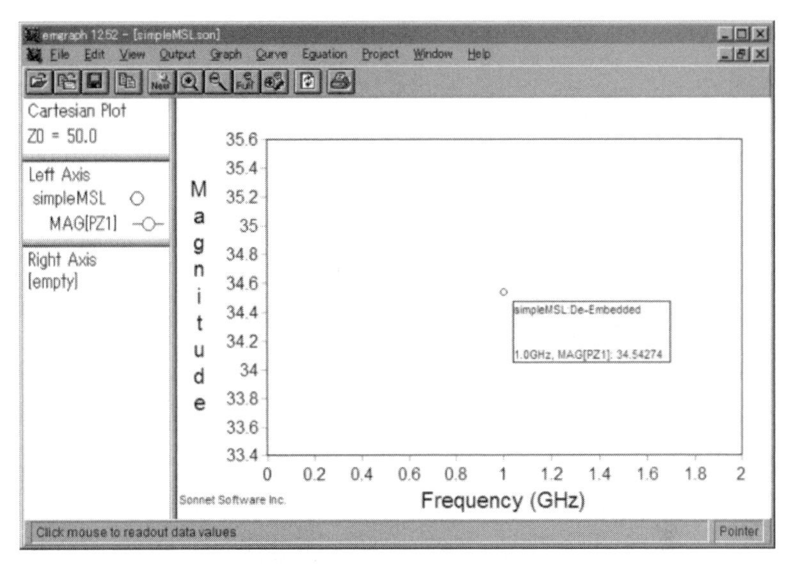

Figure 5.5 Characteristic impedance at 1 GHz calculated by Sonnet Lite.

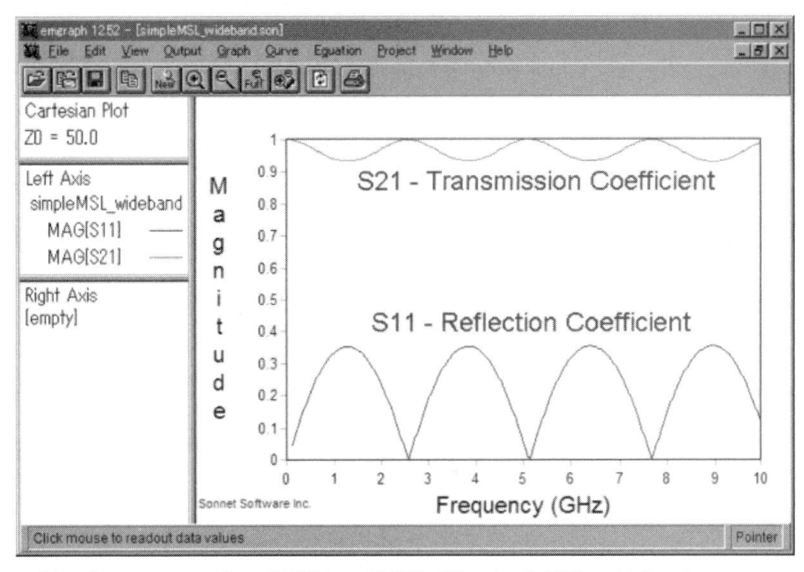

Figure 5.6 S-parameters from 0.1 GHz to 10 GHz. File: simpleMSL_wideband.son.

space, 300 million meters/second divided by 60 mm, means the line should be matched at 5 GHz. However, we see that the first matched frequency is 2.55 GHz, shown in Figure 5.6. Something strange is happening. Of course, it is the substrate.

5.2.1 Wavelength in Vacuum and Dielectrics

In the calculation of the previous section, we ignored the wavelength shortening effect of the substrate when the electromagnetic wave travels along the MSL. We see that the MSL has strong electric field between the line and the ground plane, as shown in Figure 5.2. In addition, some of the electric fields (electric lines of force) go out into space and then penetrate the dielectric substrate to head down to the ground. In any case, because a portion of the electromagnetic wave travels inside the dielectric, the velocity is slower than in free space. This makes the wavelength shorter.

The factor by which the wavelength is shortened in a dielectric is the inverse of the square root of the relative permittivity. For example, if the relative permittivity everywhere is 4.0, then the wavelength is half of the free space wavelength. In the previous MSL example, the relative permittivity of the substrate is 4.8. However, some of the electromagnetic wave is also in the air above the substrate. Therefore, we introduce the concept of effective relative permittivity. When the electromagnetic wave travels partly in one dielectric and partly in another, the effective relative permittivity is somewhere in between the two dielectrics. Exactly where in between is determined by EM analysis.

In this example, we calculated a free space resonant frequency (the frequency of the first matched point in Figure 5.6) of 5 GHz. We observed 2.55 GHz. This means the wavelength (which is inversely proportional to frequency) shortening effect is 0.51. Taking 1 over 0.51 squared, we have an effective relative permittivity of 3.8. This agrees with the calculation by Sonnet Lite, shown in Figure 5.7.

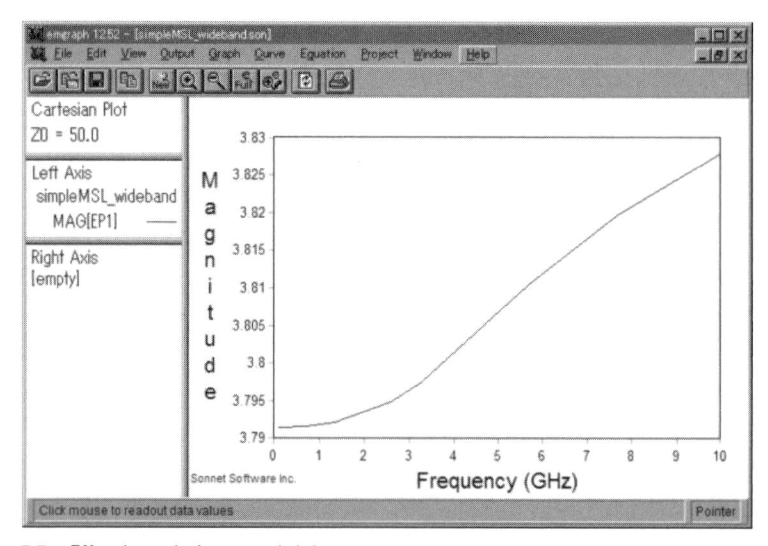

Figure 5.7 Effective relative permittivity.

The question remains, why is the reflection very small at integral multiples of one half wavelength? Basically, this is because there are two reflections. The first reflection is when the 50Ω electromagnetic wave enters the 34.5Ω line. The second reflection is when the 34.5Ω wave hits the 50Ω load at the end of the line. By the time this second reflection makes it back to the input, it is exactly out of phase with the first reflection, and they perfectly cancel each other. All the EM textbooks include an equation for the input impedance of a terminated transmission line (not given here). It turns out that if the transmission line length is an integral multiple of a half wavelength and it is lossless, the input impedance is equal to the value of the termination load. In fact, the actual characteristic impedance of a transmission line does not even matter!

5.3 Undesired Radiation from a Substrate with Ground Slit

Because the ground plane and the V_{cc} plane can decrease the inductance of a line, there is an advantage for reducing switching noise. In an actual substrate, many holes to install the parts are punched everywhere, and it is possible to create a slit right under a line in places.

Figure 5.8 is a model to simulate the effect on reflection coefficient with a slit 7 mm long and 1 mm wide at the center of the MSL ground plane shown in Figure 5.4.

5.3.1 Interpretation of S-Parameters

Figure 5.9 shows the S-parameters. Notice the reflection coefficient S_{11} in decibels (dB). We will examine the surface current distribution when there is a slit

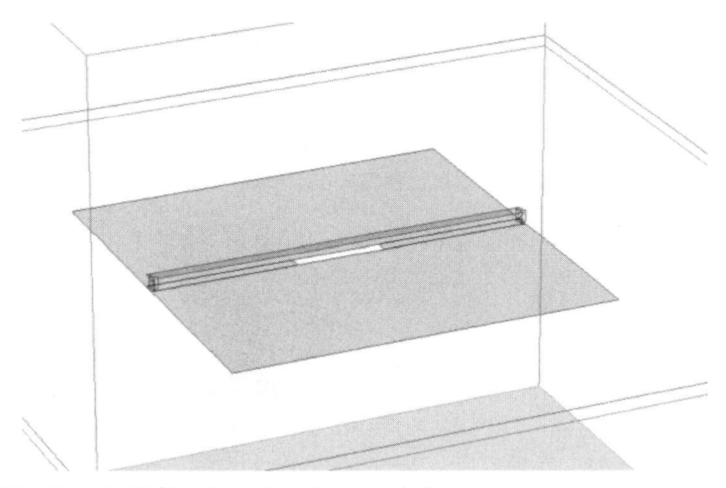

Figure 5.8 Model of MSL with a slit on the ground plane.

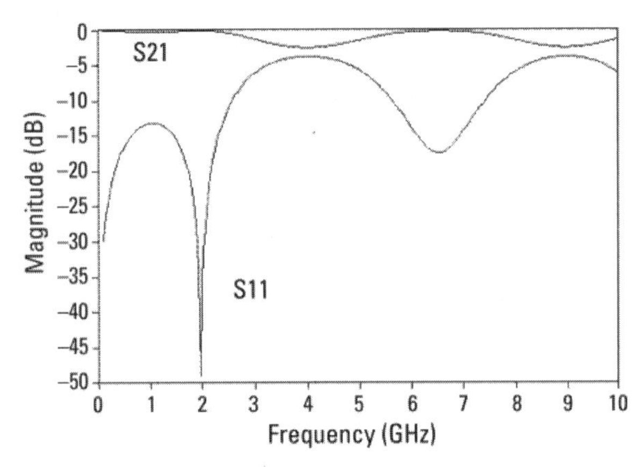

Figure 5.9 S-parameters of an MSL with a slit. File: simpleMSL_7mm_slit.son.

like this in the ground plane. Normally, when simulating an MSL with Sonnet, we use the bottom of the analysis space (Box Bottom) to play the role of the MSL ground plane. When we do this, we cannot view the ground current. So, here we create a new level under the line, and we include the ground plane with a slit by actually drawing it, just like we drew the original MSL.

In the result at 2 GHz, shown in Figure 5.10, we see that high current flows along the edge of the slit. However, in Figure 5.9, we see that the reflection is large in the regions around 4 GHz and 9 GHz.

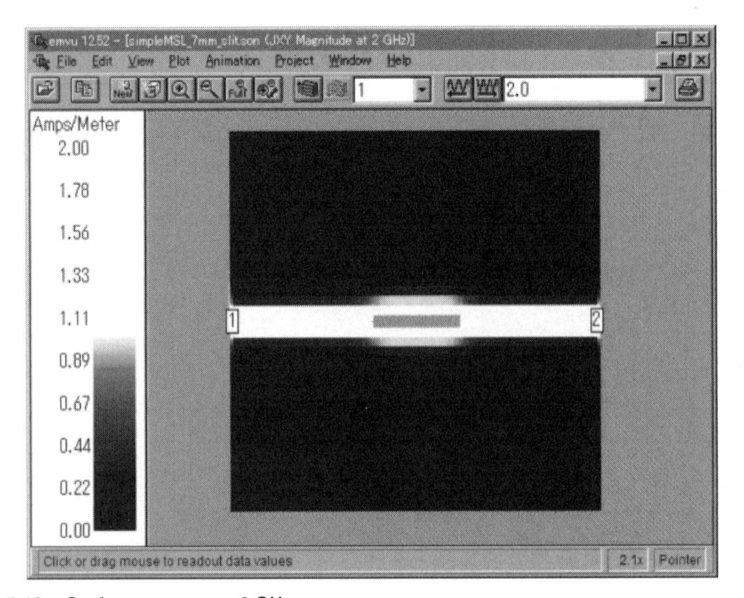

Figure 5.10 Surface current at 2 GHz.

5.3.2 Difference Due to the Location of the Slit

Next, we analyze what changes when we move the slit location. Figure 5.11 shows the S-parameters for two of locations: first, with the slit centered on the line (this is the top result) and, second, with the slit moved along the line length so that one end is at the midpoint of the line. There is almost no change below 5 GHz, as shown in Figure 5.11. The reflection coefficient has changed substantially above 6 GHz.

5.3.3 Difference Due to Slit Direction

Figure 5.12 shows the slit rotated so that it is orthogonal to the line. The ground return current is now forced to flow out and around the end of the slit. We would expect the reflection to be larger. In fact, the reflection has become extremely large above 8 GHz, as seen in Figure 5.13.

Figure 5.14 and Figure 5.15 show the surface current distribution on the ground plane. At 1.8 GHz, where the reflection is small, the effect of the current flowing out and around the end of the slit is small. At 8.4 GHz, the reflection is large, as in Figures 5.13 and 5.15, and we see a strong standing wave. The standing wave on the right side of the slit is due to the line impedance not being matched to the load. The much stronger standing wave to the left is due to the additional influence of the slit.

As we have described previously (in Chapter 2), the standing wave is generated by the superposition of a forward traveling wave and a reflecting wave. The incident electromagnetic wave at port 1, on the left, is traveling to the right. Then, due to the slit discontinuity, a portion of it is reflected and returns back to port 1. Thus, the incident forward traveling wave exists together with the reflected wave, and this creates the standing wave seen in Figure 5.15.

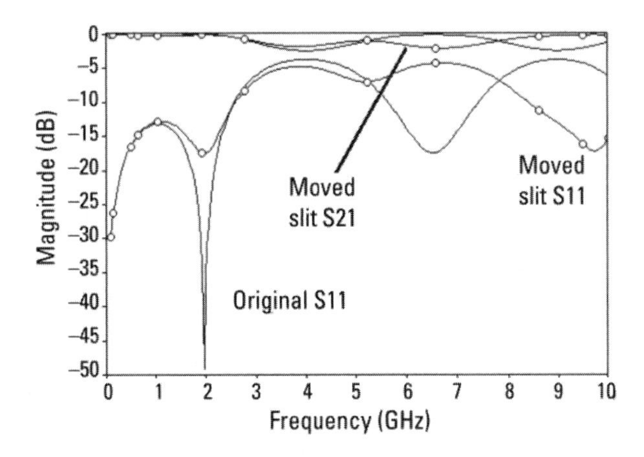

Figure 5.11 S-parameters for two slit locations.

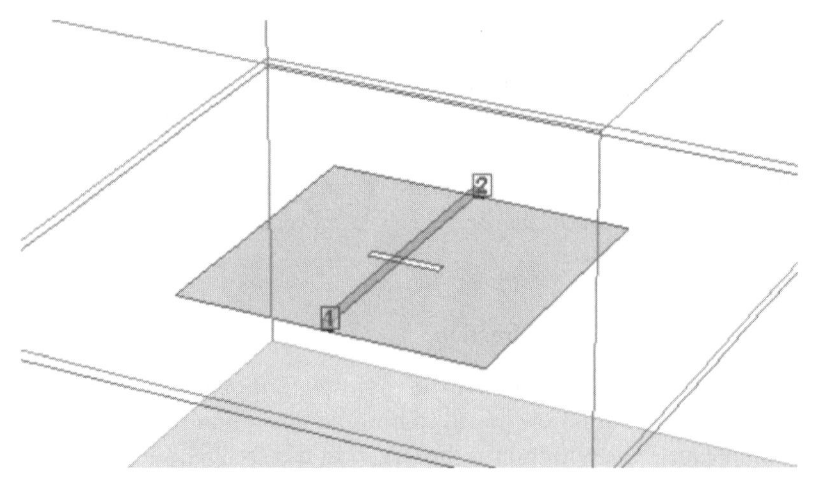

Figure 5.12 Model with the slit orthogonal to the line. File: simpleMSL_7mm_cross_slit.son.

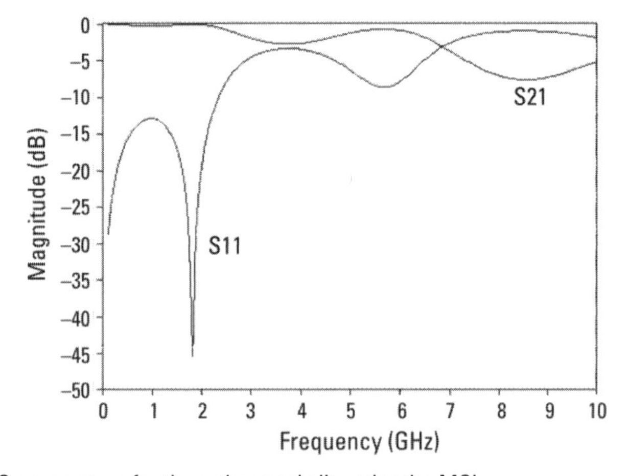

Figure 5.13 S-parameters for the orthogonal slit under the MSL.

5.3.4 Electromagnetic Field Around the Substrate

How is the electromagnetic field distributed in the vicinity of the substrate with a slit in the ground plane? Figure 5.16 is the result of simulation for the model with the slit located orthogonal to the line. It shows the magnitude of the electric field at 9.5 GHz where S_{21} (transmission coefficient) is small (results from XFdtd, by Remcom USA). The fields are plotted on the plane of symmetry of the line, exactly down the center of the line. The observer's viewpoint is slightly above the plane containing the ground plane. The black rectangle is the half of the ground plane that is on this side of the plane of symmetry.

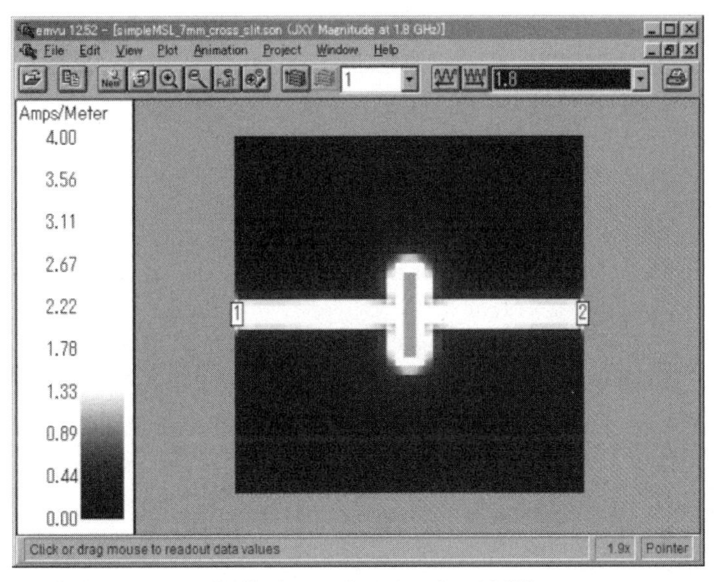

Figure 5.14 Surface current distribution on the ground at 1.8 GHz.

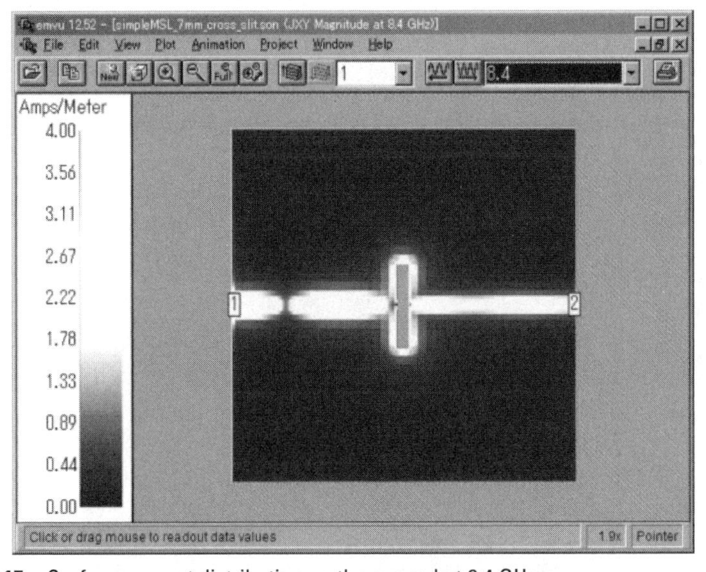

Figure 5.15 Surface current distribution on the ground at 8.4 GHz.

The high field (white) areas above the ground plane are from the intense electric field between the ground plane and the MSL. We can see a clear standing wave here. The intense electric field weakens as we move away from the MSL (toward the viewer). However, it does not go to zero. In fact, we can see that the electric field extends out from the slit and the MSL, and it wraps down, under the ground plane. So, we actually have electric and magnetic fields extending into space in all directions.

Some distance from the neighborhood of the slit, a strong area in the electric field looks like a floating cloud. Because Figure 5.16 shows the electric field strength, we can say that this is an area where the electric energy is concentrated. Moreover, because the magnetic energy exists at the same time, the electromagnetic energy propagates into space and we can imagine an undesired electromagnetic wave is radiated.

Figure 5.17 is the result of analysis of this undesired radiation. To get these results, we treat the circuit as though it is an antenna and use features of the electromagnetic software to plot the radiation pattern.

A characteristic that shows the performance of an antenna is radiation efficiency. This is the ratio of the radiated power to the input power. This quantity is easily calculated by electromagnetic software. In this case, the radiation efficiency is 1.6 percent. This means that only a small amount of power is radiated at this frequency. However, depending on the situation, for example, if you have a sensitive receiver nearby, or if there are a few hundred lines like this that are radiating all at once, the problem could become large.

5.3.5 What Happens at Frequencies of Low Reflection?

Finally, we examine what happens at 1.7 GHz, where the value of S_{11} (the reflection coefficient) is small. The reflection from port 1 is small at this frequency,

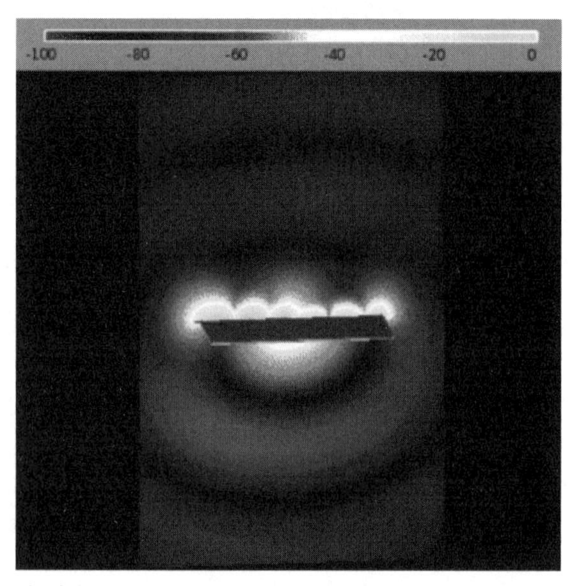

Figure 5.16 Result of simulated electric field for the MSL with the slit.

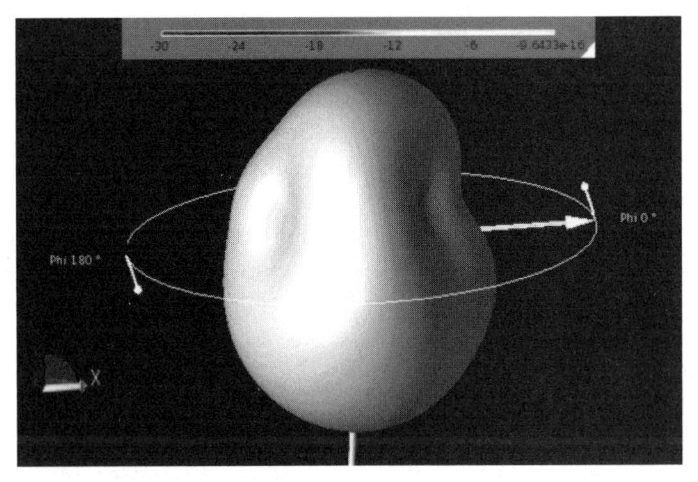

Figure 5.17 Undesired radiation at 9.5 GHz.

and S_{21} (the transmission coefficient) is almost 1. Even with the slit, the electromagnetic wave applied to the input port is almost entirely transmitted to the output port. Therefore, the electromagnetic energy radiated into space at this frequency is extremely small.

Figure 5.18 is the radiation pattern at 1.7 GHz, where S_{11} (reflection coefficient) is small. It forms an almost uniform radiator (there is a null in a direction not shown). One might at first think it is a good radiator in all directions. However, this is not the case. The radiation efficiency is an extremely small 0.02 percent. This circuit has essentially no electromagnetic energy that influences nearby circuits.

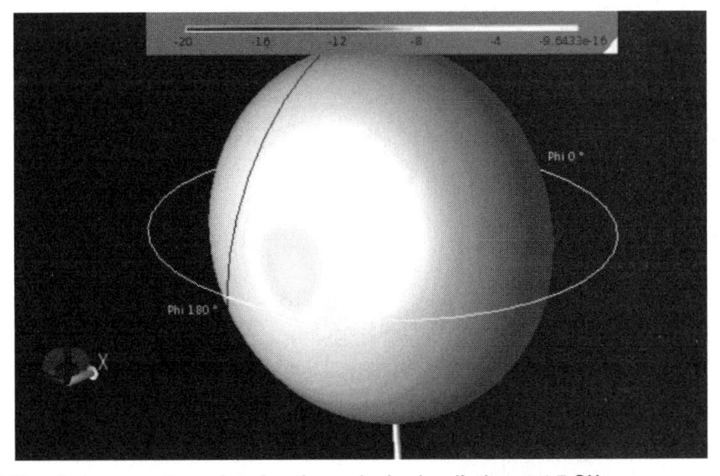

Figure 5.18 Antenna pattern showing the undesired radiation at 1.7 GHz.

5.4 Electromagnetic Shielding and Radio Wave Absorption

In Figure 5.19, the inner conductor is completely surrounded by the outer conductor, and the outer conductor is connected to ground, which we take to be zero electric potential. We can define any single point we want as zero potential. This corresponds to designating where we place the black lead of the voltmeter when we measure voltage (i.e., potential). Since the outer conductor is (for our discussion) a perfect conductor, then when we say one point on the conductor is zero potential, the entire conductor is also zero potential.

Under this condition, the influence of the electric field outside of the outer conductor does not reach the inner conductor at all. The outer conductor is called an electrostatic shield. Since the conductor also shields the interior from electromagnetic waves, it can be called an electromagnetic shield. Magnetic field that is constant with time (e.g., from a bar magnetic) penetrates such a shield easily (unless the conductor is a super-conductor). So it is not a magnetostatic shield.

An electromagnetically shielded room is a room that is completely covered on all sides by a metallic wall, usually sheet copper. The metal shielding intercepts all electrical noise from the external world. Such a room might be used, for example, to detect the slight bio-electricity that living things generate. This allows it to be used for such things as development of biosensors.

5.4.1 A Physical Quantity to Represent the Effectiveness of a Shield

To quantify the effectiveness of a shield, we can use shield effectiveness or shield efficiency (SE). This value is defined by the following expression, where the electromagnetic field before the shield is applied at a certain observation point and assumed to be ($\mathbf{E}_0$, $\mathbf{H}_0$), and an incidence wave is interrupted with the shield and assumed to be (**E**, **H**).

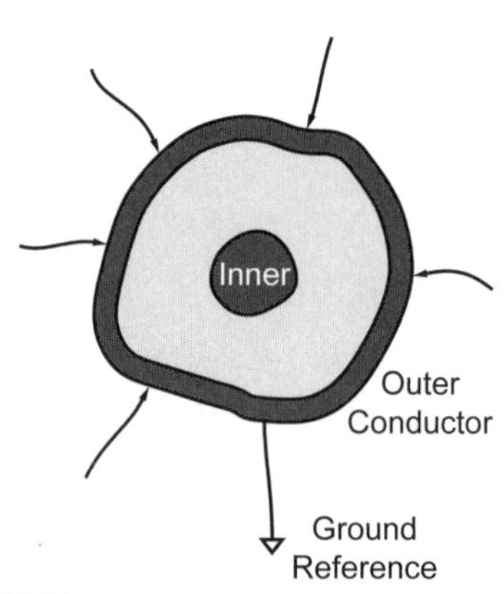

Figure 5.19 Electrostatic shield.

$$SE_E = -20 \log_{10}(|\mathbf{E}| / |\mathbf{E}_0|)$$
$$SE_H = -20 \log_{10}(|\mathbf{H}| / |\mathbf{H}_0|)$$
$$SE_P = -10 \log_{10}(|\mathbf{P}| / |\mathbf{P}_0|)$$

P and $\mathbf{P}_0$ are the power passing through a specific plane in the space before and after the shield is applied. This is also known as the magnitude of the Poynting vector, named after the professor who first derived it.

5.4.2 Effectiveness of a Magnetic Shield

For magnetostatic and low-frequency fields, a cover of magnetic material or a superconductor is needed to shield the magnetism. For instance, the shielding effectiveness for a uniform magnetic field by a spherical shell of the radius R, thickness t, and relative permeability μ_r is approximately shown by the following expression.

$$SE_H \approx 20 \log_{10}\left(1 + \frac{2\mu_r t}{3R}\right) \qquad \text{[dB]}$$

Figure 5.20 shows the appearance of a magnetic shield. The relative permeability of a magnetic material of high permeability is typically not more than 10^6. Thus, improved shielding cannot be expected when using a single layer, as shown in the left figure. Better shielding might be obtained by using a multilayer structure. In addition, when shielding from a magnetostatic field,

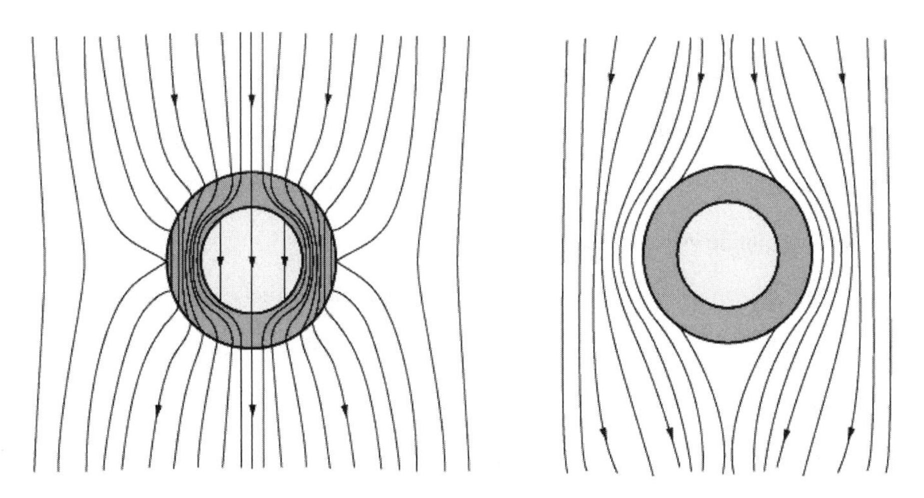

Figure 5.20 Magnetic shields, with magnetic material on the left and with superconductor on the right.

one should allow for the ability to degauss the shield. This is done by passing alternating current through the shield. This eliminates any residual magnetism. Residual magnetism results when the static magnetic field induces some "permanent" magnetism in the shield material.

A static magnetic field cannot penetrate a superconductor. This is called the Meissner effect, named after its discoverer. Note that this characteristic cannot be explained by simply invoking infinite conductivity because this effect actually ejects magnetic field from the conducting material when it transitions into a superconductive state. It was explained in 1935 by Fritz and Heinz London. Their theory allows the magnetic field to penetrate a very short distance into the superconductor. This distance is called the London penetration depth.

Figure 5.20 on the right shows the appearance of a superconducting shield—the lines of magnetic force cannot enter into the superconductor (the London penetration depth is too small to be seen at this scale). In general, a superconducting shield is used when a weak magnetic field must be observed and there is a much stronger magnetic field present.

5.4.3 The Effect of High Frequency on a Magnetic Shield

As for metals such as aluminum and copper, the direction of the electron spin on adjacent atoms is clockwise and counterclockwise. Therefore, the magnetism due to the electron spin of adjacent atoms are opposite of each other, and it is not magnetized. Though these nonmagnetic metals are not effective for magnetic field shielding at low frequency, the shielding effect can be strong for high frequency.

This is because induced current (eddy current) is generated in a direction that counters the changing magnetic field on the surface of the metal. The time-varying high-frequency magnetic field generates the current on the surface of the metal by Faraday's law of electromagnetic induction. Because of metal resistance, loss is generated. This is called eddy current loss, and a portion of the electromagnetic energy becomes thermal energy and is lost.

5.4.4 A Standing Wave in Space

Because electromagnetic energy propagates as a wave in space, space itself can be viewed as a transmission line. The ratio of electric field to magnetic field in free space is 377Ω. This is called the wave impedance, the impedance of free space, or the surge impedance.

As shown in Figure 5.21, standing waves arise when an electromagnetic wave is radiated inside a fully shielded (metal walled) room. In places, there is zero electric field and there is no coupling to an antenna (or circuit) that is sensitive to the electric field. A dipole is one such antenna, and we call it an electric field detection type antenna.

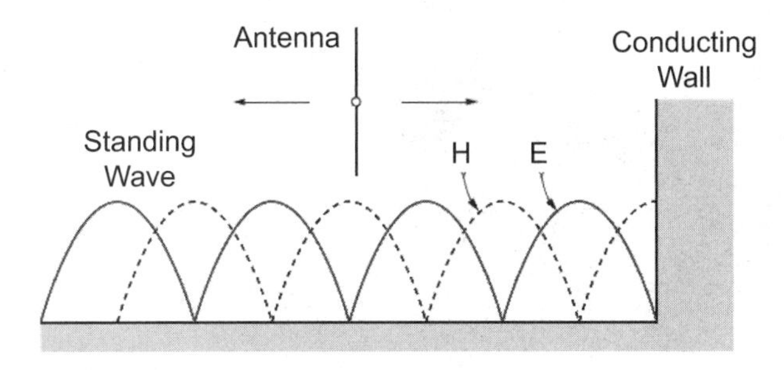

Figure 5.21 Standing wave in a room near a metallic wall.

5.4.5 Absorption of Electromagnetic Waves

Figure 5.22 shows the electric field distribution of a plane wave perpendicularly incident (i.e., going straight into) a metallic wall, to the right, which has a surface impedance of 377 Ω/square. Figure 5.23 shows the magnetic field distribution on a plane perpendicular to the electric field.

The electromagnetic wave ends at the wall to the right. The electric field and the magnetic field of these figures seem smooth; there is no standing wave. In this region, the wave acts like there is no wall there at all!

Figure 5.24 is the reflection coefficient (negative return loss in decibels) at an observation point (port) set in space in front of the wall. We see that the reflection is very small, and that the wall surface resistance of 377 Ω/□ acts as a nonreflective surface—it is a matched load.

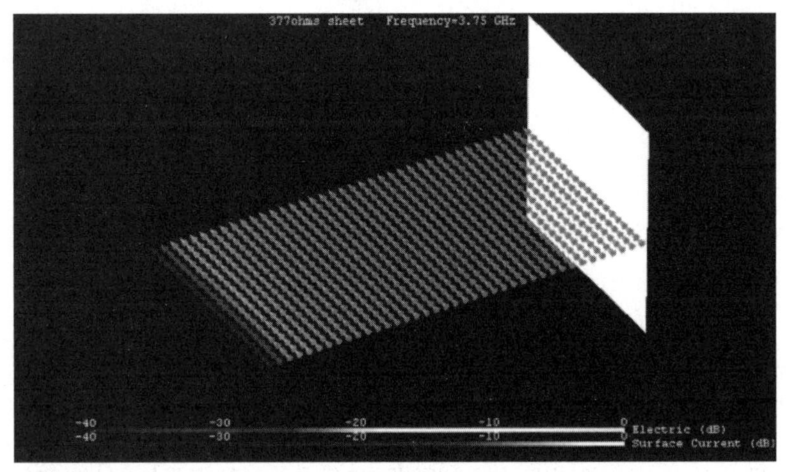

Figure 5.22 Electric field distribution of a plane wave incident on a resistive (absorptive) wall (CST MicroStripes).

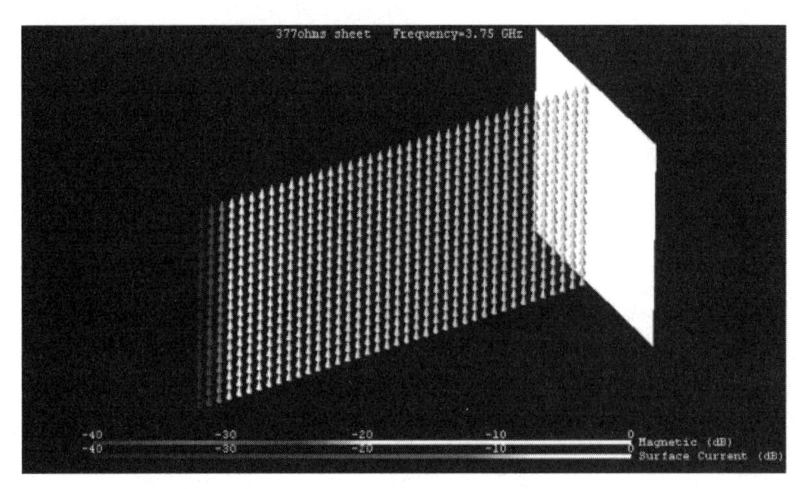

Figure 5.23 Magnetic field distribution of a plane wave incident on a resistive (absorptive) wall (CST MicroStripes).

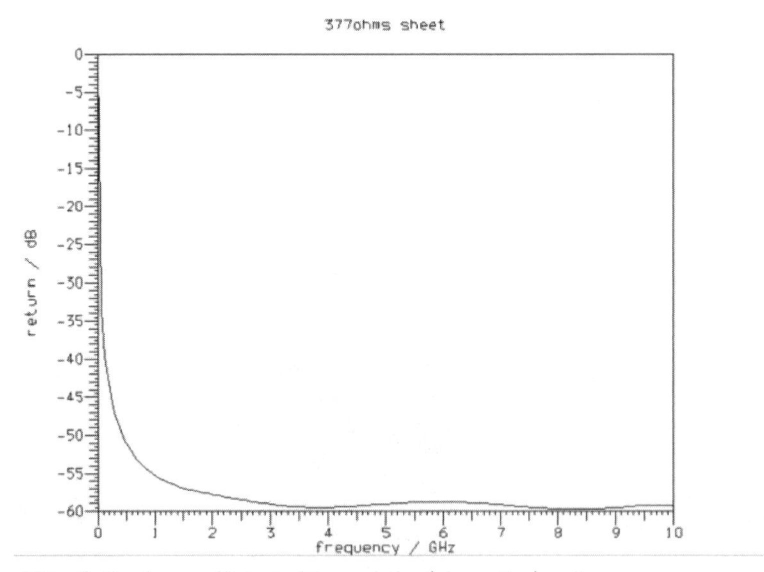

Figure 5.24 Reflection coefficient of the resistive (absorptive) wall.

The horizontal electric field, E_x, and the vertical magnetic field, H_y, are shown at an observation point 1 mm in front of the wall in Figure 5.25. The lower curve shows the electric field at 1 V/m (right-hand vertical axis) at most frequencies. The higher curve is the magnetic field at 2.65 mA/m (left-hand vertical axis). So, calculating the impedance of this wave, we have |E|/|H| equal to 377Ω.

"A transmission line named space" has a characteristic impedance of 377Ω. So we can realize nonreflection by terminating the wave with a 377 Ω/□ wall.

5.5 Confirming This Chapter by Simulation

When there is a slit in the ground plane of a microstrip line, the electromagnetic energy in the vicinity of the line might be coupled into space, depending on the frequency. In addition, when the total slit length approaches a sizable fraction of a wavelength, the electromagnetic wave along the edge resonates and strong undesired radiation is generated. Here, we capture an MSL on a ground plane with a slit and examine the S-parameters.

5.5.1 Drawing the MSL Ground Plane Including the Slit

Figure 5.26 shows the example MSL described earlier, with a ground plane slit at right angles to the line. Sonnet Lite's implementation of the Method of Moments solves for the current distribution on the surface of the metal. Thus, when we have more metal (e.g., the floating ground plane), we need more memory. To keep this problem within the memory limits of the free Sonnet Lite, we set the cell size to 2 mm for both x and y directions, and the Box Size is 64 mm × 64 mm.

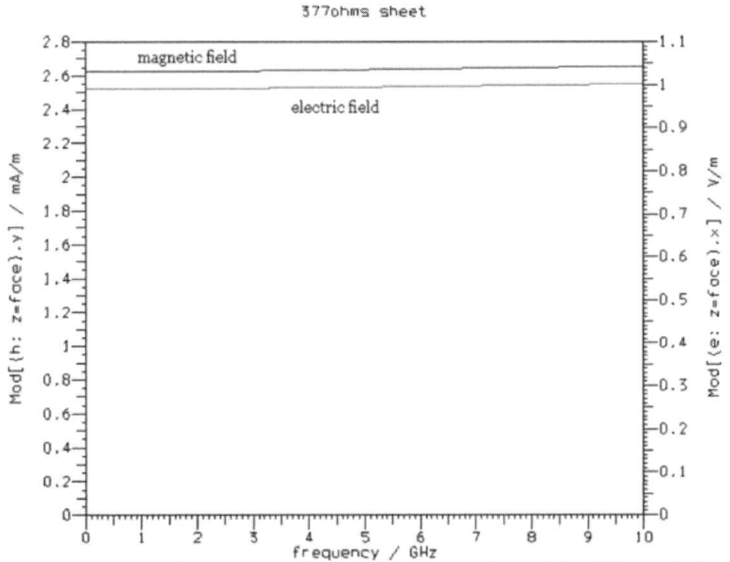

Figure 5.25 Electric field E_x and magnetic field H_y in front of a resistive (absorptive) wall.

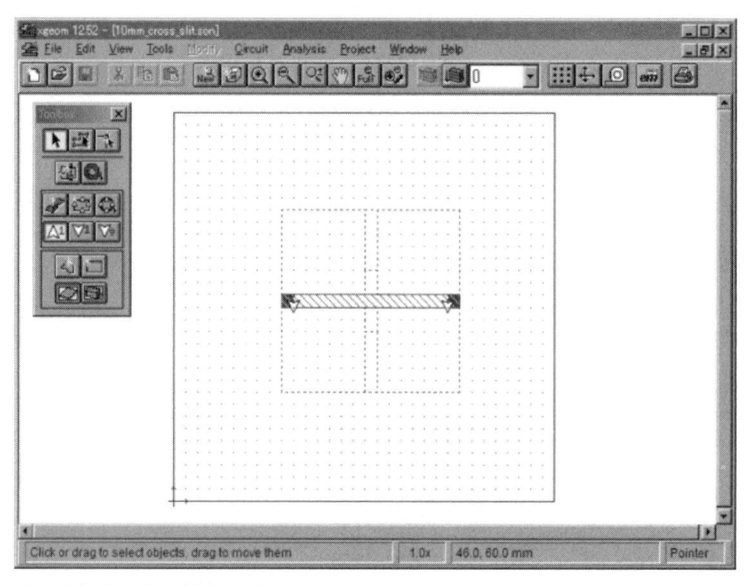

Figure 5.26 Model of an MSL with a ground plane slit at right angles to the line. File: 10mm_cross_slit.son.

We select Circuit > Dielectric Layers… and set the dielectric layers, shown in Figure 5.27. By default, Sonnet Lite starts with two defined layers (one is usually air, and the other is the substrate). We add one more layer by clicking the upper left Add button. Because the top and bottom layers are 20-mm-thick air, the relative permittivity and tanδ are left at the default lossless values.

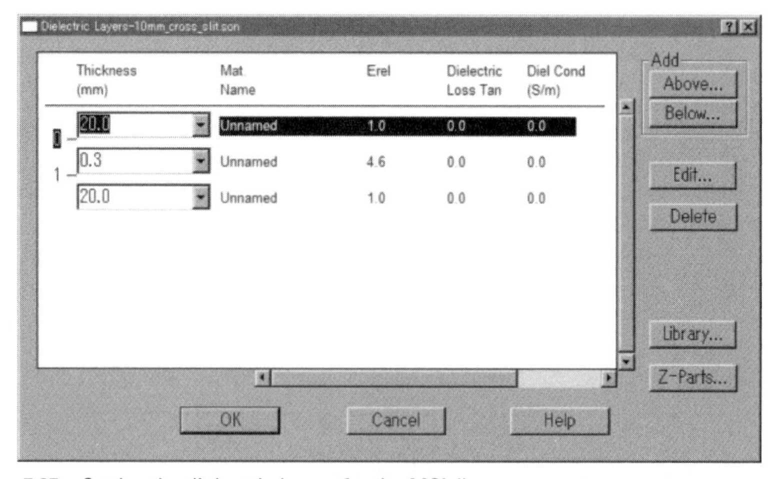

Figure 5.27 Setting the dielectric layers for the MSL line.

As the central layer is the substrate dielectric, we set the thickness to 0.3 mm, Erel (relative permittivity) to 4.6, and Loss Tan (tanδ) is left at 0 (lossless). Click OK and then we can draw the circuit.

5.5.2 Add the Ground Plane and the MSL Ports

At this time we should be viewing Level 0. This is the top side of the dielectric substrate and where we will place the MSL. Press Ctrl+D (or use the down arrow key) to move down to Level 1. Draw four rectangles exactly as shown in Figure 5.28 for the ground plane with a slit.

Next, a square covering just one cell, which will become the base of via, is drawn on the ground, like the magnified Figure 5.29. By clicking the Edge Via button on the left in the ToolBox, as in Figure 5.30, the cursor changes. Clicking on the right side of the one cell square, as in Figure 5.31, a via is drawn up one level and is indicated by an upward triangular mark. If you see a downward triangular mark, it is actually a via down to the layer below, so select it and delete it with the delete key. By selecting Tools > Add Via > Up One Level, the captured via will appear in the correct, upward direction.

Now, click the Add Port button, just above the Edge Via button, in the ToolBox. Next, click on the edge where we placed the via (as indicated by the upward pointing triangle) and port 1 is set, as in Figure 5.32. This is called a via port and forms the input port for our MSL.

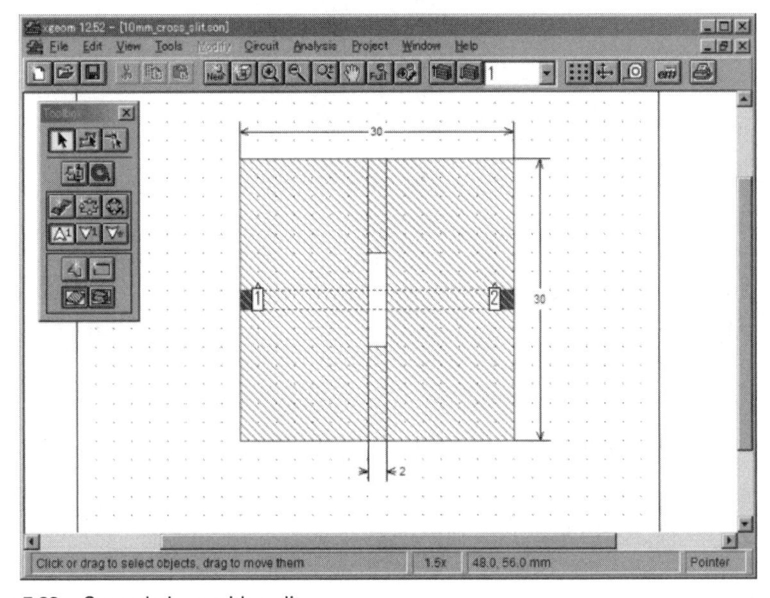

Figure 5.28 Ground plane with a slit.

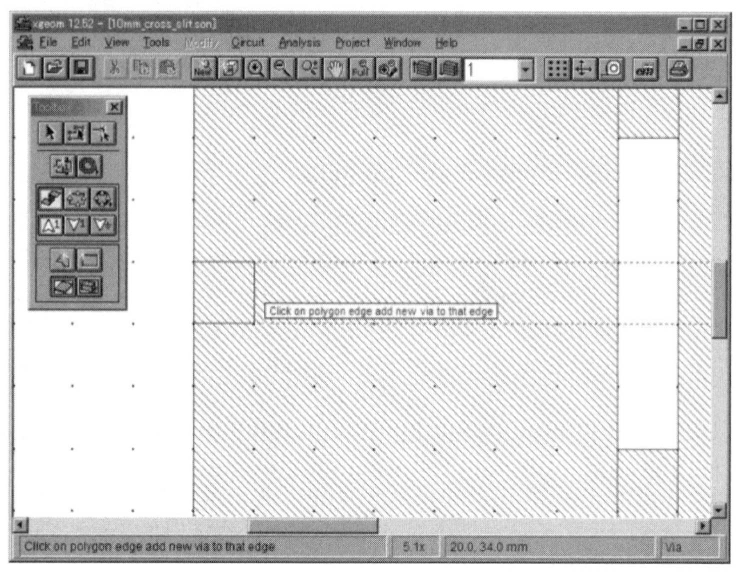

Figure 5.29 A one cell square becomes the base of a via.

Figure 5.30 Edge Via tool in the ToolBox.

Next set the same via port on the right end of the line. This time, click on the right edge of the base square, so that the via appears just inside the ground plane. If we were to click on the left edge, the via could appear just outside the ground plane and the two ports would not be identical.

5.5.3 Including the Effect of Radiation

We wish to include radiation in this model. So in Circuit > Box, set the top and bottom of the Box to Free Space, as in Figure 5.33. The thickness of the air layer is 20 mm both above and below the substrate, as in Figure 5.27. To accurately allow radiation, this distance should be at least a quarter to a half of

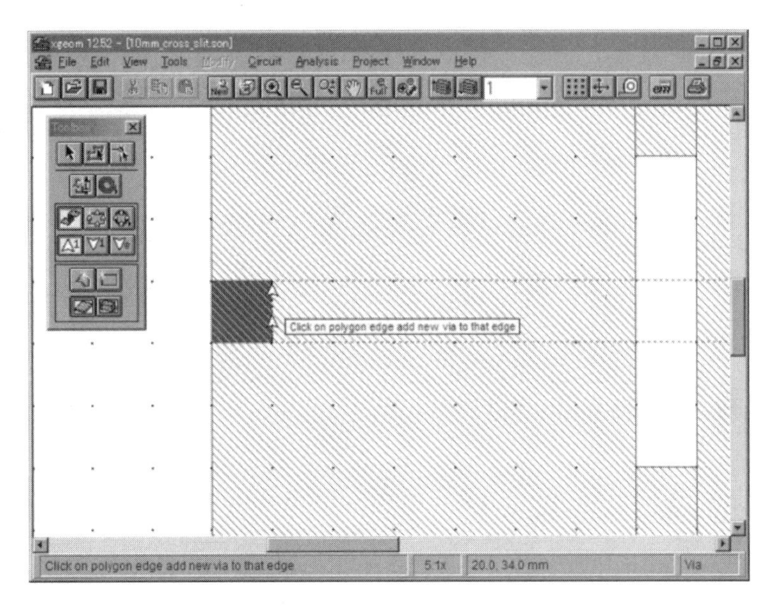

Figure 5.31 Via goes up one level.

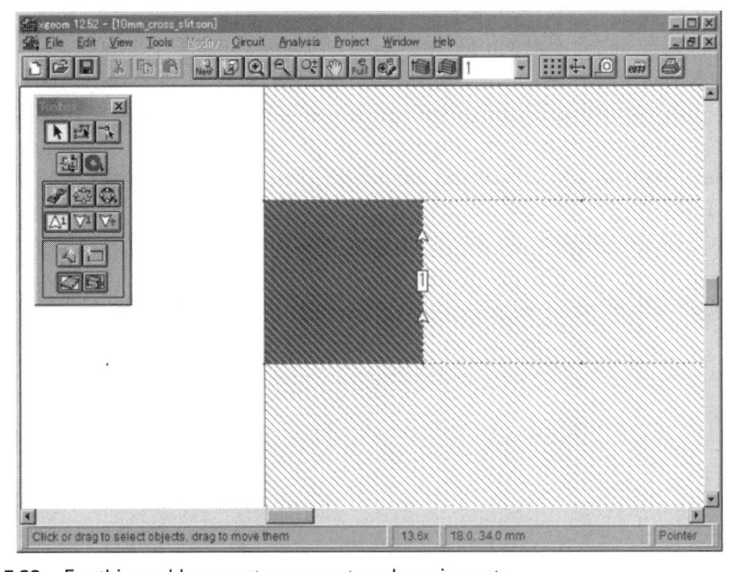

Figure 5.32 For this problem, ports are captured as via ports.

a wavelength. This is because the characteristic impedance of the wave radiated by the circuit becomes 377Ω when far enough away. The Free Space option in Sonnet sets the top cover and bottom cover to a surface impedance of 377Ω/□.

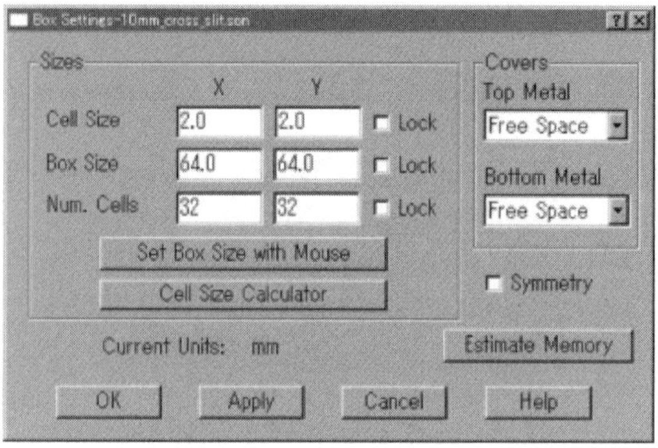

Figure 5.33 Top and bottom of Box are set to Free Space.

5.5.4 Simulate a Radiating MSL

Now let's simulate for a specific band. Select Analysis > Setup…and set the frequency range from 0.1 GHz to 4.0 GHz using the default adaptive band sweep (ABS), as in Figure 5.34. Figure 5.35 shows the resulting S-parameters. The vertical dotted line appears by selecting Graph > Marker > Add > Vertical Line Marker. You can move it using the left and right arrows to display the frequency.

The frequency of the minimum S_{11} (reflection coefficient) is 1.43 GHz.

Next, select Analysis > Setup…and change Analysis Control to Frequency Sweep Combinations and click the Add… button. You can specify a fixed frequency by selecting Single Frequency (Figure 5.36).

Now, we can set a combination of different frequency sweeps. For example, click the Add… button and set an ABS sweep. In fact, we can set any of the

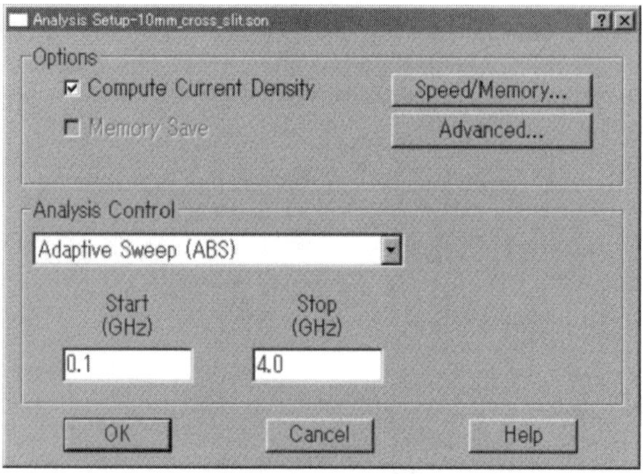

Figure 5.34 Frequency range is set from 0.1 GHz to 4.0 GHz.

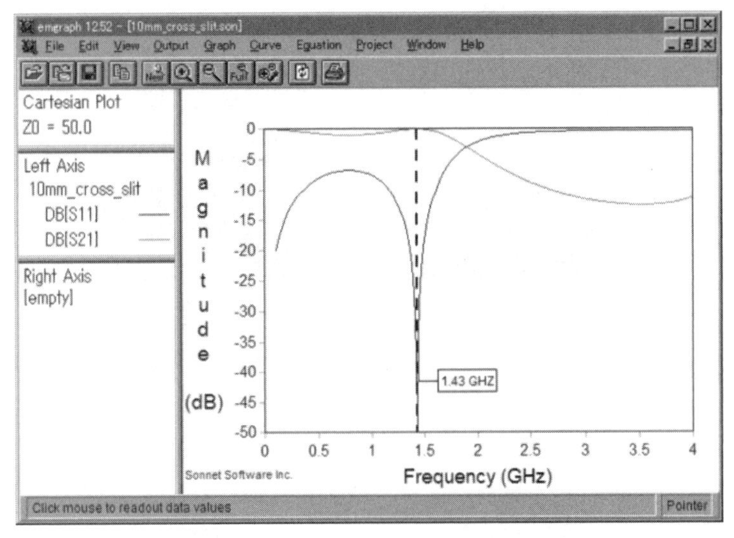

Figure 5.35 Resulting S-parameters with a vertical frequency marker.

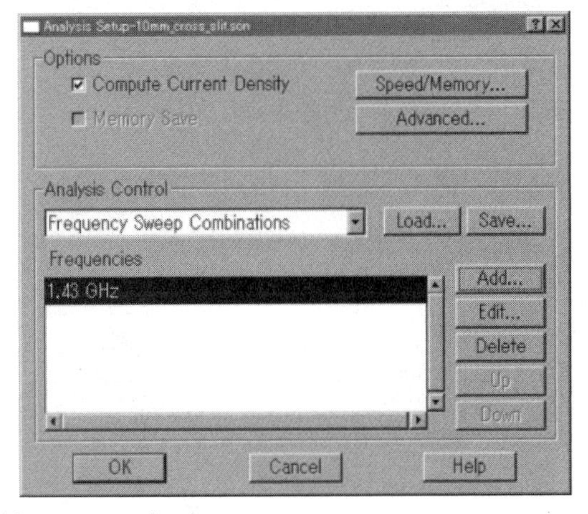

Figure 5.36 Fixed frequency selection.

combinations shown in Figure 5.37. When we have multiple frequency sweeps showing in the large window, they can be moved up and down using the Up Down button, as shown in Figure 5.38. If we move the fixed frequency sweeps to the top, our simulation will start at those fixed frequencies first. When the ABS sweep is last, Sonnet Lite will store all the special ABS interpolation information from all of the previous sweeps, and it will be able to complete faster.

Figure 5.39 shows the ground return current distribution at 1.43 GHz. At the frequency of minimum S_{11} (reflection coefficient), we see that the return

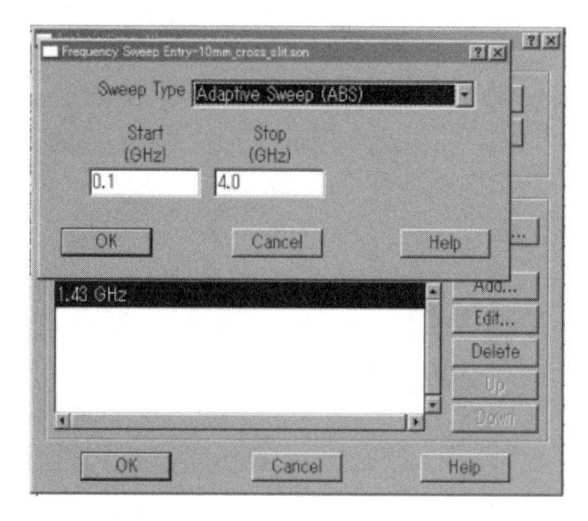

Figure 5.37 Setting an ABS Sweep under Frequency Sweep Combinations.

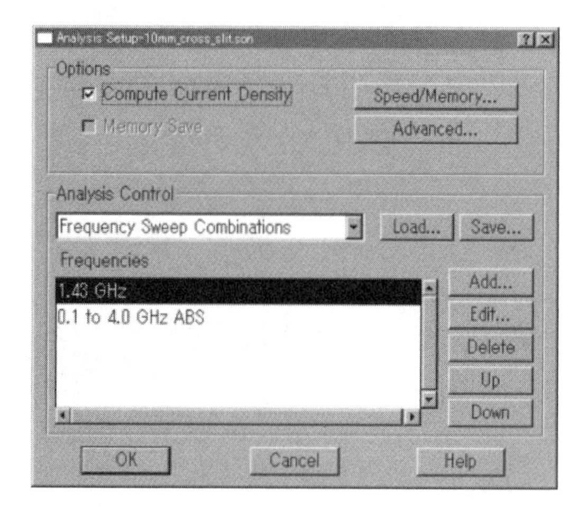

Figure 5.38 Setting fixed frequencies.

current flows most strongly along the edge of a slit—it does not spread out away from the slit.

S-parameter plots display in decibels by default. Double-click the left mouse button on the DB[S11] label located on the left side of the graph, as in Figure 5.40, and change the Data Format from Magnitude(dB) to Magnitude. Then the S-parameters are plotted on a linear scale from 0 to 1. Back in the dialog box of Figure 5.40, we can also double click the MAG[S21] in the Unselected column, or, by clicking the small triangular button to the right, move it to the Selected column. Click OK and S_{11} and S_{21} are displayed.

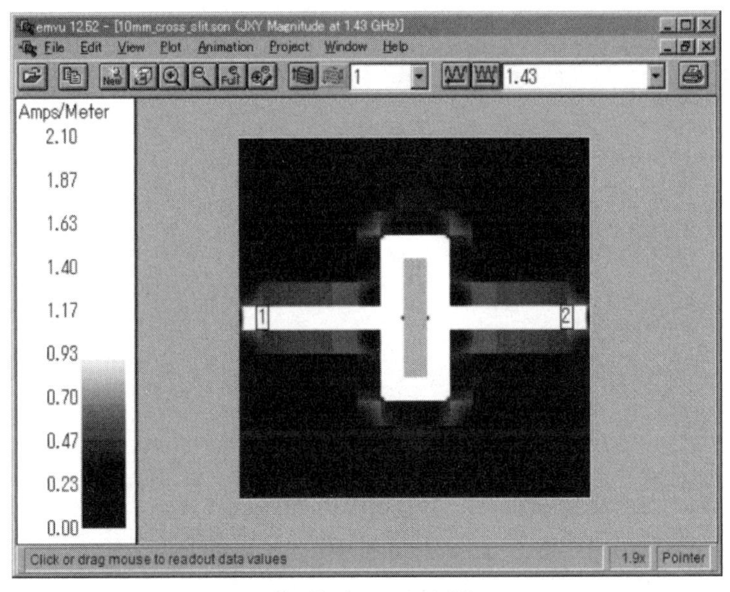

Figure 5.39 Ground return current distribution at 1.43 GHz.

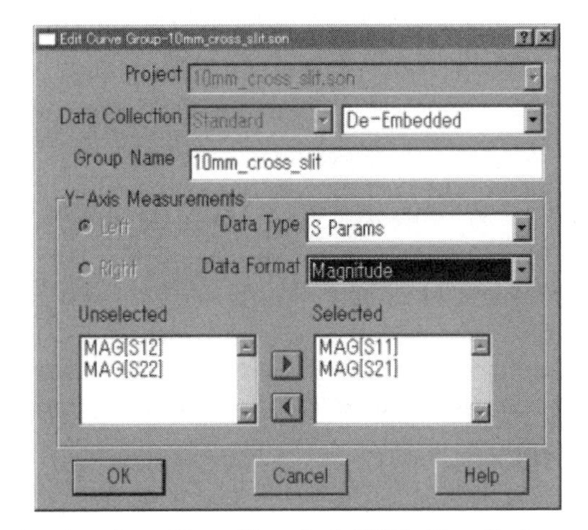

Figure 5.40 Changing data format from Magnitude(dB) to Magnitude.

In addition, we can use the magnifier + icon (Zoom In, in the icons along the top) to select a region to be magnified around 3.4 GHz, where the reflection coefficient is large. The resulting graph looks like Figure 5.41. A data marker, as shown, is obtained by selecting Graph > Marker > Add > Data Marker and clicking on the curve where you want the marker attached.

In Figure 5.41, reading from the data marker S_{11} at 3.42 GHz is 0.2424 and S_{21} is 0.968. S-parameters are a ratio of voltages. When there is no radiation

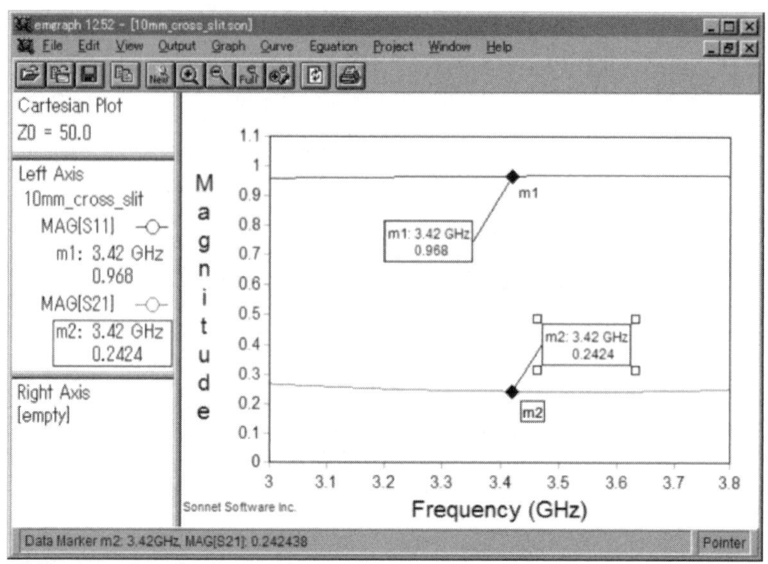

Figure 5.41 Insertion of a data marker.

or loss, the sum of the reflected power ratio and the transmitted power ratio is equal to 1, and we have the law of conservation of energy.

$$|S_{11}|^2 + |S_{21}|^2 = 1$$

At 3.42 GHz, we get 0.0588 + 0.937 = 0.9958. The power lost (to radiation and any other loss mechanisms) is 0.0042, or 0.42 percent. In this simulation, we used only lossless metals and dielectrics. Thus, this value of power loss is caused only by radiation.

Keep in mind that this amount of loss is for the situation where both ports are terminated in 50Ω resistors. Other terminations will result in different amounts of radiation—in some cases, smaller; in other cases, larger. In resonant situations, radiation can become extremely large.

5.6 Summary

1. Return current flows on the ground plane of an MSL. The return current distribution looks almost like a mirror image of the line current.
2. Electromagnetic energy can be radiated into space where the return current encounters the slit. This radiation is typically undesired.

3. The magnitude of the undesired radiation caused by the slit in the ground is strongly dependent on the position and orientation of the slit and on the frequency.
4. If there is a slit in the ground, energy will be exactly split between the load, circuit losses, and radiation.
5. High-frequency electromagnetic shielding utilizes eddy currents to absorb and reflect incident radiation.
6. A wall with a surface resistance of 377 Ω/□ absorbs (does not reflect) incident electromagnetic waves, provided the waves come from a sufficient distance (i.e., they are plane waves).

6

Understanding the Differential Transmission Line

6.1 Smaller, Better, Faster

As components become smaller and substrates become thinner, cell phones realize ever more advanced functionality, and the width of PCB lines steadily shrinks. In Chapter 3, we learned that two or more nearby lines can generate strong electromagnetic coupling (crosstalk), especially at the higher frequencies. Because microstrip lines share the same ground plane, it is easier for such electromagnetic coupling to occur, especially if the ground metal is thin (and resistive) and has lots of holes in it (inductive). When this gets to be too much of a problem, we give each microstrip line its own private ground return. This pair of lines is called a differential line or a differential pair. This is inspired by the two parallel wires that appear in Chapter 1.

In this chapter, we examine how a differential pair improves crosstalk by simulating their electromagnetic fields. In addition, we investigate whether or not the differential line is always a wonder drug, and whether or not it always achieves the expected effect.

6.1.1 What Is a Differential Transmission Line?

There are three ways in which a differential transmission line is typically structured (Figure 6.1). In addition, a pair of differential lines (i.e., two side-by-side differential pairs), as in Figure 6.2, is like Figure 6.1(c) with no shielding above or below. We indicate that current goes into the line with the positive port

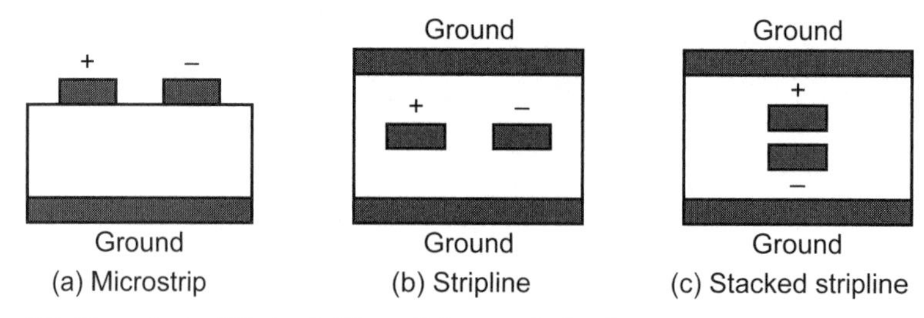

Figure 6.1 Three typical kinds of differential transmission lines.

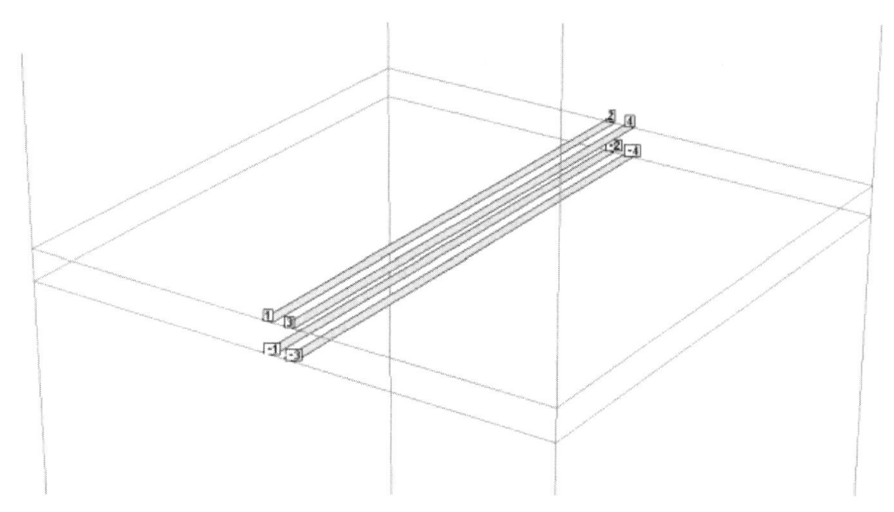

Figure 6.2 Sonnet model of a pair of differential lines. File: spl.son.

number. The return current comes back on the line with the same port number preceded by a minus sign. This type of line is useful on a two-sided substrate.

The MSL, which appears in all previous chapters, is a single-ended, or unbalanced, transmission line. These names mean that the signal line (the microstrip line) is not physically similar to the ground return (the ground plane). For a differential pair, current going out one line and coming back on the other is called a normal mode or differential mode. If two lines are close enough, the radiated electromagnetic waves from the line current and the return current cancel each other and the unwanted radiation decreases. This cancelation is most effective if the dielectric constant of the substrate is high and the two lines forming the differential pair are physically as close as possible to each other.

The structure of Figure 6.1(b), with ground planes above and below, is known as a stripline. Normally for stripline, ground return current is split equally between the top and bottom ground planes (which is why the top and bottom ground planes must be well connected together with vias). In this case, all the ground return current flows on the second of the two lines in the differential pair, so no return current should be flowing in the ground planes. The ground planes act only to shield the differential pair from outside interference, as described near the end of this chapter.

6.2 Is the Differential Line a Wonder Drug?

Our hope is that the differential transmission line reduces crosstalk. To test this hypothesis, we analyze a pair of differential pairs, shown in Figure 6.2. We also analyze two microstrip lines with the same substrate material, shown in Figure 6.3. The microstrip lines share a single ground return path (the ground plane) that is separated from the signal paths (the microstrip lines) by the thickness of the substrate. The differential pairs each have their own private ground return paths (pick one of the two lines in a pair) separated from the signal path (pick the other of the two lines in a pair) by the separation between the lines. Less separation means less crosstalk. So we now compare crosstalk.

When modeling both a differential line and an MSL, it can be difficult to decide exactly what we should compare because the two cases have different geometries. Here, we fix the distance between the lines at 1 mm and the width of lines at 1 mm as well. If it is important to have a uniform characteristic impedance set at 50Ω, then the dimensions would be different.

As we learned in Chapter 3, when we set the port numbers as shown, in Figures 6.2 and 6.3, the amount of crosstalk is directly evaluated by examining the S-parameters. Of course the crosstalk is a function of the port terminations. Since we are looking at 50Ω S-parameters, we are implicitly assuming 50Ω terminations.

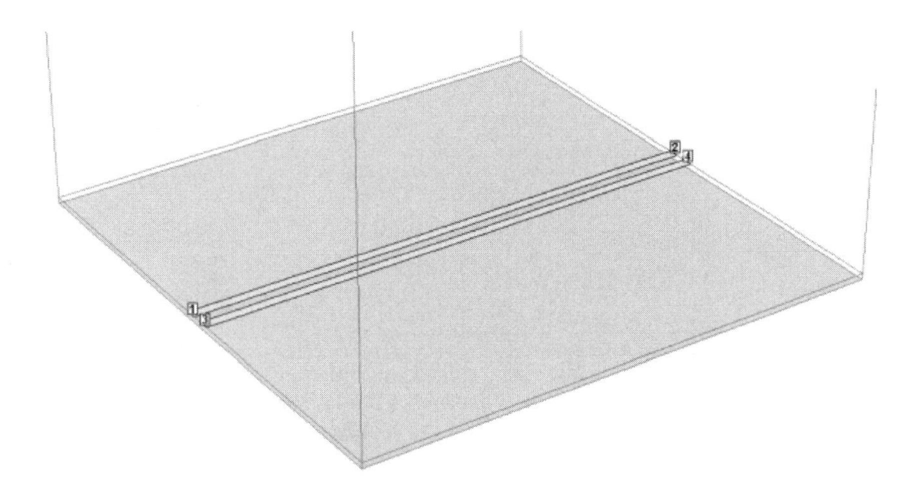

Figure 6.3 Sonnet model of coupled MSLs. File: msl.son.

S_{31} is the ratio of the transmitted wave voltage amplitude at port 3 to the incident wave voltage at port 1. This is backward crosstalk. S_{41} is the ratio of the transmitted wave voltage at port 4 to the incident wave voltage at port 1, so this is forward crosstalk.

Figure 6.4 shows the S-parameter results for the microstrip lines. Since these lines were designed to have a 50Ω characteristic impedance in order to efficiently drive 50Ω terminations, S_{11} (reflection coefficient) is small from 100 MHz to 2 GHz.

6.2.1 Interpreting the Loss Results

The transmission coefficient, S_{21}, drops from 1 (100 percent transmitted, which is the same as 0 dB) as frequency increases. We see that the transmitted wave voltage on port 2 falls away as the frequency becomes higher. This is mainly because of resistive line loss. The power lost due to resistance is proportional to the current squared. So this is called I squared R loss. The voltage drop along the line that is due to resistance is called IR drop, which is Ohm's Law.

James C. Rautio has written a definitive paper on MSL loss that finds there are three distinct frequency bands for loss (J. C. Rautio and V. Demir, "Microstrip Conductor Loss Models for Electromagnetic Analysis," *MTT Transactions*, March 2003). In the frequency range shown in Figure 6.4, the loss increases gradually due to the appearance of the edge singularity. The edge singularity is the high current at the edge of the line. This is present at high frequency, but is not seen at low frequency. In addition, the substrate dielectric also has increased loss due to loss tangent, tanδ. S_{21} (transmission coefficient)

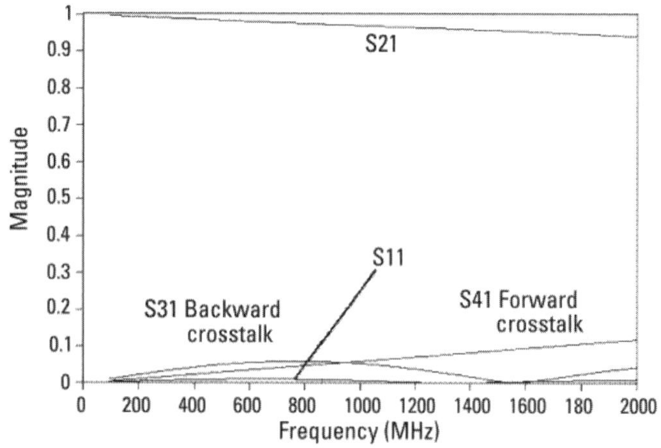

Figure 6.4 S-parameters of the two microstrip lines showing crosstalk. The vertical scale is a linear voltage scale, not decibels.

decreases gradually with increasing frequency. At even higher frequencies, resistive loss increases due to current concentrating on the surface of the metal. This is called skin effect loss.

Sonnet uses a frequency domain solver based on the Method of Moments. This means it analyzes one frequency at a time, and it assumes the input signal is a sine wave at the frequency of analysis. The wideband data shown in Figure 6.4 is obtained by full EM analysis at only a few frequencies, during which some extra EM data is saved. Then this extra data is used by the adaptive band synthesis (ABS) feature to accurately interpolate those results to hundreds of frequencies. In this case, the EM data was actually calculated at only eight frequencies.

6.2.2 Interpreting the Crosstalk Results

Up to around 1 GHz, the microstrip S_{31} (backward crosstalk) exceeds S_{41} (forward crosstalk), as seen in Figure 6.4. It is small at the frequencies we have analyzed (below 2 GHz). S_{11} (reflection coefficient) has the same general frequency variation as the backward crosstalk. The frequency where S_{11} goes to zero corresponds to an electrical line length of one-half wavelength.

The effective relative permittivity, ε_{eff}, of a single MSL, seen in Figure 6.5, allows us to calculate the wavelength shortening coefficient. This is determined from epsilon effective by the following formula:

$$\text{Wavelength shortening coefficient} = \frac{1}{\sqrt{\varepsilon_{eff}}}$$

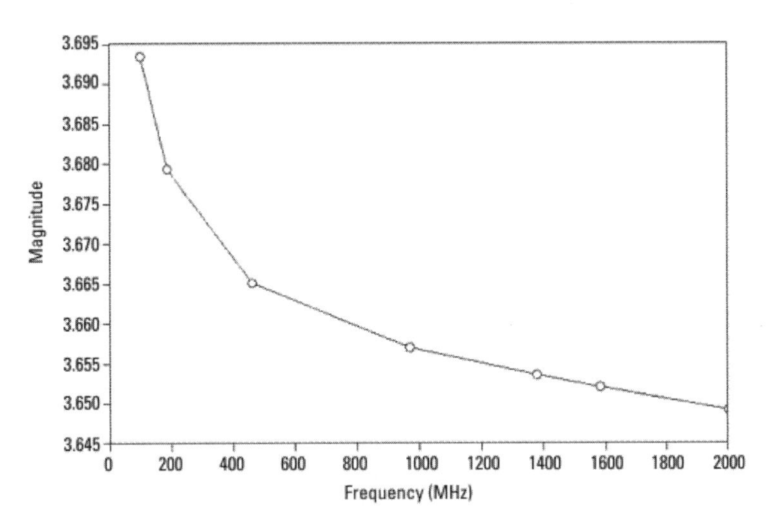

Figure 6.5 Effective relative permittivity, ε_{eff}, of an MSL.

As the physical line length is 50 mm, the frequency, f_0, at which it is one-half electrical wavelength long is calculated as follows:

$$f_0 = \frac{3\times 10^8}{2\times 50\times 10^{-3}\times\sqrt{3.66}} = 1.57 \text{ GHz}$$

Because the transmission line electrical length is one-half wavelength at this frequency, the 50Ω load impedance (on port 2) is also seen directly at port 1. There is no reflection.

S_{41} shows a monotonic increase as frequency becomes high. We compare this to the differential pair result. In Figure 6.6, S_{41} is displayed in decibels to make it easy to view differences between small forward crosstalk magnitudes. The differential pair forward crosstalk is less than the microstrip by 5–10 dB.

6.2.3 Reducing Crosstalk with a Differential Pair

Figure 6.7 shows the result of the same MSL model as Figure 6.3, using the MicroStripes (from CST) time-domain TLM method (see Chapter 9). We see the electric field vectors around the line illustrated with small cones.

We see that the electric field originating on the left line is present between the right line and the ground plane. This induces a voltage on the right line.

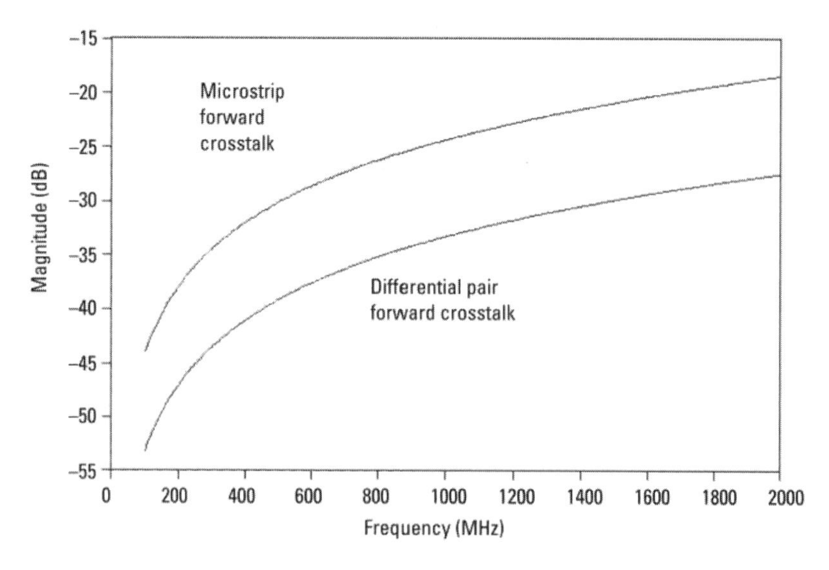

Figure 6.6 Comparing forward crosstalk for an MSL and a differential pair.

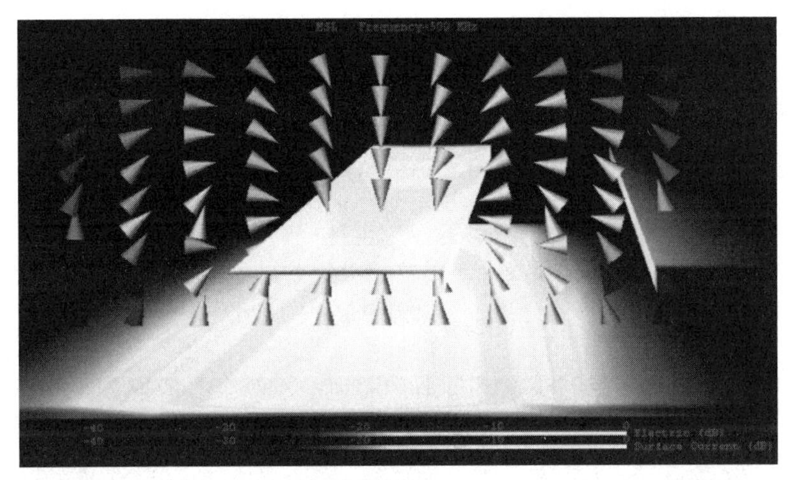

Figure 6.7 Electric field vectors around an MSL.

Coupled magnetic field (not shown) also induces current in the right line. This generates crosstalk.

Figure 6.8 is a display of the time averaged Poynting vector power (W/m^2). The Poynting vector (named after the person who first derived it) is the product of the electric and magnetic fields. It points in the direction perpendicular to both the electric and magnetic fields (which are themselves also at right angles to each other). This is the direction of power flow. The magnitude of the Poynting vector, shown in Figure 6.8, shows energy transferring to the right line.

Figure 6.8 Time-averaged Poynting vector power around the MSL.

The electric field vectors between the differential pair, shown in Figure 6.9, are approximately –40 dB. At the right side of the MSL in Figure 6.7, it is between –30 dB and –20 dB, so it is stronger than the differential pair line.

The electric field between the left differential pair is strong. This generates an electromagnetic field that couples to the adjacent (right) pair but is weaker. Figure 6.10 shows the magnetic field vector that couples the left pair to the right pair.

Corresponding to Figure 6.8 for an MSL, Figure 6.11 is the time-average Poynting vector power for the differential pairs. To visualize the distribution

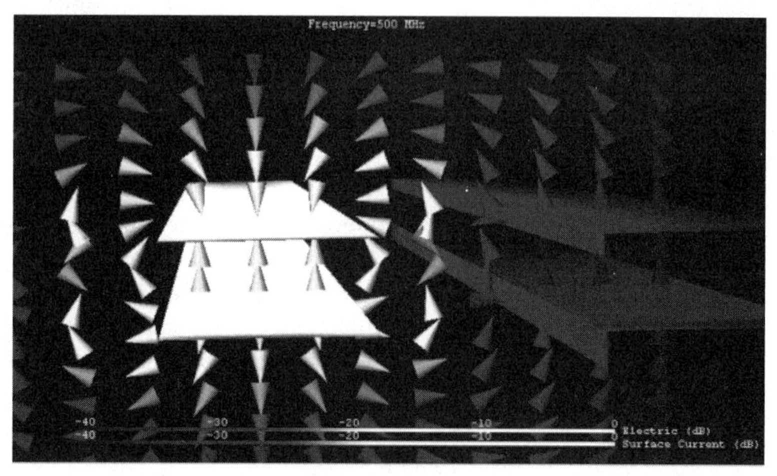

Figure 6.9 Electric field vectors around the differential line.

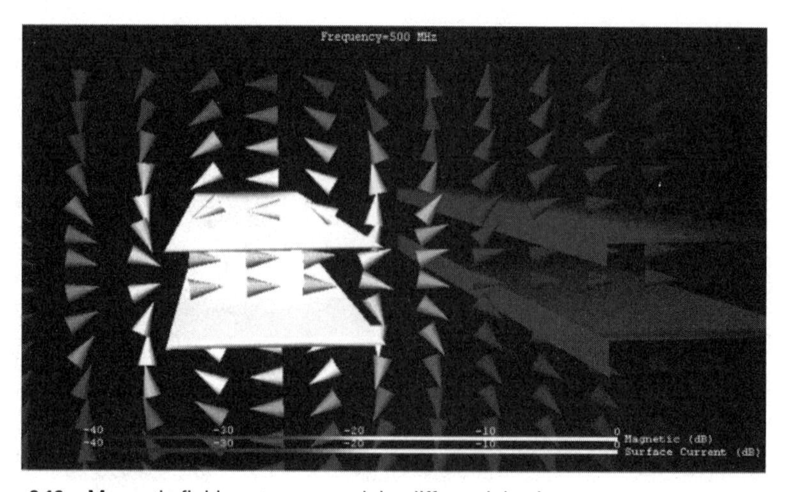

Figure 6.10 Magnetic field vectors around the differential pairs.

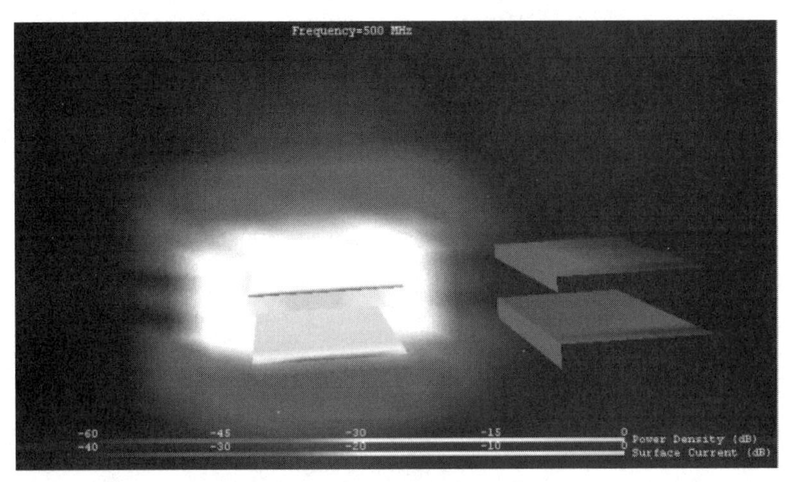

Figure 6.11 Time-averaged Poynting vector power around the differential line.

easily, the scale is set to –60 dB. This make it easy to view the small crosstalk power. Note that the power flow around the right line is limited to the edge.

From above, the differential line is excellent in its ability to concentrate electromagnetic field to within the paired lines. We see that the electromagnetic energy coupled to the adjacent line is lower than the adjacent MSLs, which share a common ground.

6.2.4 The Case of the Differential MSL

We calculate crosstalk for two closely spaced differential MSL pairs (four lines, two per pair), in Figure 6.12. We set the port numbers to +1, –1, +2, –2, +3, –3, +4, and –4; the separation between each line is set to 1 mm, and the width of each line is also 1 mm. Ports are terminated by 100Ω loads by changing the Sonnet default of 50Ω to 100Ω (Sonnet Lite cannot do this). Figure 6.13 shows the forward crosstalk. We see that it is slightly lower than the differential (stacked) pair, shown in Figure 6.6. The strange glitch at 1.65 GHz is a classic problem; we explain it later in this chapter.

6.3 Is the Differential Line Perfect?

Things can go wrong. Let's figure out how.

6.3.1 Relation Between Normal Mode, Common Mode, and Radiation Noise

For the differential pair, the normal, or usual, mode is the odd, or balanced, or push-pull mode. This is because the input signal is applied between the two

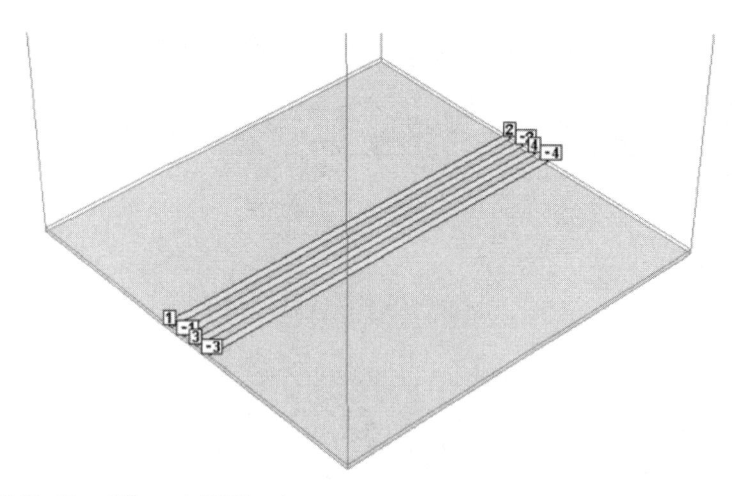

Figure 6.12 Two differential MSL pairs.

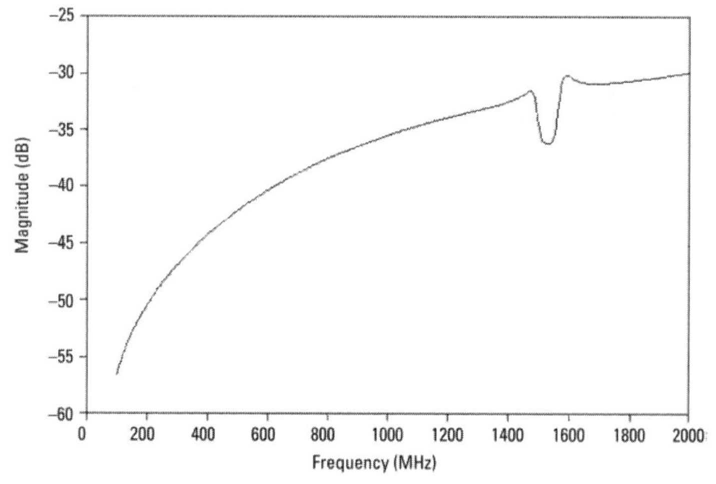

Figure 6.13 Forward crosstalk between differential MSL pairs. A glitch is seen at 1.65 GHz. This is explained later.

input terminals. If a crosstalk or other external signal couples equally to the two lines, that noise does not appear at the load because the noise voltage on each line is equal.

For the common mode, both lines are connected in parallel (or "push-push"), and the signal is split equally between the two. The signal current then returns via the ground plane. Now, the problem is that noise that couples from nearby lines, or from external sources, acts just like the signal on the lines and can cause problems. Thus, we must make sure our most sensitive circuits are resistant to common mode excitation.

6.3.2 Investigation of the Common Mode

Figure 6.14 is a differential MSL pair. It is based on the line and the dielectric of Figure 6.3. Because we want to include radiation from the edges of the ground plane, the ground plane is included in the drawing. An additional small piece of metal is attached at two line ends to set ports 1 and –1 in Figure 6.15.

Figure 6.16 shows the current on the two lines that form the differential pair. The magnitude of the current is the same on each line, but the directions (not shown) are opposite. When clicking on the line, the value of surface current density is displayed at the bottom left of the screen. The difference in

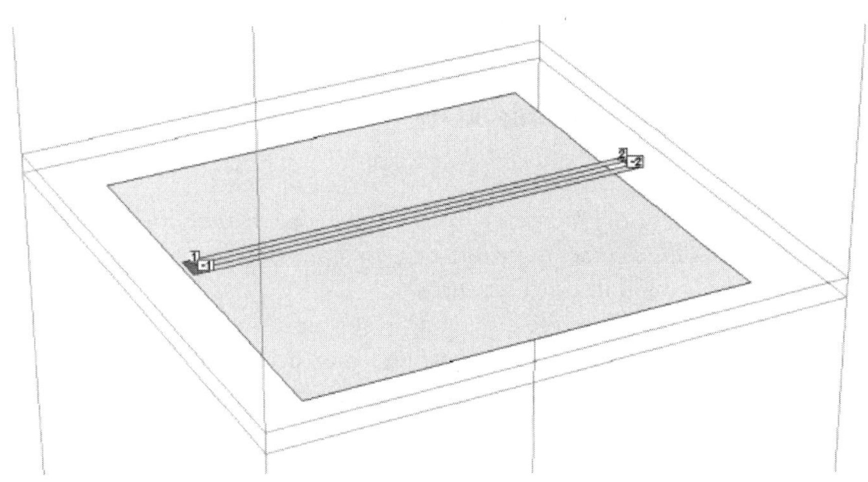

Figure 6.14 Differential MSL pair.

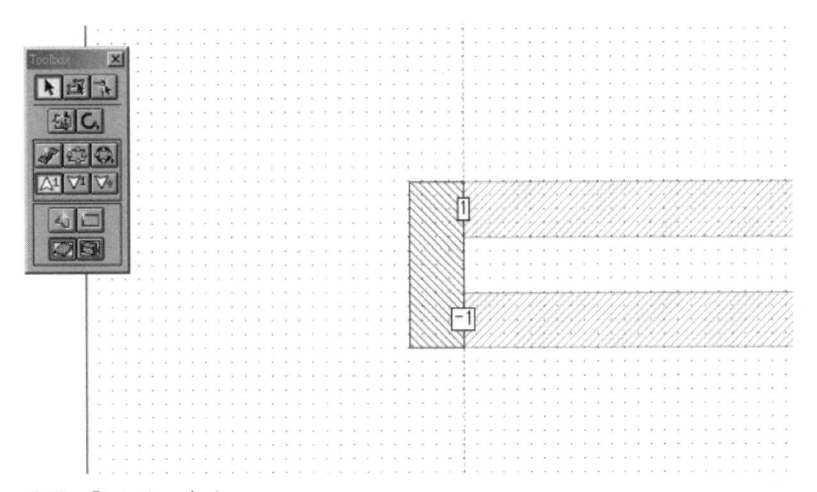

Figure 6.15 Ports 1 and –1.

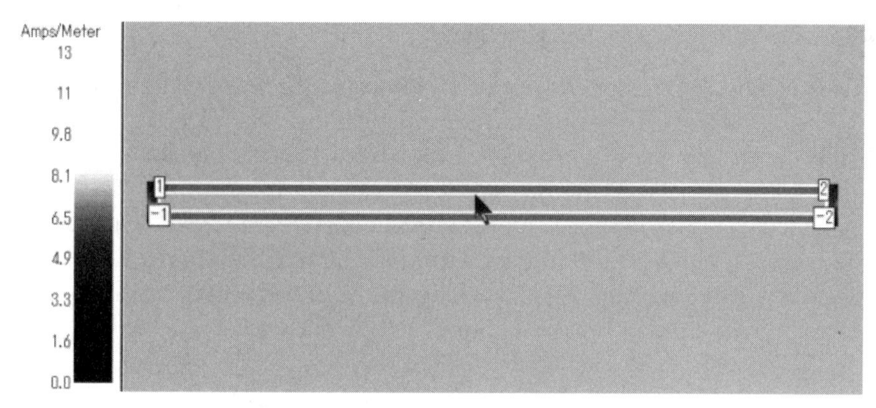

Figure 6.16 Surface current distribution on the differential pair.

magnitude between the two currents between the two lines is almost exactly zero. So it is clear that only the normal mode is flowing along the entire length of the line.

Next, we move the differential pair to the edge of the substrate, shown in Figure 6.17. This represents a worst case. In a practical situation, the nearby Sonnet box wall would be other circuitry.

In order to clearly see the difference in surface current density, we adjust the upper limit of the display scale, shown in Figure 6.18. Table 6.1 lists the differences calculated from the values by clicking on various locations and reading them; all are about 0.46 A/m.

From the above, we can see the common mode component takes on a much larger value when we place the line on the edge of the substrate, next to

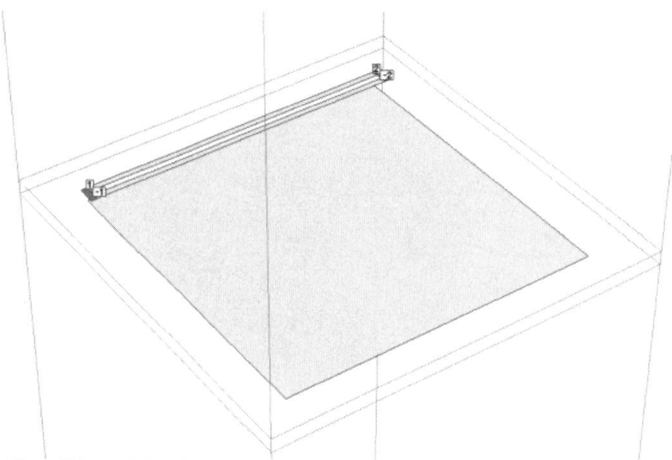

Figure 6.17 The differential pair moved to the edge of the substrate, next to the Sonnet box wall.

distance from port [mm]	10	20	30	40
surface current on upper line [A/m]	14.38	14.42	14.46	14.49
surface current on lower line [A/m]	13.92	13.97	14.00	14.03
difference [A/m]	0.46	0.45	0.46	0.46

Table 6.1 Difference in Current Magnitude Between the Two Lines as Calculated from the Values Obtained by Clicking on the Lines

Figure 6.18 Surface current distribution on the differential pair.

the Sonnet box wall. Almost no common mode component (indicated by any difference in current magnitude on the two lines) is seen when we locate the differential transmission line at the center of the substrate.

6.4 Radiation Problems with Differential Lines

A properly designed differential line has very low radiation. However, even if properly designed for lower frequencies, it might still radiate at higher frequencies. We next describe how that happens.

6.4.1 The Electric Field Observed at 3m from the Circuit

Figure 6.19 displays the electric field intensity for a differential pair located in the center of a substrate. This is the electric field distribution at 500 MHz. It shows how the electric field suddenly weakens a short distance from the line. Consequently, in this model, electromagnetic wave radiation into free space is close to nothing.

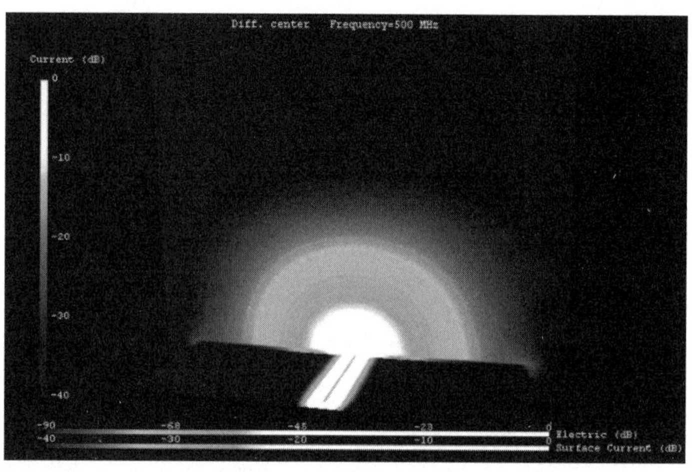

Figure 6.19 Electric field intensity for a differential pair.

Figure 6.20 is a comparison of the differential pair in the center of the substrate and one that is on the substrate edge, based on data obtained using Sonnet to get the electric field 3m from the circuit. For most of the frequency range, the electric field is strong for the line on the substrate edge. Assuming that the main cause of radiation is the common mode component and assuming that its far-field region is similar to that of a dipole antenna, 3m is far enough to be in the far field. In the far field, the electromagnetic waves are effectively propagating in free space, so Figure 6.20 amounts to undesired radiation.

Radiation is caused by the acceleration of charges, as in a microscopic dipole antenna, shown in Figure 6.21. When dividing a conductor into tiny segments, the current, *I*, flowing in any given segment is the time rate of change of the charge, *q*, in that segment:

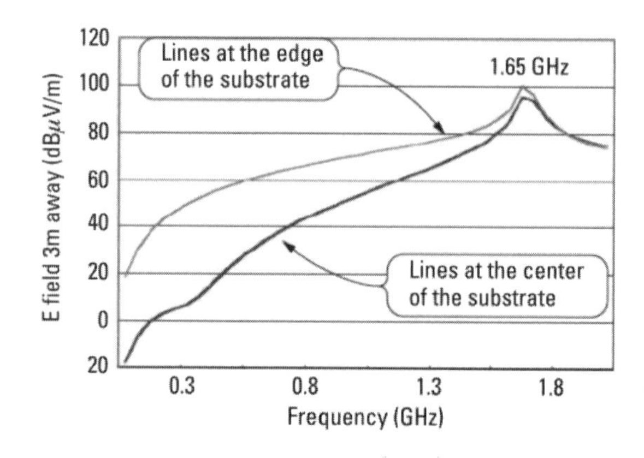

Figure 6.20 Electric field intensity at 3m from each circuit.

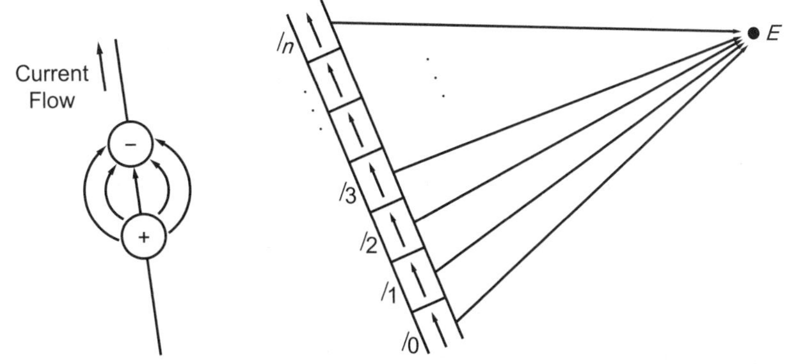

Figure 6.21 Viewing the entire circuit as a group of tiny (in the limit, infinitesimal) dipoles generating a radiated electric field.

$$I = \frac{dq}{dt}$$

In other words, when the infinitesimal dipoles of Figure 6.21 are joined together, we have a current that causes radiation.

The electric field, E, observed in free space is related to this current:

$$E \propto \frac{dI}{dt}$$

From this expression, we see that the electric field intensity at the observation point 3m away is proportional to the time rate of change of the current. The higher the frequency, the larger the rate of change, and the stronger the radiated electric field becomes. This is the reason that the curves in Figure 6.20 are increasing.

When all the current flows in the same direction, as in Figure 6.21, we have a common mode. In the normal mode, the differential pair can be represented by two rows of the tiny dipoles. When the distance between the pair of lines is small, the radiation from closely spaced and oppositely directed segments cancel each other and electromagnetic waves are not radiated.

This differential pair has a matched termination of 100Ω. If we change the termination to 50Ω, the electric field 3m away (peak value) increases by 5 dB. Thus, an improperly terminated line results in increased radiation.

6.4.2 Magnetic Field Around a Line

In Figure 6.20, a peak appears at around 1.65 GHz, and when looking at the surface current on a line, in Figure 6.22, there is minimum current at the center of the length of line, and we see that it is one half wavelength long. The reso-

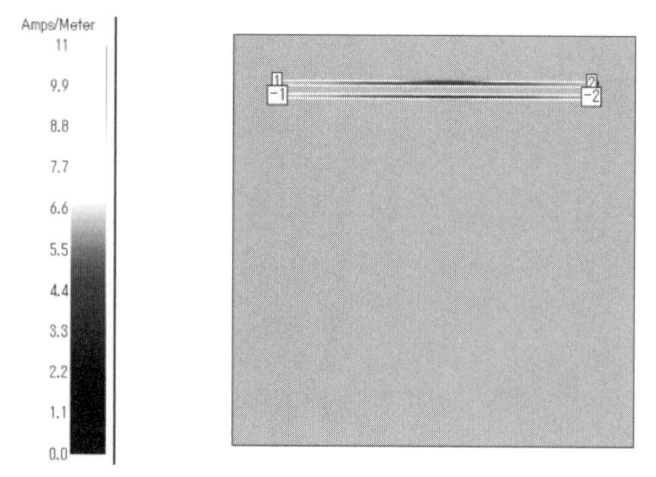

Figure 6.22 Surface current distribution on a differential pair at 1.65 GHz.

nant frequency of a 50-mm line length in free space is 3 GHz. Thus, we need to consider the wavelength shortening effect explained in Section 6.2.

While a differential pair transmission line can suppress unwanted radiation, why do we see strong radiation at 1.65 GHz?

Figure 6.23 is the current distribution on the ground plane. We see strong current along the edge. For example, the edge current is weak at 1.55 GHz, shown in Figure 6.24, which is 100 MHz below the resonant frequency. From this, we can see that radiation occurs from the edge of the ground plane when its size is about a half wavelength (or multiple thereof). This is the explanation

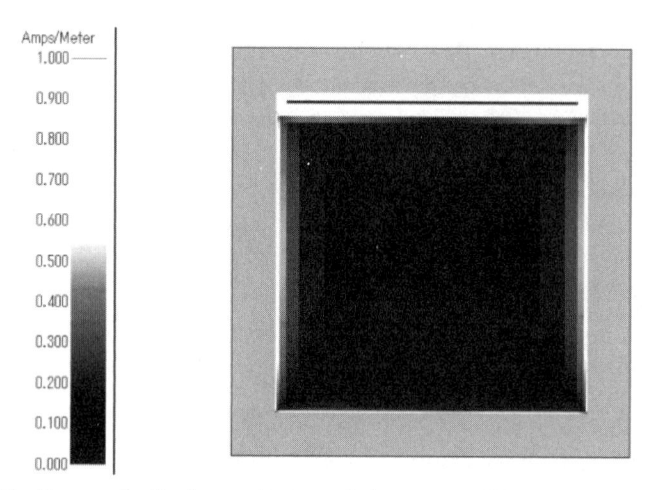

Figure 6.23 Current distribution on the ground plane at 1.65 GHz.

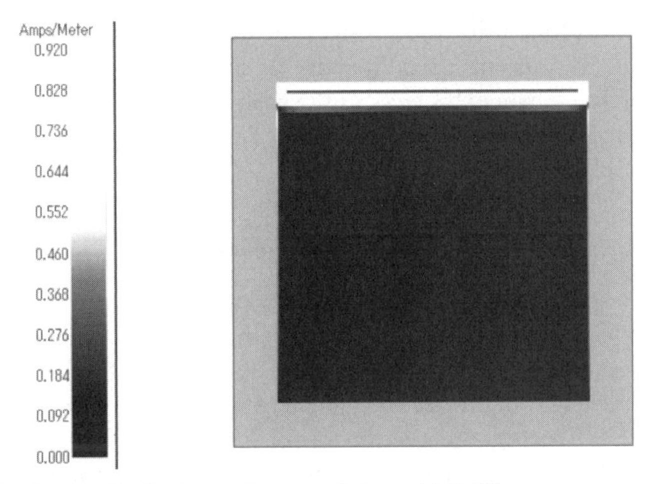

Figure 6.24 Current distribution on the ground plane at 1.55 GHz.

for the strange glitch in Figure 6.13. So, whenever you see a glitch, be sure to consider the size of the ground plane.

Figure 6.25 shows the magnetic field vector in free space. We can imagine the magnetic lines based on Ampere's right-handed screw rule: "When rotating a right-handed screw in the direction of magnetic field, the direction of the screw becomes that of the current that is generating that field." This is often visualized with the thumb of our right hand pointing in the direction of the current and the fingers curling in the direction of the magnetic field.

Note that strong current is observed on the edge of the ground plane. The direction of the magnetic field is parallel to the metallic surface, and the higher

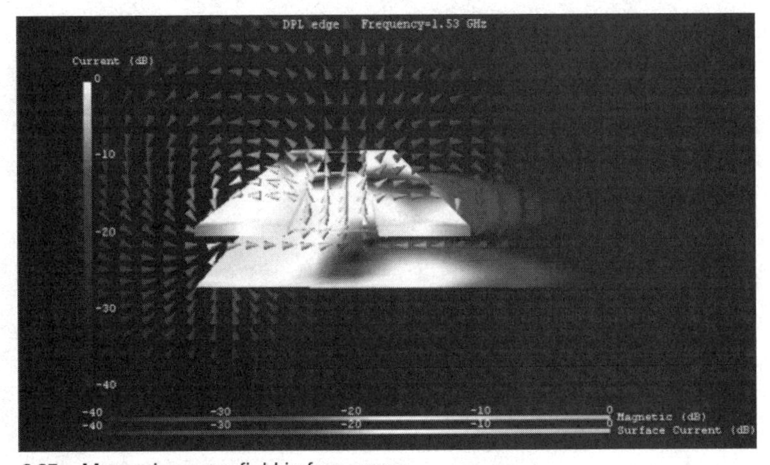

Figure 6.25 Magnetic vector field in free space.

the magnetic field intensity, the stronger the current flowing on the metallic surface. Thus, the current along the edge of the ground occurs because the ground plane is resonant. This gives us the peak of unwanted radiation at 1.65 GHz, where it is one half wavelength in size. There is also a resonance at every integral multiple of a half wavelength.

6.4.3 Field Loops and Electromagnetic Interference (EMI)

Figure 6.26 displays the electric field intensity over a large area at a higher frequency (7.49 GHz). This frequency generates the highest electric field at 3m of all frequencies up to 10 GHz.

The wavelength of an electromagnetic wave at 7.49 GHz is 40 cm, and the dark stripes that indicate a weak area of field intensity appear every half wavelength (20 cm). When viewing it with a time animation feature, we can see that the waves move from the circuit to free space.

The electric field vectors below the line head directly down to the ground surface, as in Figure 6.9. If all electric field vectors are terminated around the line, there is no radiation and we have a pure transmission line. However, electric field vectors loop out and spread to free space, as in Figure 6.26. They do not return to the circuit, and they flow out into space, one by one. This helps us visualize the mechanism by which an electromagnetic wave is radiated.

Of course, the magnetic field must also be present. We could view the Poynting vector power described in Section 6.2 to see the radiated power flow out from the circuit and become radiation. This sort of thing is the principal cause of EMI.

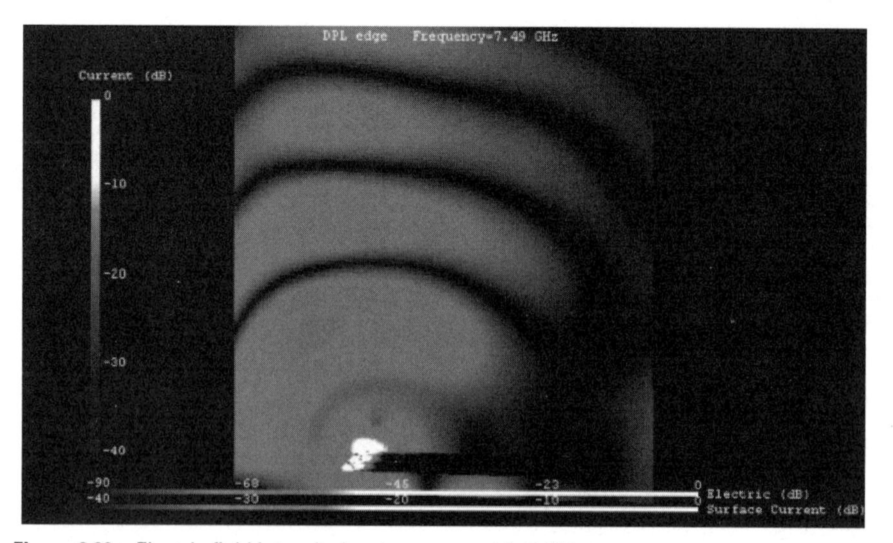

Figure 6.26 Electric field intensity in a large area at 7.49 GHz.

6.4.4 Far-Field Radiation

Figure 6.27 is the far-field radiation pattern, something that is more commonly used to evaluate transmitting antennas. When a differential transmission line is in the center of the substrate, the total radiation power as calculated by the EM analysis is 7.8 μW. This is a very small amount, but even at this level we must still be careful if our circuit has hundreds, or even thousands, of these lines.

On the other hand, when a differential transmission line is on the edge of a substrate, the radiated power increases to 72.4 μW, about a tenfold increase, as shown in Figure 6.28. In general, a differential transmission line has the advantage of having low EMI; however, when common mode components are involved, EMI can increase, so careful design is advised.

Awareness has spread widely that common mode noise is an important cause of EMI. Because undesired radiation takes the form of an electromagnetic wave radiated from a circuit, we can design the differential transmission line, as well as other transmission lines, using an electromagnetic field simulator. In this way, we can quickly make the trade-offs we need in order to meet our requirements.

When the ground plane width is narrow and there is a metallic housing or other circuitry near a differential pair in the microstrip of Figure 6.1(a), then unwanted current is induced because the electromagnetic field generates coupling between the various portions of the circuit. One form of this is common

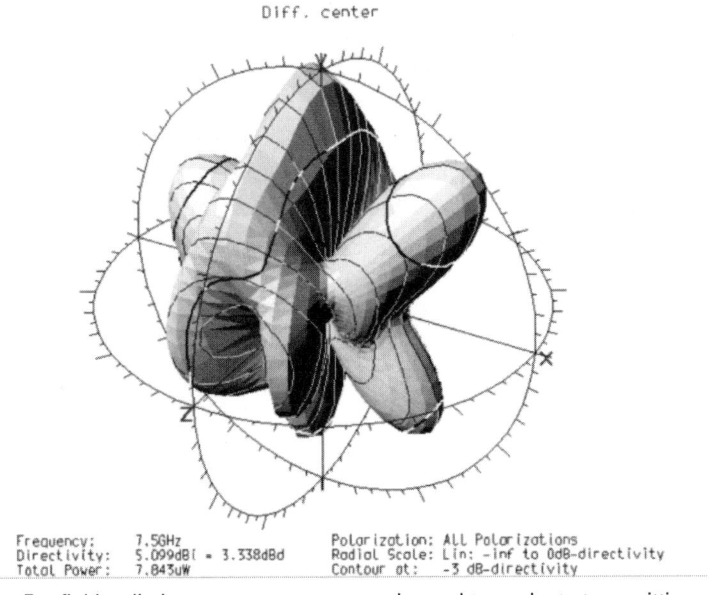

Figure 6.27 Far-field radiation patterns are commonly used to evaluate transmitting antennas. In this case, we evaluate a differential pair in the center of a substrate.

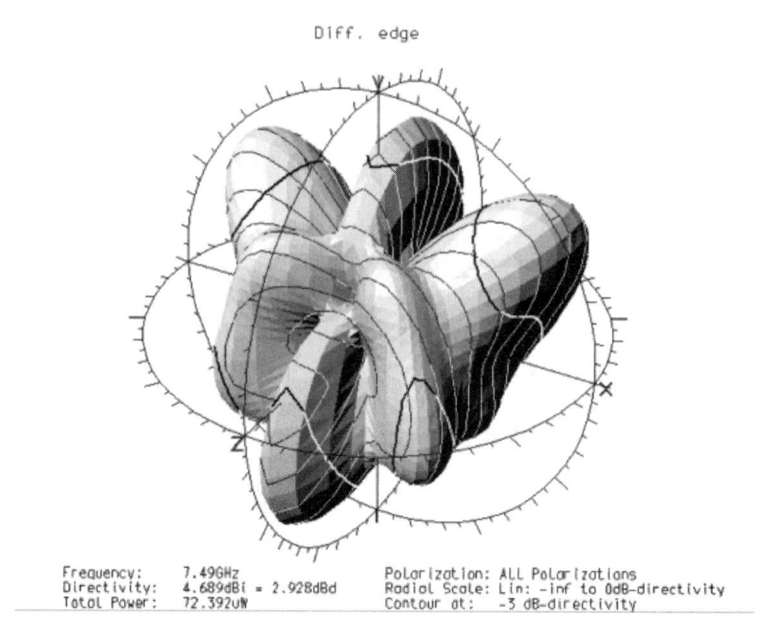

Figure 6.28 Far-field radiation pattern of a differential pair at the edge of a substrate.

mode current. The electromagnetic field generated by the signal line current and ground return current do not perfectly cancel, so the far field radiation can become large. We must also always keep in mind that the common mode current sometimes generates strong currents along the edge of the ground plane at high frequency. This too can be an important cause of EMI at high frequency.

The differential transmission line structures like Figures 6.1(b, c) have a characteristic that the common mode component is not easily excited because of the shielding effect of the two grounds. However, when two separate differential pairs approach each other, the mutual electromagnetic coupling can become strong, especially if the line terminations are not well matched.

6.5 Confirming This Chapter by Simulation

Ports of the differential transmission line described earlier in this chapter are set as numbered pairs, for example, 1 and –1, in Sonnet Lite. For a microstrip line, it is common to use box wall ports. These are ports on the edge of the substrate. Such ports use the box sidewall as their ground reference. When we want to see if strong current is flowing on the edge of our ground plane, we must model a ground plane floating inside the Sonnet box. In that case, we cannot use box wall ports. We must use a port that is placed inside the Sonnet box, rather than at the box wall. We demonstrate this technique by modeling a two-port differential transmission line.

6.5.1 A Microstrip Line Differential Pair

When assigning the dielectric substrate material, select FR-4 from the library included in Sonnet and use the default tanδ = 0.025. We also set the copper thickness to 30 μm (conductivity 58,000,000 *S*/m) as the line metal.

First, we model a line in order to set its dimensions so that the characteristic impedance is 50Ω. We set the line width to 1 mm and vary the dielectric thickness, and we see approximately 50Ω with a substrate thickness of 0.55 mm, as in Figure 6.29.

Next, we place two lines separated by 1 mm, as in Figure 6.30, and simulate the signals leaking to the next line. This is the crosstalk. Strictly speaking, when lines are close enough to couple, their characteristic impedance changes, but we ignore that for now. This circuit, Figure 6.30, is the same as Figure 6.3 of this chapter, so we simulate from 100 MHz to 2,000 MHz using the ABS sweep. Make sure you obtain the result of Figure 6.4 when you plot it with a voltage magnitude scale, rather than a decibel magnitude scale. Note that we are still using box wall ports for the initial simulations.

6.5.2 A Stacked Differential Pair

Figure 6.31 shows the bottom level of a pair of stacked differential lines, which requires an additional dielectric layer of the same size. The signal lines are one layer above and appear identical to the microstrip lines of Figure 6.30. The negative ports, shown in Figure 6.31, are attached to the lower lines, which form the ground return for the upper lines. The structure is the same as Figure 6.2. The line widths and the line separations are all 1 mm.

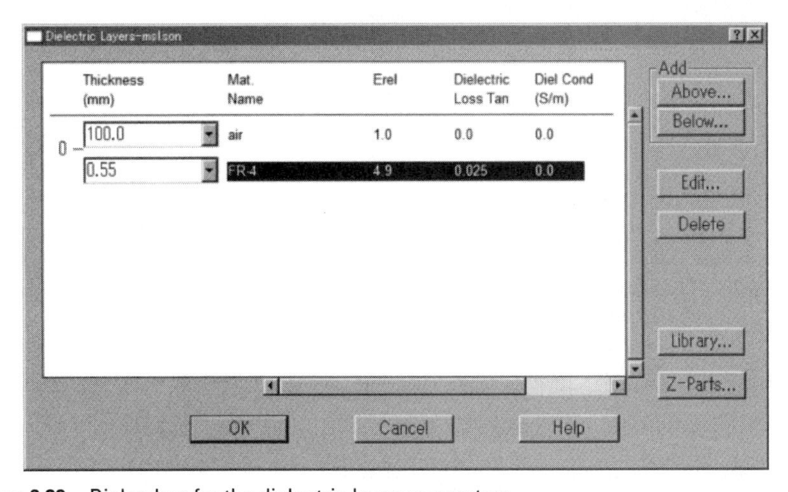

Figure 6.29 Dialog box for the dielectric layer parameters.

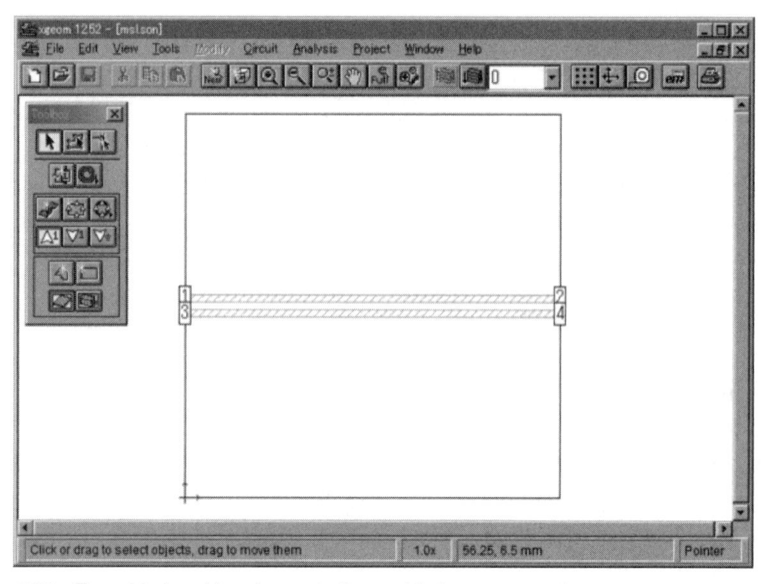

Figure 6.30 Two side-by-side microstrip lines with 1-mm separation.

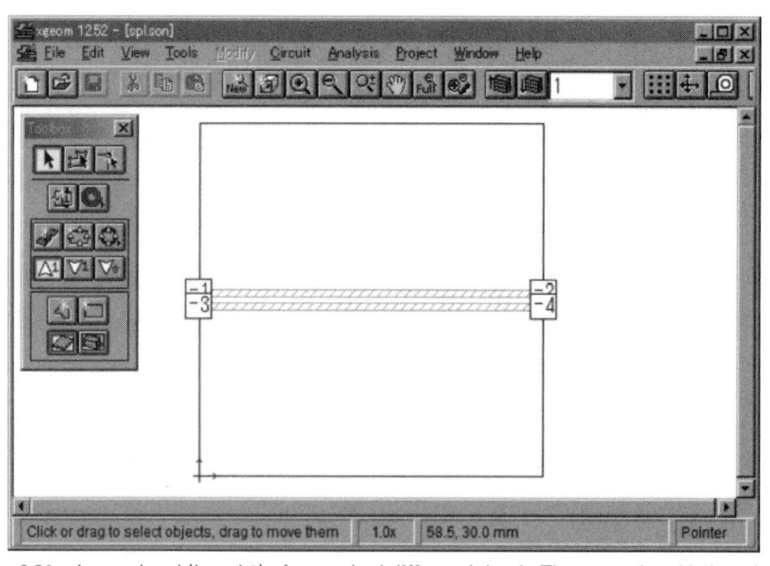

Figure 6.31 Lower level (Level 1) of a stacked differential pair. The upper level is Level 0 and appears identical to the microstrip lines of Figure 6.30.

For a single differential pair over a finite ground plane, shown in Figure 6.32, on a substrate of the same size, we set a 1-mm line width and separation, with the substrate thickness at 0.55 mm. In Sonnet, differential ports are realized with pairs of 1, –1 and 2, –2, as in Figure 6.32. After setting ports in the

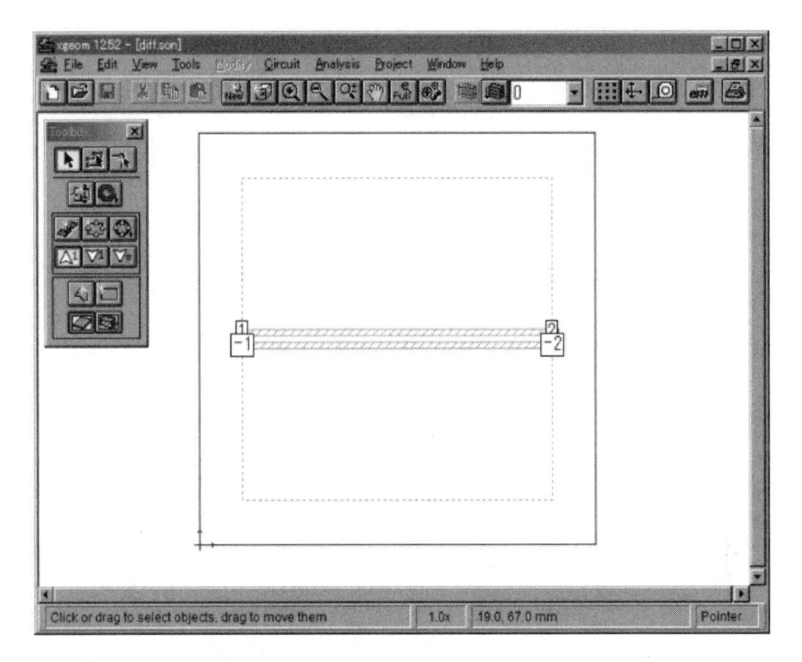

Figure 6.32 A single differential pair over a finite ground plane. File: diff.son.

same manner as the regular box wall ports, double click on the ports and set the ground return port numbers to –1 and –2.

Notice that to the left of ports 1 and –1, and to the right of ports 2 and –2, there is a small rectangular piece of lossless metal. This small piece of metal takes the place of the box wall that was there in previous analyses. It adds a small amount of inductance to the results, but that is so small it is not of concern for this analysis. The finite size ground was added as a large square on the next level down. It is indicated by the dashed line going around the circuit near the edge of the box. When you view diff.son in Sonnet, just press the down arrow key on your computer to view the ground plane.

6.5.3 Evaluating Radiation

Figure 6.33 shows the cell size and box size settings. We want to evaluate radiation from the circuit so we set the Top and Bottom of the box to Free Space. Figure 6.34 shows the settings for the dielectric layers. Select FR-4 from the Sonnet library and set 0.55 mm as the thickness.

Figure 6.35 shows the Sonnet layout for a differential pair transmission line moved to the edge of the ground plane. We specify the frequency range where we see strong current on the edge of the ground plane. Because the ABS sweep determines analysis frequencies automatically (starting with the

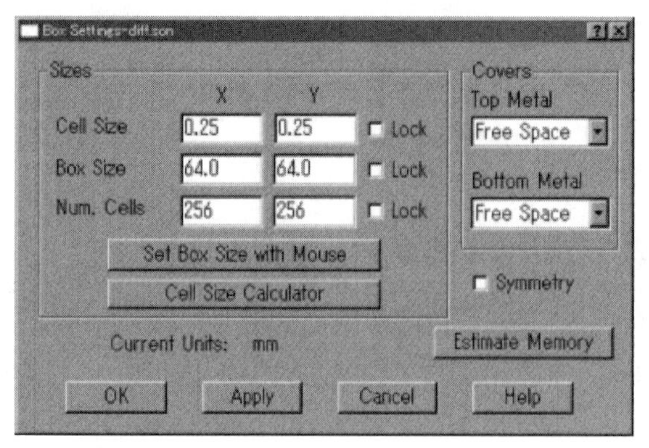

Figure 6.33 Cell size and box size settings.

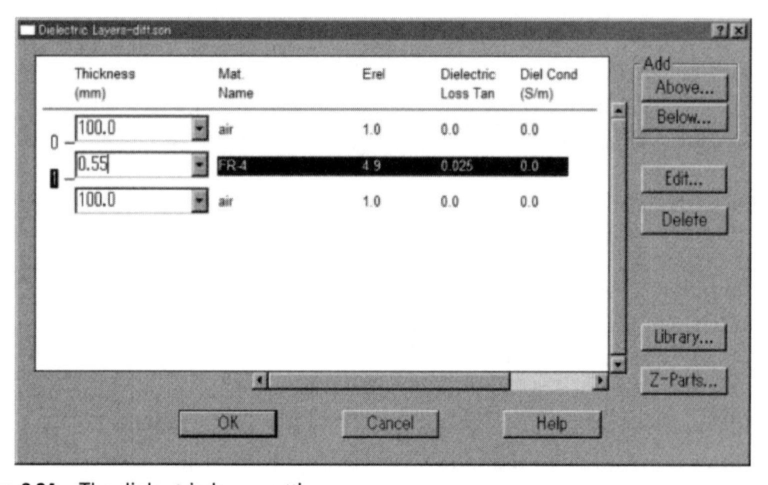

Figure 6.34 The dielectric layer settings.

minimum and maximum frequencies that we specify), here we set a linear frequency sweep, shown in Figure 6.36.

6.5.4 Reducing Memory Requirements

Setting the frequency step to 50 MHz means there are more EM analysis frequencies and the simulation takes longer. By checking Compute Current Density in the upper left of Figure 6.36, Sonnet saves the current distribution at all frequencies.

As this ground plane is wide, the required memory is also large (memory scales with the area of circuit conductor). Figure 6.37 shows the screen displayed after clicking Speed/Memory... button at the upper right in Figure 6.36. When setting the slider bar to center or to the far right, larger subsections are

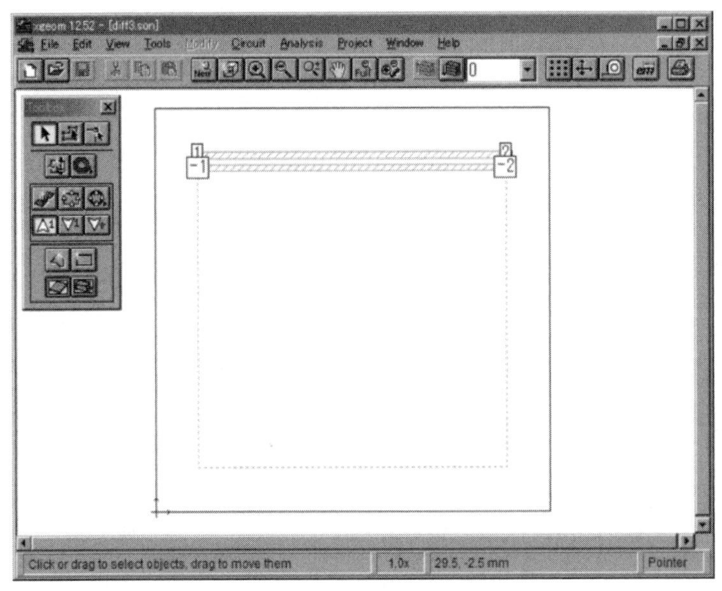

Figure 6.35 Two differential line pairs at the edge of the ground plane. File: diff3.son.

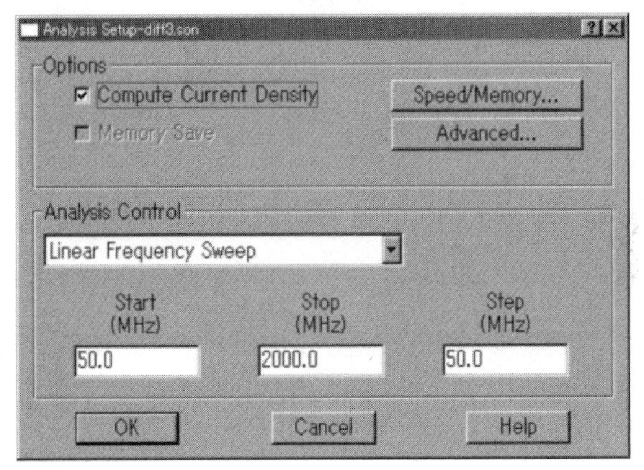

Figure 6.36 Linear frequency sweep set for a frequency range where we expect strong current on the edge of the ground plane.

used and accuracy decreases slightly, but it fits within the 16-MB Sonnet Lite limit.

The surface current distribution at 50 MHz shows that at low frequency, the surface current on Level 1 (the ground plane) is strong only directly under the line, shown in Figure 6.38.

With the current displayed, repeatedly press the right arrow key on the keyboard, and the displayed frequency increases. At 1,600 MHz (1.6 GHz), we now see strong current on the left and right edges of the ground plane. Looking

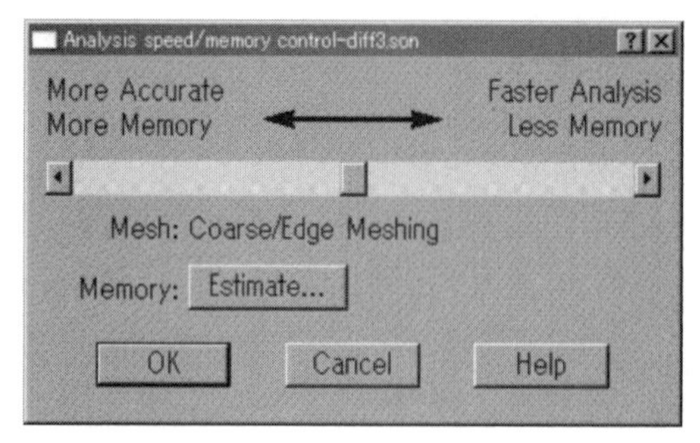

Figure 6.37 Memory usage control.

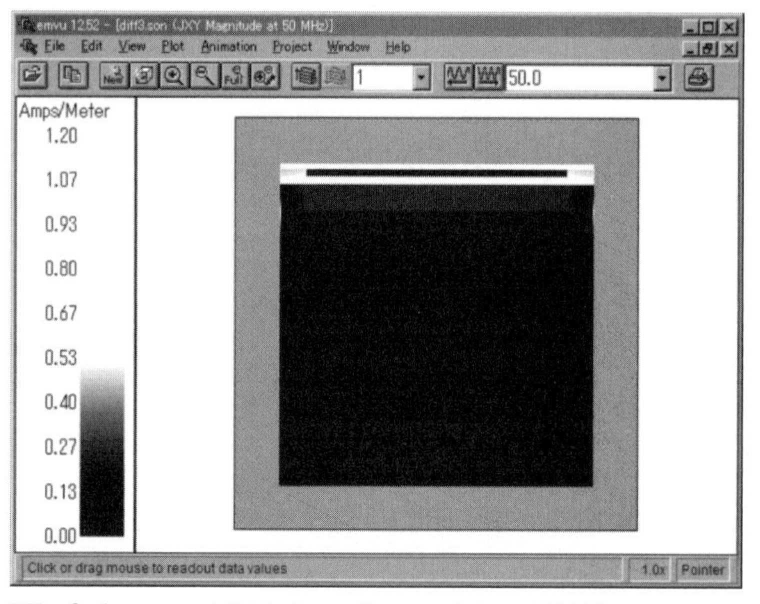

Figure 6.38 Surface current distribution on the ground plane at 50 MHz.

at even higher frequencies, like 1,650 MHz in Figure 6.39, we see that the surface current distribution is spreading to the interior of the ground plane.

The current concentrating on the ground edge of one plate has no nearby oppositely directed current to cancel its radiation. When this current is strong, as in Figure 6.20, the electric field at 3m can be a major cause of EMI.

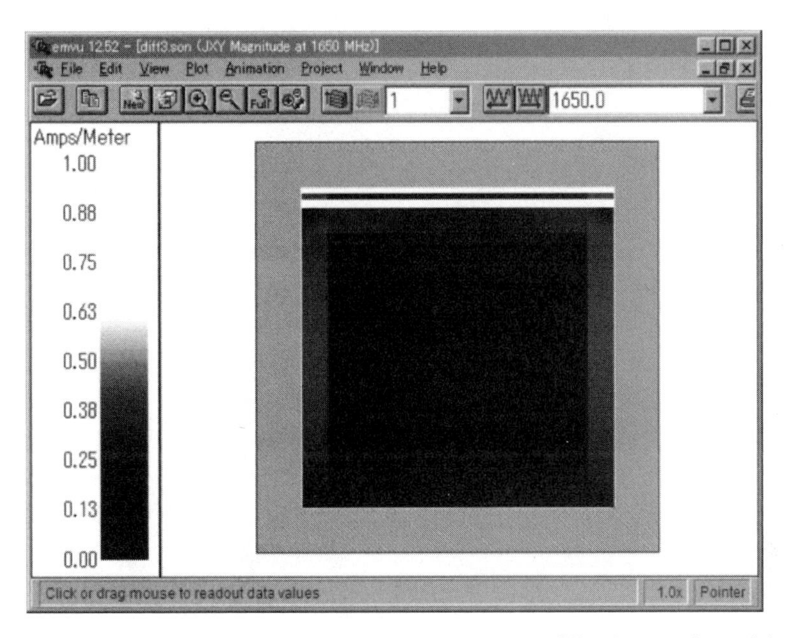

Figure 6.39 Current distribution on the ground plane at 1.65 GHz shows substantial edge current.

6.6 Summary

1. Properly designed shielded differential transmission line structures (e.g., stripline) are not prone to radiation and EMI because of the shielding effect of two grounds.
2. The crosstalk in a differential transmission line is smaller than in a similar microstrip line.
3. The differential transmission line tends to concentrate the electromagnetic field around the line, and thus less electromagnetic energy is coupled to other nearby lines than with similar microstrip line.
4. When positioning a differential transmission line on the edge of a substrate, the common mode component becomes a problem. This is not seen when the line is in the center of the substrate.
5. When positioning a differential transmission line on the edge of a substrate, a resonance occurs when ground plane is about an integer multiple of a half wavelength in size and transmission line common mode current couples to the ground plane. In this situation, electromagnetic waves are radiated.

7

Electromagnetic Compatibility Design Is Commonsense High-Frequency Design

7.1 Taking Out the Mystery

In Chapter 6, we learned the merits of the differential pair. But, depending on the layout, we found that sometimes the electromagnetic field around the substrate is radiated into free space. In addition, a conducting enclosure, or a finite size ground plane that couples to high-frequency circuits, can resonate at many frequencies. The electric energy supplied to a device is the most significant source that can excite these resonances; however, it is entirely possible that the electric energy from external devices can get in through an aperture (e.g., cooling vents) and excite resonances. If waves can get in, then they can also get out and radiate. Have you ever been awakened at night by a cell phone that was placed too close to a clock radio? In this chapter, we examine the role that electromagnetic field simulators play in achieving good electromagnetic compatibility (EMC) and electromagnetic interference (EMI) performance.

7.1.1 What Is EMC?

Electromagnetic compatibility is the characteristic that an artificial system can achieve its desired performance without radiating excessive electromagnetic energy and polluting the electromagnetic environment or being affected by the electromagnetic environment.

A simpler description is that EMC is an environmental issue dealing with electromagnetic waves. Or, here is an analogy with the heat of a thermos bottle: something that does not radiate electromagnetic waves (heat) is resistant to being affected by electromagnetic waves. Something that does not absorb electromagnetic waves (heat) is resistant to emitting electromagnetic waves. This is electromagnetic compatibility! (It turns out that heat is also an electromagnetic wave.)

The same situation exists with a good antenna that can receive electromagnetic waves well, as it can also transmit electromagnetic waves well. That is, a transmitting antenna can also be used as a receiving antenna as it is. The electromagnetic term for this characteristic is reciprocity, a critical concept for EMC.

The important system concepts for high-frequency EMC design are the three elements shown in Figure 7.1. The basic elements of EMC are that electromagnetic energy is radiated from and received by antennas (even if it does not look like an antenna), there are high frequency noise sources both inside and outside a circuit, and this unwanted noise energy is transported over various conductors, or even through free space, as a transmission line to arrive at an unintended "antenna." The basic problem is that the components of a system might perform in a manner that is inconsistent with its intended function.

In order to meet the needs of miniaturization and weight reduction, the degree of circuit integration is becoming very high. Because circuit components are designed so that they work with lower power and lower voltage, they become increasingly sensitive to the influence of electromagnetic energy radiated by adjacent electrical and electronic systems. The range of applications of artificial systems using electrical and electronic equipment has become completely embedded in our lives. Electromagnetic environmental problems in our society(e.g., electromagnetic interference to a cardiac pacemaker) must be carefully considered.

Thus, EMC is the property that the electromagnetic environment and our artificial systems can peacefully coexist. It is extremely important to eliminate as best as we can susceptibility to the electromagnetic environment and

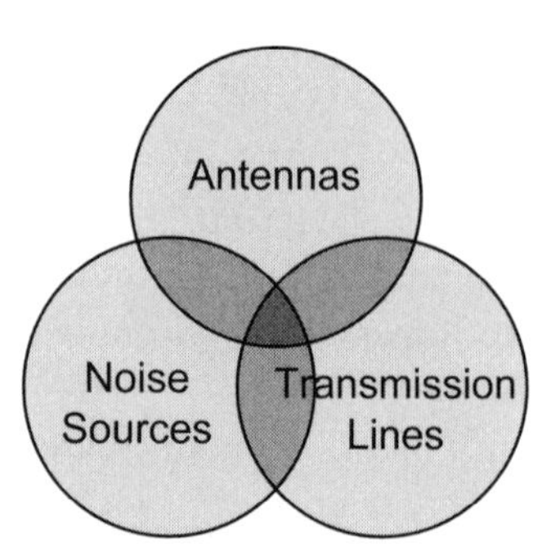

Figure 7.1 The three base elements of EMC.

enhance immunity against the electromagnetic environment for our electrical and electronic systems.

7.1.2 Modeling EMC Problems

In EMC problems, it is important that we can simulate complicated shapes because we must investigate not only radiation from the signal source but also the radiation caused by the resonances of all of the circuitry, including ground planes and radiation from apertures, of a multilayer printed circuit. The following tasks must be accomplished:

1. Analyze the current distribution of the entire structure and dig into the critical portion(s) qualitatively.
2. Obtain S-parameters of critical transmission lines and analyze the crosstalk between each line quantitatively.
3. Run an electromagnetic field simulation of the near and far fields of a printed circuit.
4. When placed in a product, the printed circuits should be simulated as they are in the final product design.

As for the electromagnetic susceptibility of a printed circuit in a housing, we must simulate the detailed distributions of electric and magnetic fields inside the housing at the resonant frequencies of the housing. In Table 7.1, we summarize simulation cases associated with EMC that the authors have performed.

7.2 Electromagnetic Waves Penetrating Through an Aperture

We consider a simulation for a printed circuit mounted in a housing. In this problem, we look at two cases. One is radiation from a circuit inside a housing and its effect inside and outside the housing. The second is an external electromagnetic field penetrating through an aperture of the housing (e.g., a slot for inserting DVDs) to couple to the internal circuit. We next describe the simulation for the second case.

As shown in Figure 7.2, we place a multilayer printed circuit substrate that consists of a signal line, dielectric layers, a ground layer, and a Vcc (power supply) layer, and we simulate the induced current generated by external electromagnetic field coupling to the circuit through an aperture in the housing, We use MicroStripes from CST for EM analysis. It is based on the TLM method.

Critical physical dimensions are as follows:

- Upper dielectric thickness: 2.0 mm, relative permittivity: 4.8;

Analyzed Problem	Evaluated Quantity	MoM	TLM	Meas.
Multilayered printed circuits	Current distributions	○	○	
	Field distributions		○	
	Radiation	○	○	
	S-parameters	○	○	
Housing with aperture	Cavity resonance	○	○	○
	Aperture currents	○	○	
	Field distributions		○	
Aperture impedance	Aperture impedance	○	○	
Circuits inside a housing	Induced housing current		○	

Table 7.1 Simulation Cases Associated with EMC

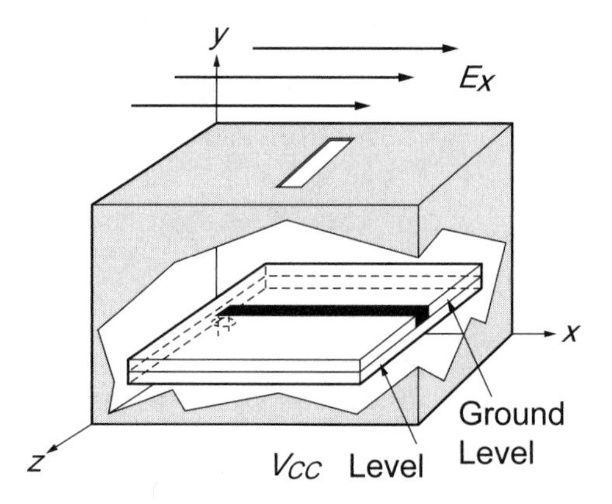

Figure 7.2 Model of a multilayer printed circuit substrate inside a housing.

- Lower dielectric, thickness: 1.0 mm, relative permittivity: 4.8;
- Signal line width: 1.0 mm;
- Substrate size: 260 mm (*x*), 180 mm (*z*);
- Housing size: 300 mm (*x*), 150 mm (*y*), 200 mm (*z*);
- Aperture: 10 mm (*x*), 100 mm (*z*).

The thickness of the metallic plates that form the housing and ground are assumed to be zero. The substrate is placed in the housing 50 mm (*y*) above the bottom face.

Starting at the top, the printed circuit consists of a signal line, a ground, and a *Vcc* (power supply) layer. The signal line is terminated on both ends with 50Ω loads; one end is connected to the ground layer and the other end is connected to the Vcc layer. We analyze the case where the aperture is placed at the upper center in the housing and the straight side is in the *z* direction.

7.2.1 Exploring the Properties of the Frequency-Domain Response

We connect the four corners of the ground plane to the housing sidewall with conductors 10 mm wide. We also connect the midpoints of the long sides of the ground plane to the housing sidewall as well. The result, Figure 7.3, shows the current induced in the circuit from an electromagnetic wave incident from above on the housing.

The MicroStripes TLM method is based on the time domain. To analyze a structure, it launches an impulse into the structure. Then the impulse response is converted to the frequency domain by a discrete Fourier transform, one form of which is the well-known fast Fourier transform (FFT). We see the frequency domain data plotted here. We set the observation point (output point) at the center of the line. Using this approach, we can easily obtain very broadband data. This allows us to examine the properties over a wide range.

Comparing this result with that of the same circuit, only with a floating ground plane (no straps connecting the ground to the sidewalls), the induced EMI current is about 5 dB lower when the ground plane is connected to the sidewall of the housing up to around 1 GHz. But at 700 MHz, 800 MHz, and 1 GHz, the peaks are higher by 20 dB, as in Figure 7.3.

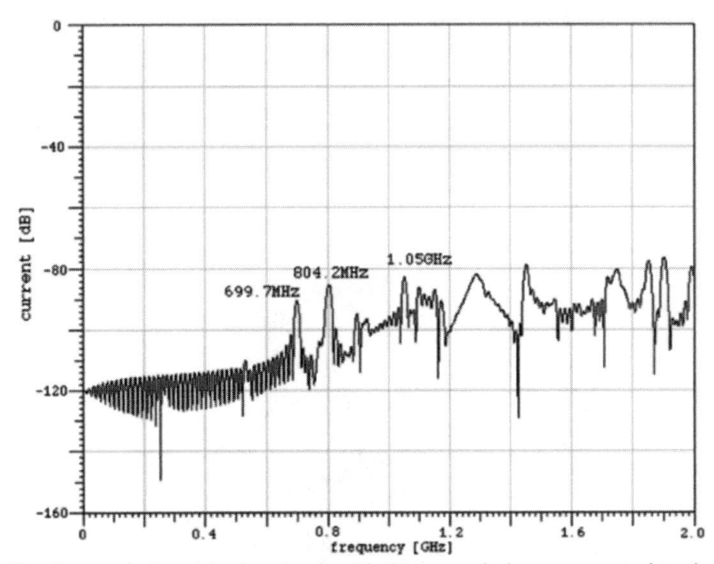

Figure 7.3 Current induced in the circuit with the ground plane connected to the housing sidewall in six places.

7.2.2 Half-Wavelength Resonator

In order to examine the cause of these current peaks, we analyze the current distributions inside the housing at each frequency. Figure 7.4 shows the current density distribution (perspective view) at the first peak, 699.7 MHz.

Looking at this result, strong current is seen, especially along the edge in the *z* direction of the ground plane and at both terminated ends. Examining the current distribution in more detail, we see that the high current continues along the whole 220-mm length, including the total length connecting to the sidewall and the edge of the ground layer.

Judging by these results, we infer that somehow we have made a half-wavelength resonator. Note that each layer is filled with dielectric, and we might expect some degree of wavelength shortening effect. However, because most of the resonant current flows on the edge of the ground plane, the electric lines of force traveling in the dielectric are limited.

Consequently, the effect of the filling dielectric is low, and the wavelength shortening has little influence. The observed resonant frequency of 699.7 MHz, where the current peak is seen in the simulation, is close to the resonant frequency of 682 MHz calculated for a free space resonator 220 mm long.

7.2.3 Magnetic Field Distribution Between Layers

Figure 7.5 shows the magnetic field distribution just above the ground plane at 804.2 MHz, where we see the second peak. In this case, strong current is not seen at the shorting points on sidewall, so we conclude that the resonance must have some other cause.

When examining Figure 7.5 in detail, the area enclosed by the centrally located portion at about 120 mm in the *z* direction over the length of 260 mm

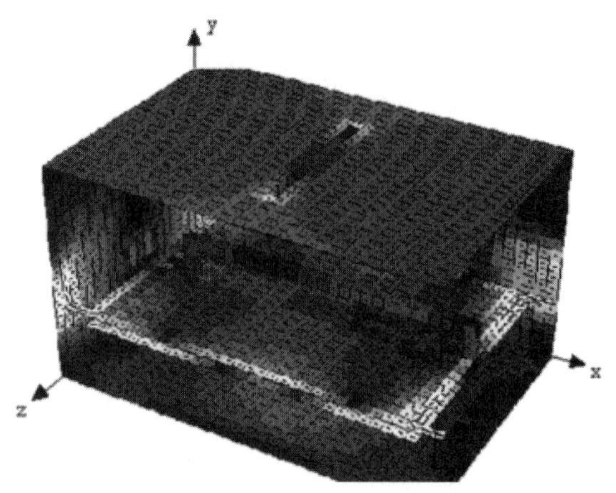

Figure 7.4 Current density distribution at the first peak frequency, 699.7 MHz.

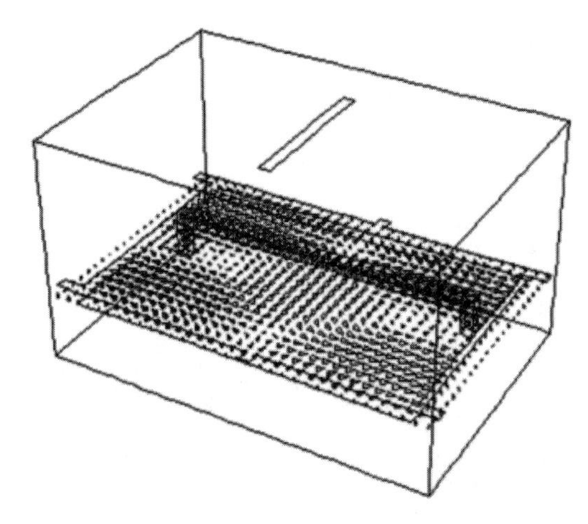

Figure 7.5 Magnetic field distribution between layers at 804.2 MHz.

in the *x* direction shows the electromagnetic field distribution to be similar to the TE_{102} mode of a cavity resonator, where, in this case, the cavity is formed by the printed circuit layers.

As for this resonance, which has strong fields in the dielectric, we see that the effect of the filling dielectric is high, because most of the electric lines of force exist in the dielectric between layers. If the entire space is filled with a dielectric of relative permittivity 4.8, the wavelength shortening effect is $1/\sqrt{4.8} = 0.46$, and the resonant frequency corresponding to this situation is 804 MHz, close to the analysis result.

7.3 Housing Resonances

When putting a high-frequency circuit inside a metallic housing, the housing sometimes becomes a cavity resonator at some frequencies. Because of this problem, we must examine the resonant modes of a hexahedral (rectangular) cavity resonator in detail. In this case, we actually built and measured the resonance of a metallic housing before putting a multilayer printed circuit inside.

Figure 7.6 is a prototype metallic housing with the dimensions of 300 mm (*x*) × 150 mm (*y*) × 200 mm (*z*). Aluminum sheet metal 1.5 mm thick is screwed on as a top cover, and it is sealed with copper foil conductive tape on the inside. The aperture at the center is 100 mm × 10 mm. This represents a slit for DVD insertion.

The measuring setup is shown in Figure 7.7. We used a double-ridge guide antenna to launch a plane wave towards the top of the metallic housing. We measured S_{11} (reflection coefficient) using a network analyzer.

Figure 7.6 Prototype metallic housing including a slit that represents a DVD insertion slot.

The horn antenna needs to be far enough from the housing so that the wave incident on the box approximates a plane wave. However, when we perform the actual measurement with the horn antenna at a large distance, the measured result was unstable. This is likely due to multiple reflections from other nearby objects that were also illuminated by the horn antenna. To eliminate the influence of these extra reflections, we fixed the antenna location 30 mm above the box. Not having a perfect plane wave means our illumination is different from what we would normally use for EM analysis, but the frequency of resonances will see little effect.

Figure 7.8 is the measured S_{11} showing that the incident wave reflects back from the top of the housing and returns to the antenna over most of

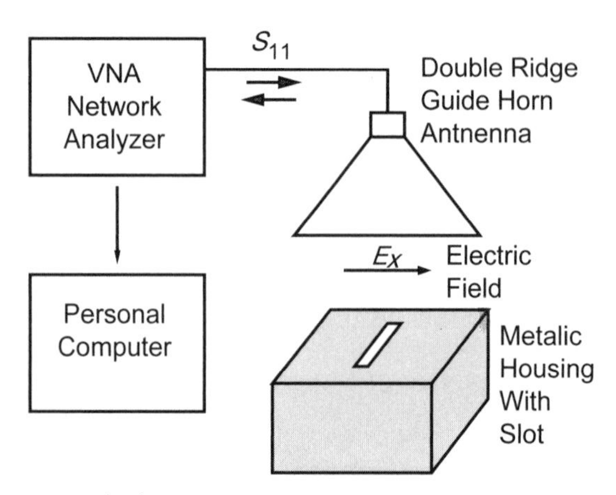

Figure 7.7 Measurement setup.

the frequency range. However at several specific frequencies, S_{11} dips, and the reflection is reduced at, for example, 1.02, 1.32, 1.60, 2.19, and 2.44 GHz.

What does it mean that there is less electromagnetic energy reflected back to the antenna at these frequencies? It does not seem natural that electromagnetic waves would be absorbed only at very specific frequencies by any resistance in the sides of the housing. Absorption like that would be broadband.

Narrowband absorption at very specific frequencies could be due to resonances in the housing. Cavity resonators have resonant modes that are examined in detail in the next chapter. These modes are resonant and absorb power only at very specific, narrowband frequencies. We can imagine that these frequencies are where the dips of S_{11} are observed, with one dip for each resonant mode.

There are an infinite number of resonant frequencies in a sealed conducting housing. In the case of a hexahedron (rectangular prism), the frequencies are given by

$$f_{mnq} = c_0\sqrt{\left(\frac{m}{2a}\right)^2+\left(\frac{n}{2b}\right)^2+\left(\frac{q}{2c}\right)^2}$$

where a, b, and c are, respectively, the x, y, and z axis dimensions of the hexahedron; c_0 is the velocity of light (the velocity of the electromagnetic wave); and m, n, and q, are mode numbers in the x, y, and z axis directions, as in Figure 7.9.

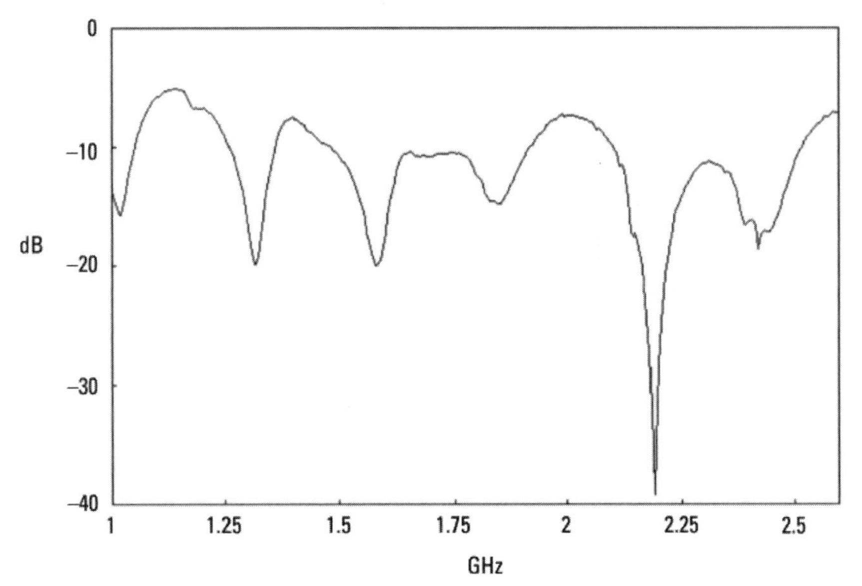

Figure 7.8 Measured S_{11} shows box resonances.

When we put this formula in a spreadsheet, like Excel, we can calculate the resonant frequencies of all modes by inputting each case of *m*, *n*, and *q* as desired. However, in this case, we do not have a perfectly sealed cavity because the top cover has a slit. In addition, we assume the excitation is an electromagnetic wave traveling in external free space and incident from above. So we should examine more carefully whether or not the cavity resonator theory is applicable.

7.3.1 Analyzing Resonant Cavity Modes

Figure 7.10 shows the simulation result of the electric field, E_x, at the observation point when the illuminating plane wave electric field is in the *x* direction. At each frequency of 1.10, 1.32, 1.64, 2.16, and 2.40 GHz, we see peaks in the observed (inside the housing) electric field, which indicate that the housing resonates. The frequencies, indicated along the bottom by small triangular marks, are the measured resonant frequencies for the metallic box of Figure 7.6. We discuss these resonant modes a little more later in this chapter.

The peak at around 1.1 GHz is a resonance that is significantly lower than the predicted 1.25-GHz resonant frequency of the TM_{011} mode, the lowest natural frequency inside the housing. The mode number indicates how many half wavelengths the fields go through in the *x*, *y*, and *z* dimensions, respectively. So, the TM_{011} mode has constant electric field in the *x* direction, and shows one half wavelength along the other two directions.

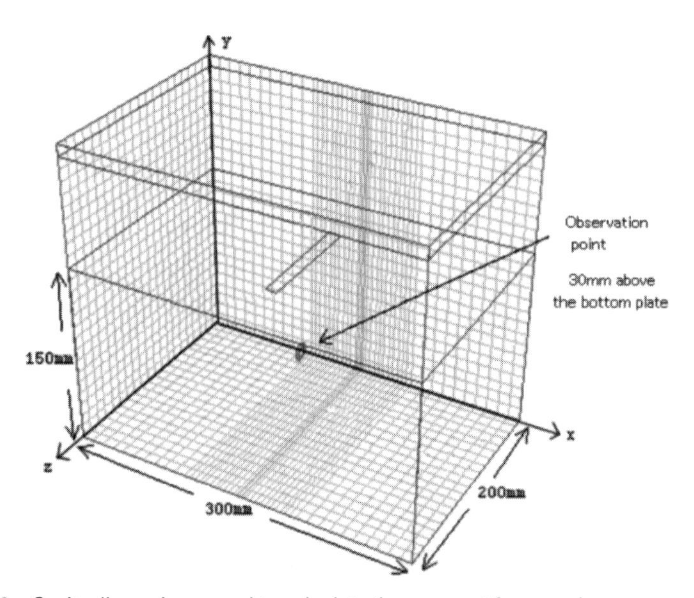

Figure 7.9 Cavity dimensions used to calculate the resonant frequencies.

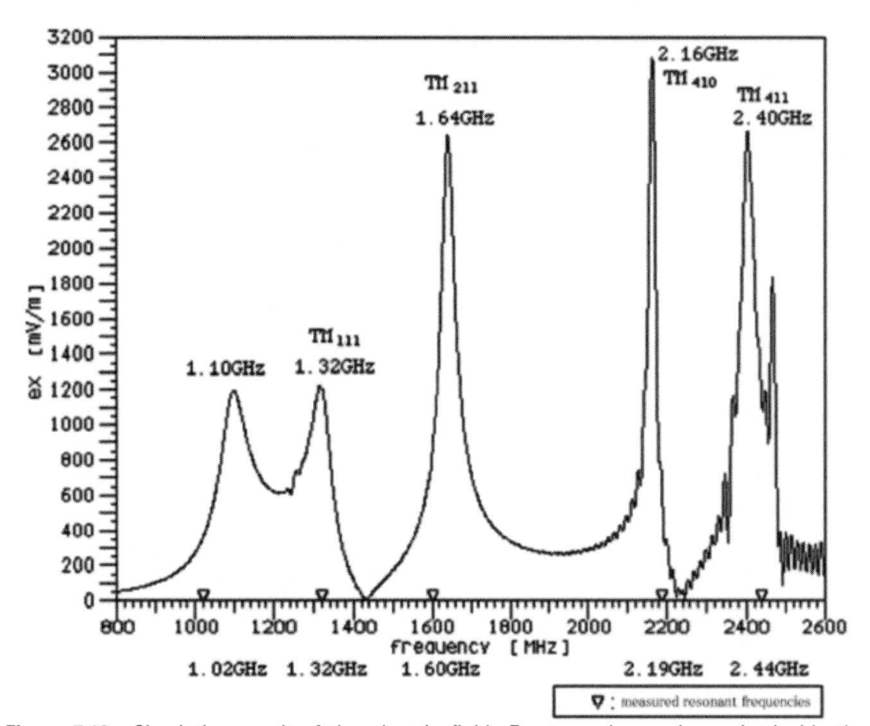

Figure 7.10 Simulation result of the electric field, E_x, at an observation point inside the housing.

The difference is large enough that it looks like our resonant cavity theory, discussed earlier, does not seem to work for this, the lowest resonance. When we examine the fields more closely, we will find that this first resonance is actually the slit resonating as a slot antenna. That mystery is solved, but now we have another one. Why don't we see that first cavity resonance? Perhaps we are trying to observe it in a location where the fields of that mode are zero? Or perhaps we are not even exciting that mode. We discuss this later.

As for each resonant frequency higher than 1.3 GHz, we find that they correspond to the resonant frequencies of the cavity resonator, 1.35 GHz (TM_{111}), 1.60 GHz (TM_{211}), 2.24 GHz (TM_{410}), and 2.36 GHz (TM_{411}), which are determined by the size of the housing. In these cases, the resonant frequencies are not as strongly influenced by the aperture.

Figure 7.11 shows, as a case in point, the magnetic field vectors of the y–z plane ($x = a/2$) for the TM_{211} mode resonant at 1.64 GHz, shown in Figure 7.11(a), and the electric field vectors of the y–z plane ($z = c/2$) in the same mode, shown in Figure 7.11(b), with cones.

When we illuminate the box slit at this resonant frequency, a strong electric field appears inside the housing. In the electromagnetic solution of this case, we can see that the electric field vectors distribute concentrically around the aperture, shown in Figure 7.11(b).

7.3.2 A Mode That Is Not Excited

Based on the theory of the cavity resonator, the resonant modes calculated previously with spreadsheet software should agree with our measurements fairly closely. However, we find that some resonances in this housing can be influenced by the slit.

Because the aperture is in the center, it cuts any current that might flow in the x direction on the top cover of the housing. Thus, the dominant TM_{011} mode simply cannot occur. Consequently, the observed cavity modes are TM_{111}, TM_{211}, TM_{410}, and TM_{411} in that order. We also find that the modes TM_{112}, TM_{113}, and so on are also suppressed by the slit-cutting current that must flow if we are to have those modes.

It might be possible to turn this mode suppression to our advantage. We can easily imagine that it is possible to shift the resonant frequency of the housing at critical operating frequencies of a printed circuit board, for example, by adjusting the position of the slit and the size of the housing.

7.4 A PCB Inside a Housing

We must also consider whether or not the information obtained by simulating an empty housing is applicable to the case of a housing with a printed circuit board inside it. So we put a circuit, shown in Figure 7.12, in the middle of the housing, 75 mm above the bottom, and simulated the current induced in the circuit by an electromagnetic field illuminating the box from the outside.

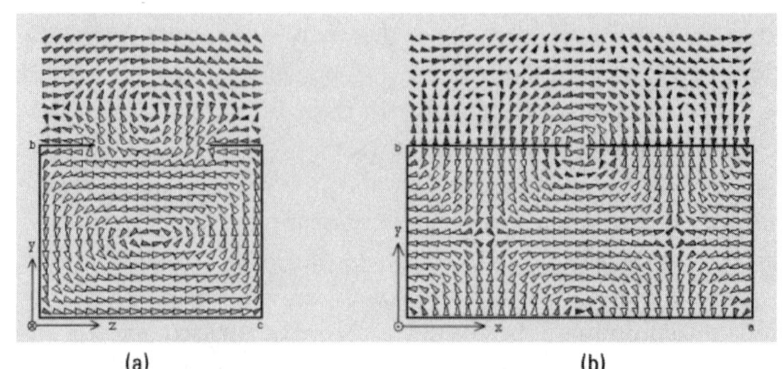

(a) (b)

Figure 7.11 (a) Magnetic field vectors of the y–z plane ($x = a/2$) for the TM_{211} mode (1.64 GHz); (b) electric field vectors of the y–z plane ($z = c/2$) in the same mode.

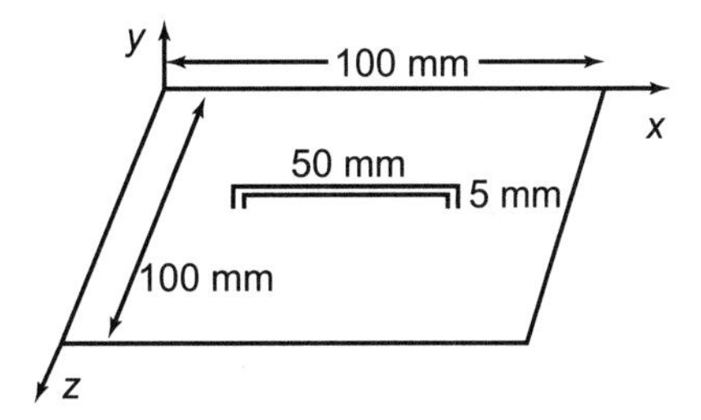

Figure 7.12 Printed circuit substrate in the middle of the housing.

Figure 7.13 shows the MicroStripes TLM model. The circuit transmission line is 50 mm long and with a 5 mm separation from the ground plane (100 mm × 100 mm). Both ends of the line are terminated with 50Ω loads.

The induced current at any location in the transmission line can be output by the simulation. Figure 7.14 shows the current at the center of a line. The resonances at 1.34 GHz (TM_{111}), 1.61 GHz (TM_{211}), 2.17 GHz (TM_{410}), 2.37 GHz (TM_{411}), and so on correspond to the resonant modes, and the induced current in the transmission line of the printed circuit board are seen as the peaks in the graph. In other words, the induced current shows peaks at frequencies that correspond to each resonant frequency of the housing. By the way, the peak at 1.46 GHz is not caused by a housing resonance. In this case, it is because the edge of the ground plane is one half wavelength on each side, 100 mm × 100 mm. Viewing the electromagnetic fields, and in particular the current, is extremely useful in determining the source of any resonances we might observe.

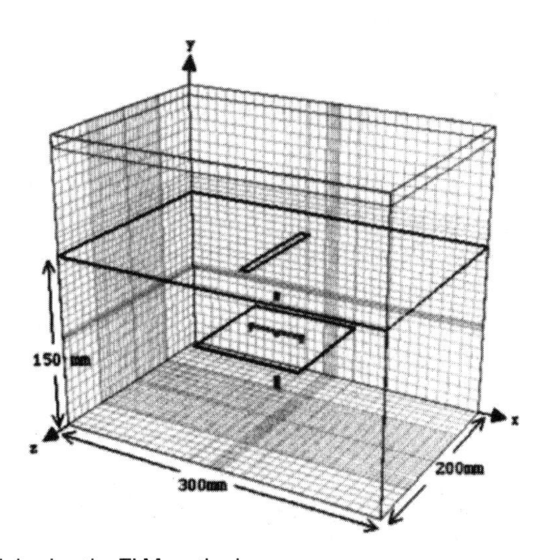

Figure 7.13 Model using the TLM method.

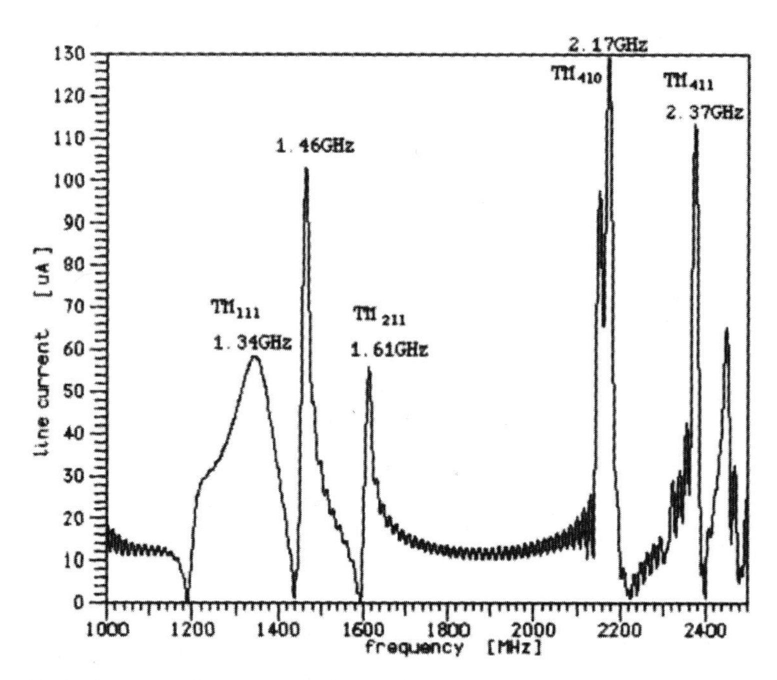

Figure 7.14 Current at the center of a line.

Figure 7.15 shows the simulation result of the electric field in the *x* direction 37.5 mm above the top of the substrate. This is the center of the free space in between the substrate and the top cover (with the slit) of the housing. We see that the frequencies of peak electric field have very good match with the frequencies of the electric field peaks in Figure 7.14.

7.5 The Microprocessor Unit Heat Sink and Radiation Noise

The heat sink of a microprocessor unit (MPU) has a size that can work as an antenna at several hundred megahertz to several gigahertz, as shown in Figure 7.16. When simulating, we put the heat sink on a MPU and it is excited directly. Figure 7.17 shows the strong electric field distribution at 3.9 GHz.

Figure 7.18 is the electric field observed just above the heat sink, which models the heat sink as electrically floating, unattached to ground. This shows peaks of the electromagnetic field at around 4 GHz and 7.5 GHz.

The electric field observed just above the heat sink when the four corners of the heat sink are electrically connected to the ground is shown in Figure 7.19. Comparing with Figure 7.18, we can see that the electromagnetic field level is reduced below 2.5 GHz. However, the peaks of the electromagnetic field near 4 GHz and 7.5 GHz remain.

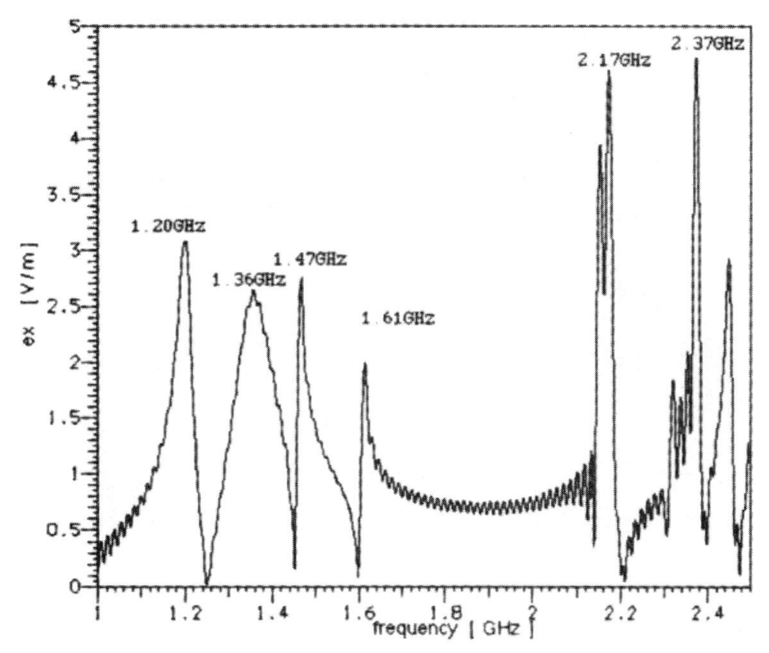

Figure 7.15 Electric field in the *x* direction 37.5 mm above the top of the substrate.

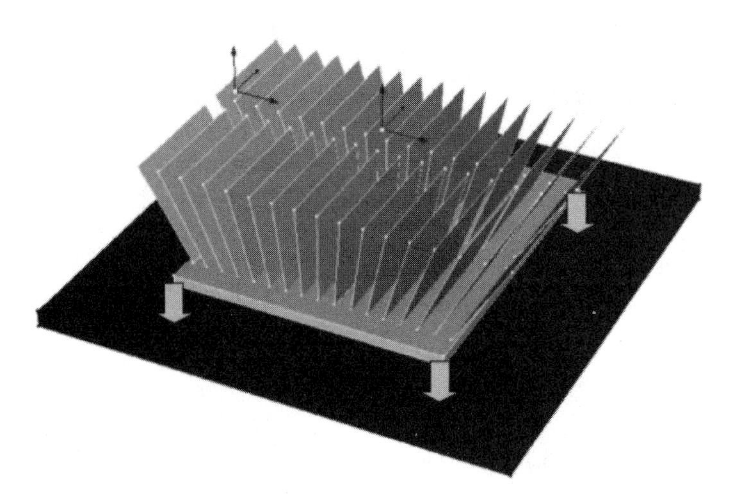

Figure 7.16 Heat sink of a microprocessor unit.

We see that we must change the size and shape of the heat sink in order to change the peak resonant frequencies. Figure 7.17 shows that strong common mode currents flow on the surface of the radiator fins, which means that they play the unintended role of an antenna at 3.9 GHz.

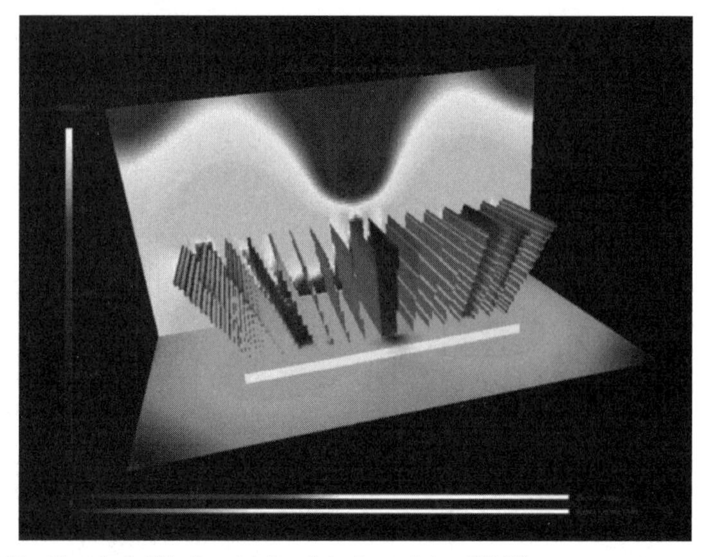

Figure 7.17 Electric field in the vicinity of the heat sink at 3.9 GHz.

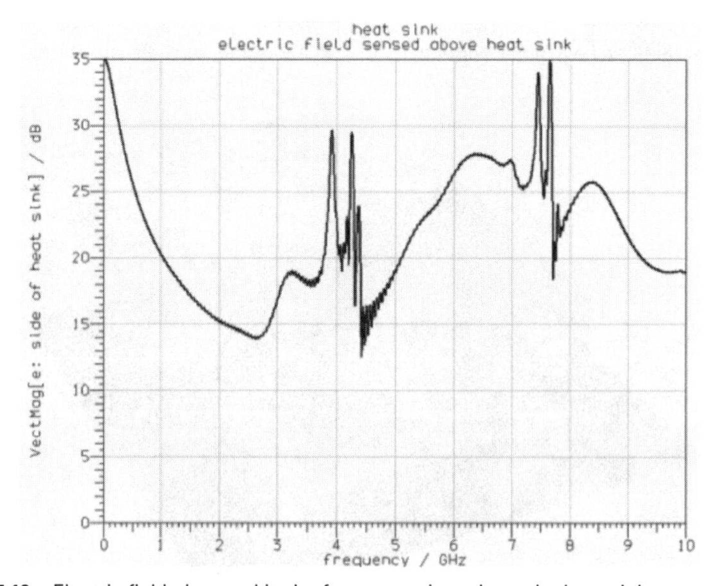

Figure 7.18 Electric field observed in the free space just above the heat sink.

7.5.1 A Ventilation Slit Becomes an Antenna

Figure 7.20 shows ventilation slits on the back of a desktop personal computer. This is the current distribution on the metallic surface at 2.4 GHz. The variation in the current distribution looks like standing waves aligned by the slits.

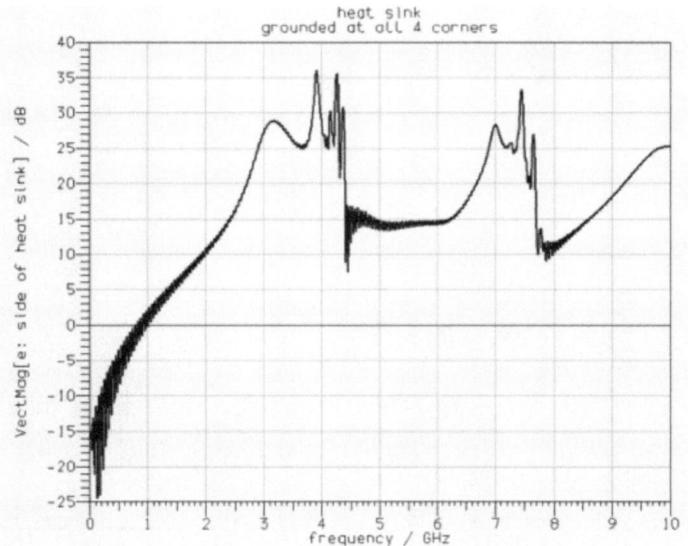

Figure 7.19 Electric field observed just above the heat sink whose four corners are connected to the ground.

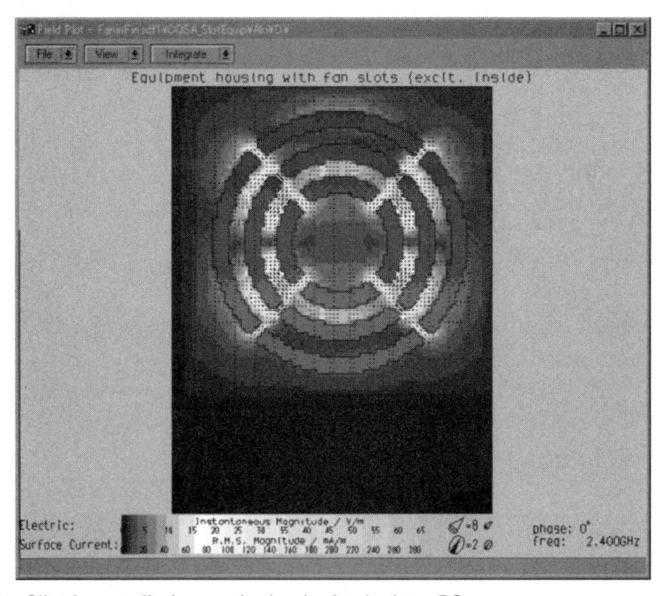

Figure 7.20 Slits for ventilation on the back of a desktop PC.

Inside the typical personal computer, we have a motherboard, peripheral boards for graphics, a power supply, and flat cables. It is possible that all of these items could cause radiating electromagnetic noise.

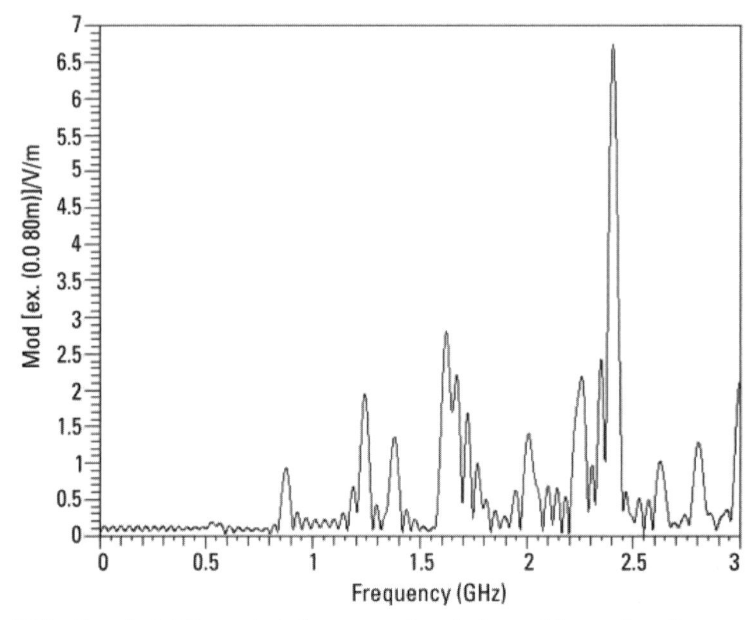

Figure 7.21 Electric field intensity in free space 8 cm in front of the cooling slits.

For simulation, we excited the slits from the inside with an electromagnetic wave to examine the electromagnetic noise radiated outside of the housing through the slits. The plot in Figure 7.21 is the result of simulating the electric field intensity in free space 8 cm in front of the slits. Much to our surprise, we see an unexpected peak at around 2.4 GHz.

In addition to desktop personal computers, thin notebook personal computers also have electromagnetic energy trapped inside a metallic housing. Just like with desktop PCs, the electromagnetic field can be radiated to free space through the cooling slits.

7.6 Troubleshooting Radiation Noise Problems

We are not finished with a project when we can get the prototype to successfully operate for 15 or 20 minutes. We have more and more cases where, in the last stage of development, we find we cannot meet requirements for radiation noise.

In conventional design flows, we start problem solving by measuring the prototype for radiation noise. However, with the incredible size of systems being developed, like a hundred operating cell phones in every city block embedded in an ocean of electromagnetic radiation from other electronic devices, the complexity becomes gigantic. Due to development schedule limitations, we must obey the following constraints:

1. Design of the circuit board mounting bay cannot be changed.
2. Design of the boards cannot be changed, either.

If there is no problem in operating a product whose radiation noise is just over required maximum levels, we might manage to ship it anyway. However, if we do not understand the problem, we cannot use this knowledge to counter a similar but potentially worse problem later. Next, we learn concrete steps for problem solving by exploring the case of a high-speed Internet router.

7.6.1 Troubleshooting Procedures

The approach that we consider given these constraints is, first of all, to identify the source of any strong electromagnetic radiation. If we use a micro loop antenna to measure the local electromagnetic field in many places and at many frequencies, we might find the source of the radiation. If we actually do this experimentally, this method can take a lot of time. However, if we use an electromagnetic field simulator, it is possible to set observation points at any location in space and to easily obtain data over a wide range of frequencies.

The problem is how faithfully we model the product. When we need to model a large-scale system precisely with all the lines, components, printed circuit boards, and wiring harnesses, it can take too much time. Thus, we divide the problem into three portions:

1. Component level;
2. Module level;
3. System level.

We gradually work our way from the fine-scale aspects to the entire system. On the component level, we focused on the grounding of an MPU heat sink. Below, we investigate electromagnetic radiation from ventilation slits of various sizes. Next, on the module level, we consider circuit substrate modules and the resonance that occurs together with the housing that encloses them. Finally on the system level, we examine the interactions among power cables, modules, and a backplane. We simulate the system after taking countermeasures, and we evaluate their effect quantitatively.

7.6.2 Size of Air Vents

We focus our attention on the air vent by investigating radiation originating from a printed circuit board inside a housing and escaping outside. Figure 7.22 shows hexagonal air vents that are modeled, including the housing metal thick-

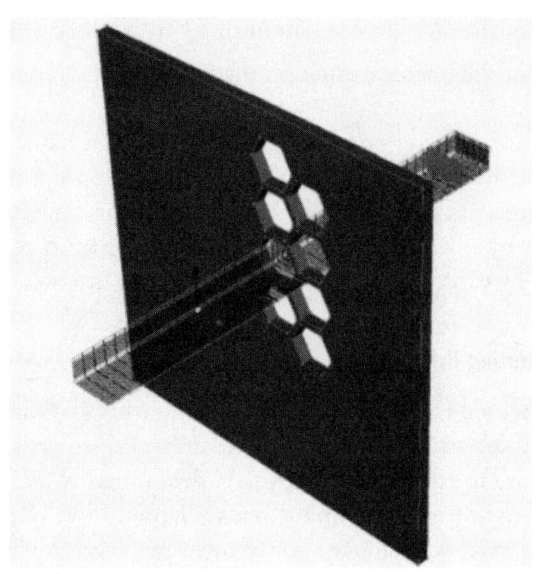

Figure 7.22 Hexagonal air vents.

ness. Rather than investigate the area of the ventilation holes, we investigate the thickness of the housing metal, comparing 0.060- and 0.200-inch-thick metal.

As for excitation, we launch a vertically polarized plane wave toward the right back side, as shown in Figure 7.22, and observe the intensity of the electric field in the same direction on the other side of the plate in front of the air vent. This is a model of the air vent only; it is not a housing. It is only a plate. The electric field strength increases with frequency, as seen in Figure 7.23.

What we can see from this result is that simply reducing the thickness of the housing metal (in which the air vents are punched) from 0.200 to 0.060 inch can deteriorate the effect of electromagnetic shield by 20 dB. And we can expect to gain a similar effect by changing the shape and size of the air vent. However, if the slots are large enough (with respect to wavelength), the slots become a resonant slot antenna at specific frequencies. It is then possible that we will see sharp radiation increases at particular frequencies.

7.6.3 Investigating Problems at the Module Level

The module in Figure 7.24 has hexagonal air vents on both sides. Figure 7.25 shows a model of the printed circuit board module, which is imported from the model used for thermal analysis.

The shapes and material attributes are assigned for the optical module (optics) at the rear of the substrate and the ASIC (application specific **integrated circuit**) portion in front. However, except for the power lines and a

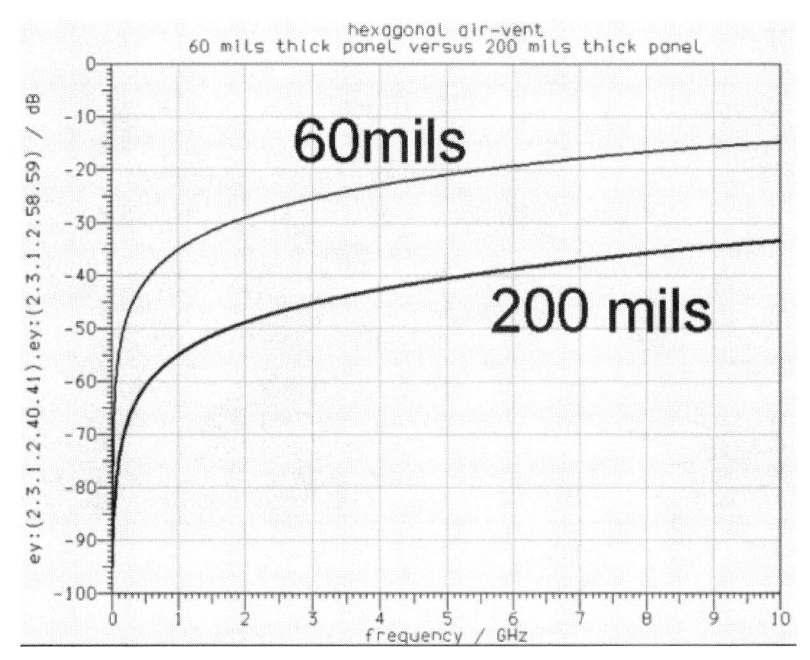

Figure 7.23 Electric field strength observed in front of the air vent.

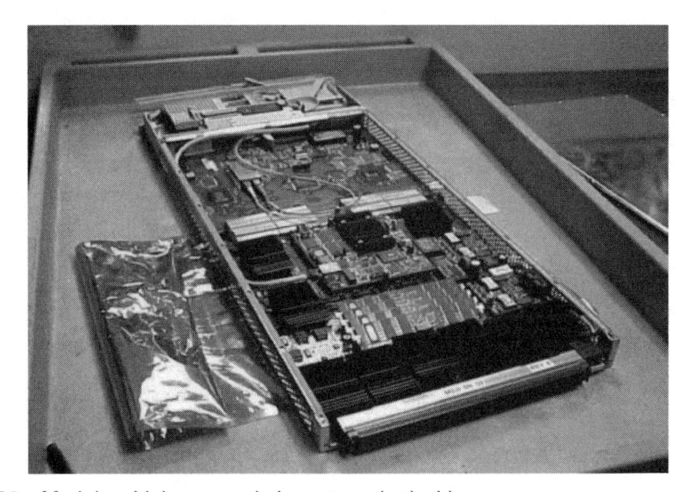

Figure 7.24 Module with hexagonal air vents on both sides.

few of signal lines, wiring between portions of the module are not modeled in detail. This is because of the constraint that the circuit design cannot be changed. However, the design team had performed an EMC evaluation so that the lengths of all transmission lines are minimized so as to allow high operating frequencies. Reconsideration of the layout was not an option.

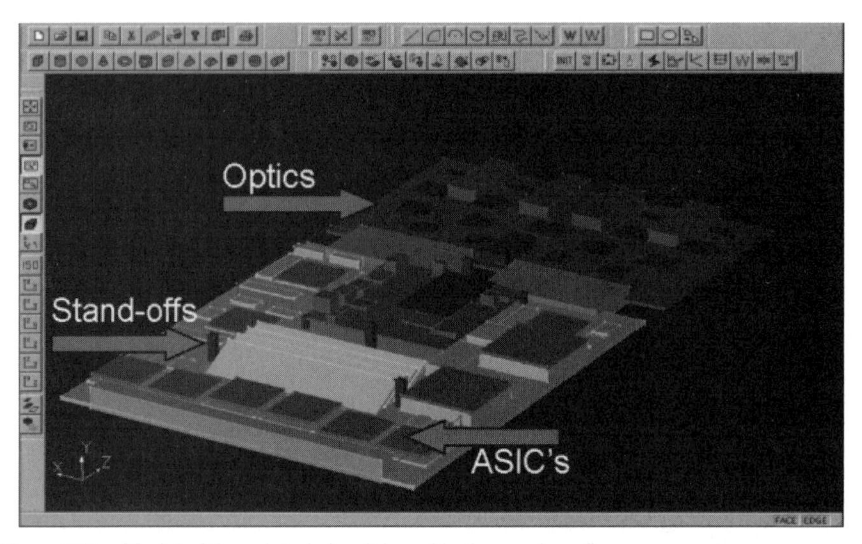

Figure 7.25 Model of the printed circuit board in the previous figure.

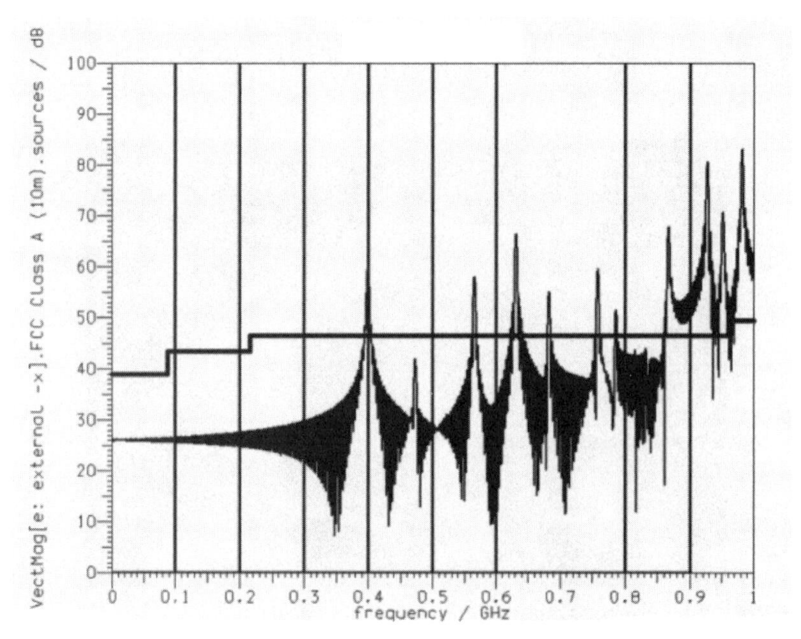

Figure 7.26 Electric field intensity at an external observation point in free space.

At first, it was suspected that the modules generated unnecessary radiation through the power supply and input/output (I/O) connectors. There is an I/O connector in the front of the module, and a crystal oscillator for a clock is located next to this connector. To test this suspect, we set an excitation source between the ASIC and the substrate with a wire and apply 3.3V of voltage source (3.3V is a logic level of this ASIC).

7.6.4 Slit Between the Module and the Cover

In the solvers using finite difference time domain (FDTD) and transmission line matrix (TLM) methods (both described in Chapter 9), we can conveniently obtain broadband frequency responses. We do this so that we know the particular frequency where a peak in the frequency response appears, and we can examine the electromagnetic fields and the current distributions at these frequencies. Here we specify several observation points in free space in advance. These points are between the module and the cover, and a few points are external to the module. We determine the frequencies where the response peaks appear.

The electric field intensity at the observation point in external free space shows a peak at 400 MHz, seen in Figure 7.26, which corresponds to the clock frequency of this ASIC. Then we examine the electromagnetic field around the module and the current distribution on conductor surfaces at this frequency.

Figure 7.27 (displayed in the same orientation as Figure 7.25) shows the result. When examining this data in detail, we find that the strongest electromagnetic field is between the module and the cover. Between them, four spacers (standoffs) are also seen in Figure 7.25. Then, when we modify the model so that these spacers are tightly connected to the cover, the peak at 400 MHz completely disappears, as shown in Figure 7.28.

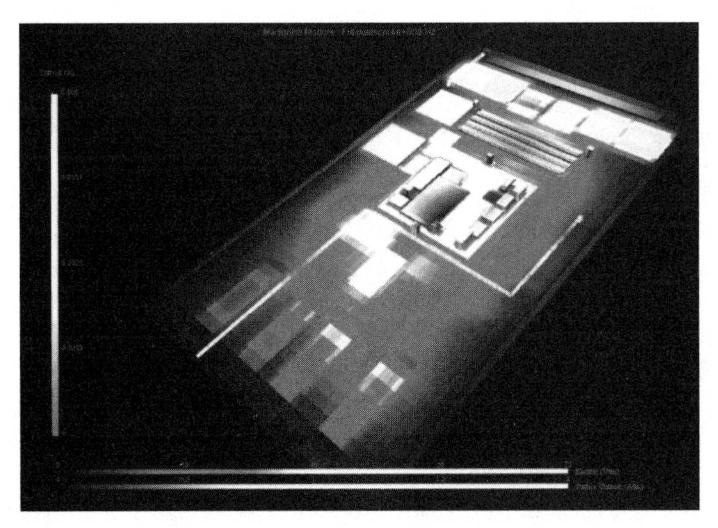

Figure 7.27 Electric field around the module and the surface current distribution on the substrate.

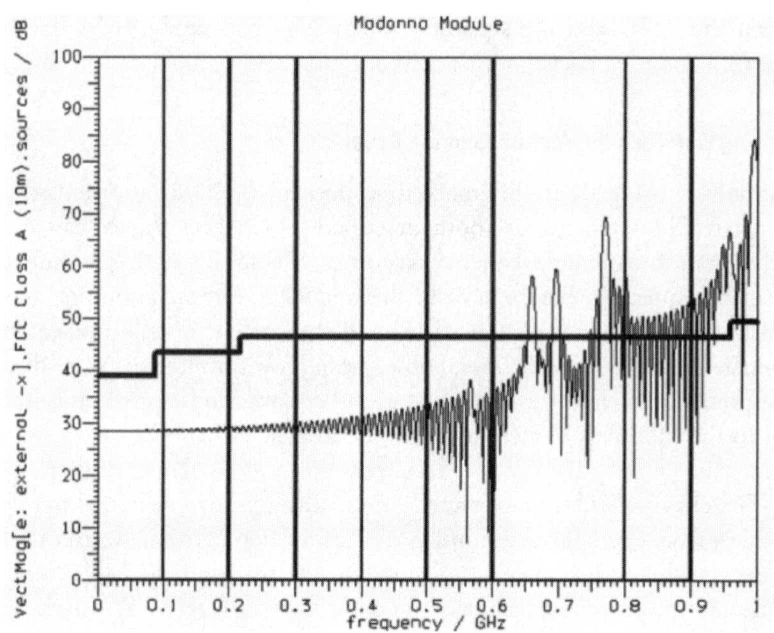

Figure 7.28 Electric field intensity at the observation point in free space; spacers are now tightly connected to the cover.

The electromagnetic field around the module and the current distribution on the conductor surfaces at 400 MHz, seen in Figure 7.29, show that the strong currents and fields of Figure 7.27 are gone.

7.6.5 EMI from the Power Cable

A communication system like a high-speed router for the Internet has several modules. Figure 7.30 is a model of 10 modules and a backplane. The thin vertical line at the left is a power cable. Figure 7.31 shows a perspective interior view. Power cables often cause EMI problems, so let's examine the simulation results.

Figure 7.32 shows the current distribution on the module and backplane surface at 465 MHz, where the peak electric field observed in free space around the router system appears. It shows a strong surface current; however, there are three spots of strong current at the 922-MHz peak, shown in Figure 7.33, and the distribution is different from Figure 7.32.

When displaying the far field radiation pattern, we can see the effect of these differences. As shown in Figure 7.34, we see that at the lower frequency, radiation is about the same in all directions. However at the higher frequency, it tends to radiate strongly in a particular direction, as shown in Figure 7.35. This trend agrees well with measurement. It is almost as though when it is multiple wavelengths in size, it has become some kind of random antenna array, with beams and nulls pointing off in all sorts of random directions.

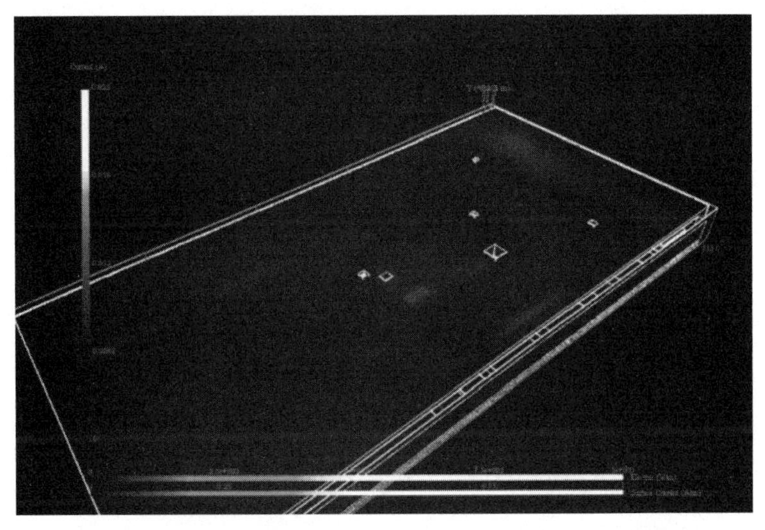

Figure 7.29 Electromagnetic field around the module and the current distribution on the conductor surfaces at 400 MHz.

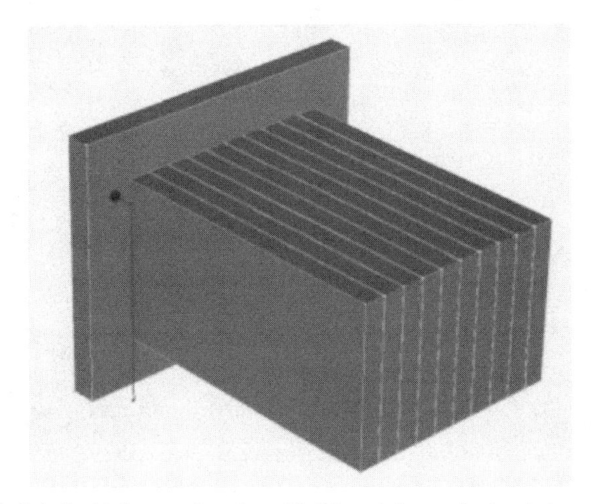

Figure 7.30 Model of a high-speed router with 10 modules and a backplane.

7.6.6 The Effect of Gaskets

In this simulation model, the modules and the backplane were assumed to be perfectly connected. But actually there is a gasket between them, and we found that we needed to model the gap occupied by the gasket. The simulation result of the radiated electric field is significantly different, shown in Figure 7.36.

The lower radiation level curve is for our original model, where the module and the backplane are connected. The higher radiation curve includes the

Figure 7.31 Perspective view of the interior.

Figure 7.32 Current distribution of the module and backplane surface at 465 MHz.

gaskets, which facilitate radiation due to the tiny, dielectric-filled slit formed by the gaskets. Below 700 MHz, including the effect of the gaskets is important.

7.6.7 Effects of Lossy Dielectric Material

Sheets of material with dielectric loss are sometimes used to absorb electromagnetic energy. It can take a long time to figure out the best positioning experimentally because measurements to determine effectiveness must be carried out over many locations and many frequencies. On the other hand, when using

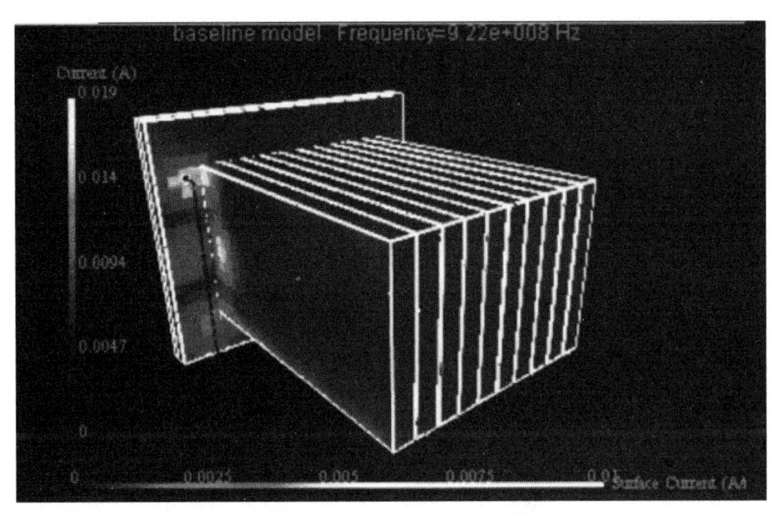

Figure 7.33 Current distribution of the module and backplane surface at 922 MHz.

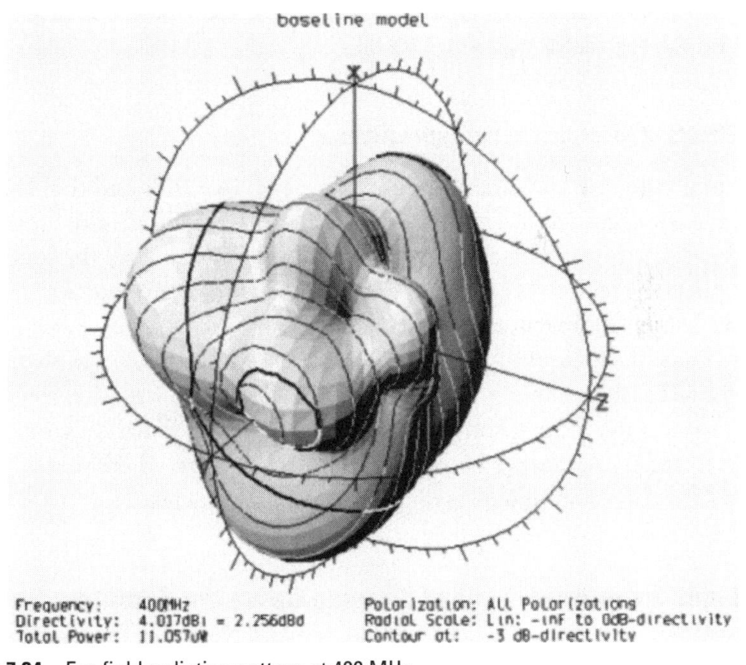

Figure 7.34 Far-field radiation pattern at 400 MHz.

an electromagnetic field simulator, the effect of each configuration can be easily quantified with a single analysis. For this particular project, we found that an absorbing sheet had the largest effect when placed on a sidewall inside the module. We achieved approximately 10 dB in the range from 400 to 800 MHz.

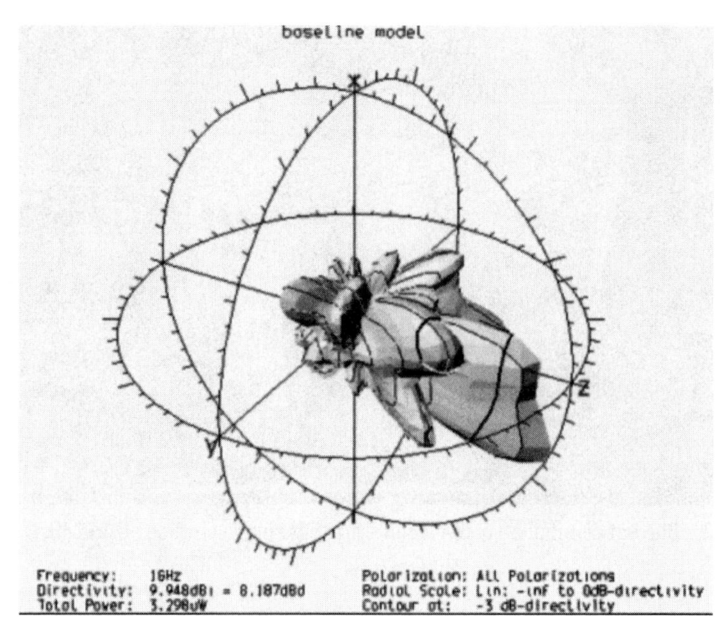

Figure 7.35 Far-field radiation pattern at 900 MHz.

7.6.8 Effects of Shielding of the Entire Module

When simulating the backside of each module (the I/O end) and the shielding around the connector, as in Figure 7.37, we can expect significant reductions in radiation. The arrow indicates a power supply cable, and now all the backplane connectors are shielded. We include cylindrical ferrite beads, and we simulate the effect of bypass capacitors to ground as well.

Analysis of the resulting model shows substantial reduction in the radiated electric field, shown in Figure 7.38. The higher radiation curve is the model without gaskets for comparison (Figure 7.36). The lower radiation curve is the result after adding the ferrite beads and bypass capacitors, where we confirm a radiated field reduction of approximately 30 dB at frequencies up to 1.2 GHz.

7.7 Radio Wave Absorber and Keeping Down the Radiation Noise

In Chapter 5, we learned that magnetic materials, or, even better, superconductive materials, provide shielding from static magnetic fields at lower frequencies. A magnetic material of high permeability that absorbs the surrounding magnetic field also has a high shielding effect.

There is a product called a noise suppression sheet that uses a soft magnetic metal powder. The process of manufacture is first dispersing the soft mag-

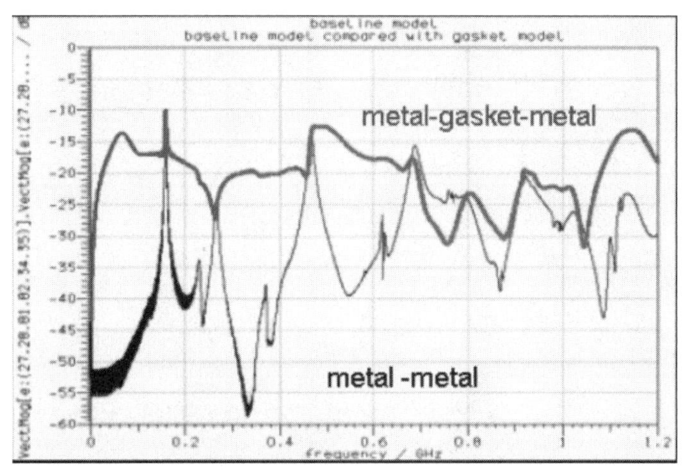

Figure 7.36 Simulation result of the radiated electric field showing the effect of a gasket.

Figure 7.37 Shielding around the connector.

netic metal powder in a bonding material (a polymer) using a solvent to make a paste. Next, the sheet is created by coating up to the desired thickness.

Figure 7.39 is a plot of the relative permeability of a typical noise suppression sheet. The real part of the relative permeability shows a high value of broadband permeability, which gives it the desired magnetic shielding effect. Here, however, we pay attention to the imaginary part of the relative permeability. The ratio of the imaginary part to the real part represents magnetic loss tangent, tanδ, which is separate and distinct from the dielectric loss tangent. If the imaginary part is large over a wide frequency range, it can absorb more magnetic energy.

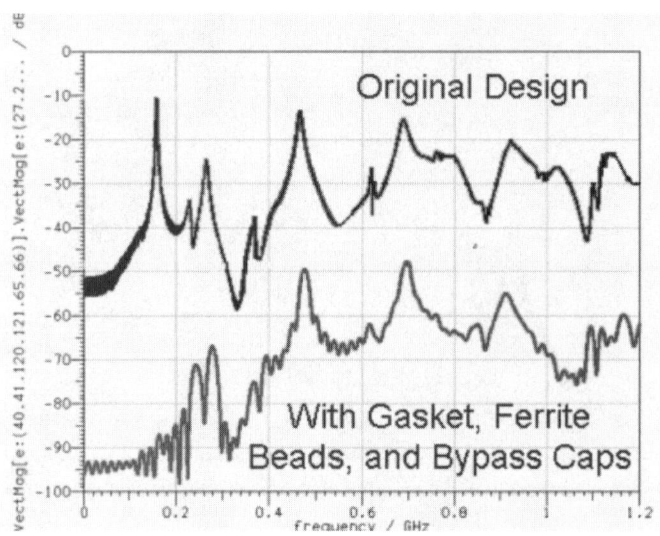

Figure 7.38 The lower curve shows the reduction in the radiated electric field when shielding is included.

Electromagnetic waves are attenuated in a lossy medium. For a lossy magnetic material, the loss results from the imaginary part, μ'', of the relative permeability. In this case, the absorbed power, W/m^2, of the electromagnetic wave is

$$P = \frac{1}{2}\omega\mu''|H|^2$$

From this formula, the effective range of the noise suppression sheet with the characteristics of Figure 7.39 can be predicted. The values of the real and imaginary parts cross at around 600 MHz, and they both abruptly get smaller at higher frequencies. This is a general characteristic caused by the magnetic saturation of magnetic materials, so we can see that it is difficult to keep high values at higher frequencies.

Because the ability to absorb the magnetic field, *H*, deteriorates when the real part of the magnetic permeability is small, we see that it is desirable to develop materials that keep large values of the real part at frequencies as high as possible.

7.7.1 Measurement of a Noise Suppression Sheet

We can measure the effect of a noise suppression sheet using two micro loop antennas (magnetic field probes). Figure 7.40 shows a method that positions them parallel to and on the same side of the sheet. Figure 7.41 illustrates two probes perpendicular to and on either side of the sheet.

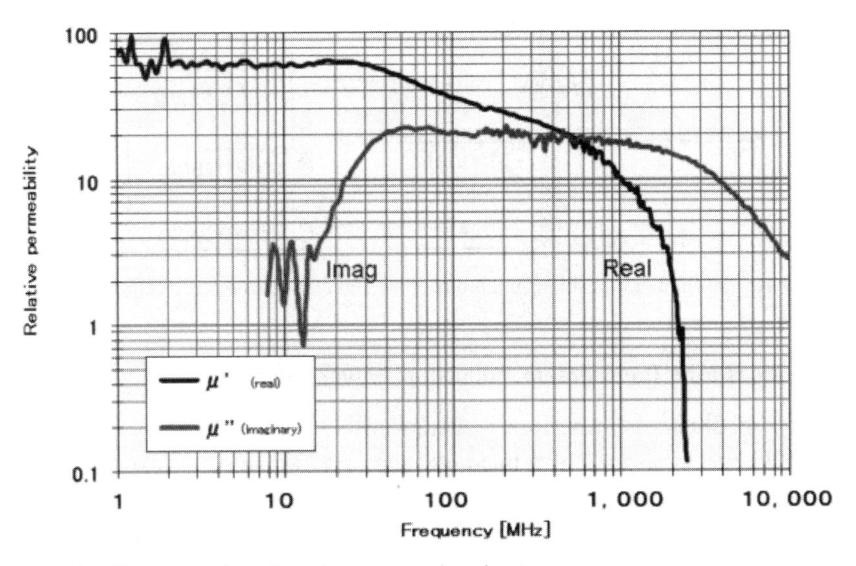

Figure 7.39 Characteristics of a noise suppression sheet.

The example in Figure 7.40 is proposed as a method to measure R_{da} (intra-decoupling ratio). The center-to-center distance of the two probes is 6 mm, and the sheet (50 mm × 50 mm) is located 3 mm from probes.

Figure 7.41 is a method to measure R_{de} (interdecoupling ratio), and the center-to-center distance of two probes in this case is also 6 mm.

Once the probes are in place, we measure their S-parameters with a network analyzer and obtain S_{21}, the transmission coefficient from input port 1 to output port 2. Denoting the value of S_{21} (in dB) without the sheet as S_{21R}, R_{da} and R_{de} are obtained by

$$R_{da} \text{ (or } R_{de}) = S_{21R} - S_{21m} \quad \text{dB}$$

When these decoupling ratios are high, it means that electromagnetic energy absorption is stronger. We can use these values to compare the effectiveness of noise suppression sheets.

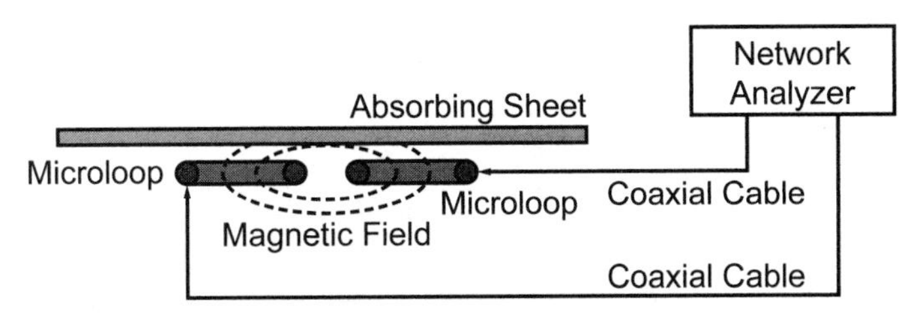

Figure 7.40 Method that positions the probes parallel to the absorbing sheet.

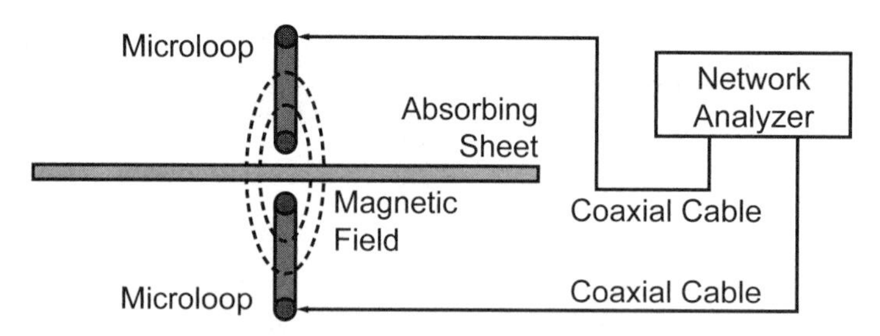

Figure 7.41 Method that positions the probes perpendicular to the absorbing sheet.

7.7.2 Measuring Transmission Attenuation Power Ratio

Figure 7.42 shows a connection to a transmission line of characteristic impedance 50Ω that is covered by a magnetic absorption coating. With this connection, we can measure the two-port S-parameters with a network analyzer, just like we did before. R_{tp} (transmission attenuation power ratio) is defined by the measured values of S_{21} and S_{11}.

$$R_{tp} = -10\log\left\{10^{(S_{21}/10)} / \left(1-10^{(S_{11}/10)}\right)\right\}$$

As shown in Figure 7.42, because the absorption sheet contacts the line, the characteristic impedance of the line is no longer exactly 50Ω. Because of

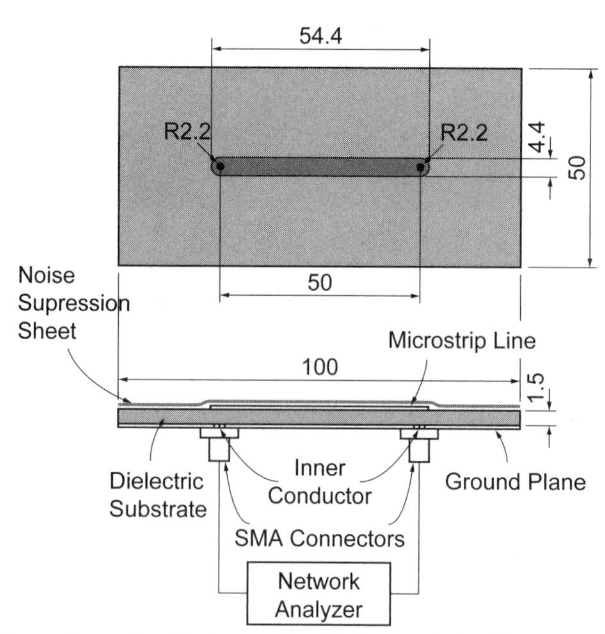

Figure 7.42 Measuring method that uses a length of transmission line to contact the absorbing sheet.

this, depending on materials, the reflection coefficient, S_{11}, of the input port might increase. The previous expression takes this into account in calculating the net attenuation.

The strong electromagnetic field generated around the upper surface of the microstrip line (especially at the edges) is very sensitive to the exact positioning of the absorptive sheet. If there is a very thin air gap between the sheet and the transmission line, the S-parameters can change significantly. Thus, in the actual measurement, we use a polystyrene foam plate that is more than 10 mm thick and apply it under pressure on top of the absorptive sheet.

7.7.3 Effects of Radio Wave Absorbers

We learned about the mechanism of radio wave absorbers in Chapter 5. The electromagnetic waves from a wireless LAN are reflected by steel racks and metallic desk partition panels. Figure 7.43 is a model drawn with a CAD tool. It is 3m × 3m × 2.1m.

Figure 7.44 displays the root mean square value of the electric field when putting an antenna at an access point on the ceiling. The bottom of the vertical dipole antenna is set at the height of 2m in the center of desk D. The striped patterns on each desk partition show standing waves. The signal is around –25 dB at the weakest point. When we move away from the access point, we might not be able to connect, depending on the position of the antenna.

The standing waves are generated by the electromagnetic waves reflecting from the desks and partitions, so in some cases we can improve it by installing

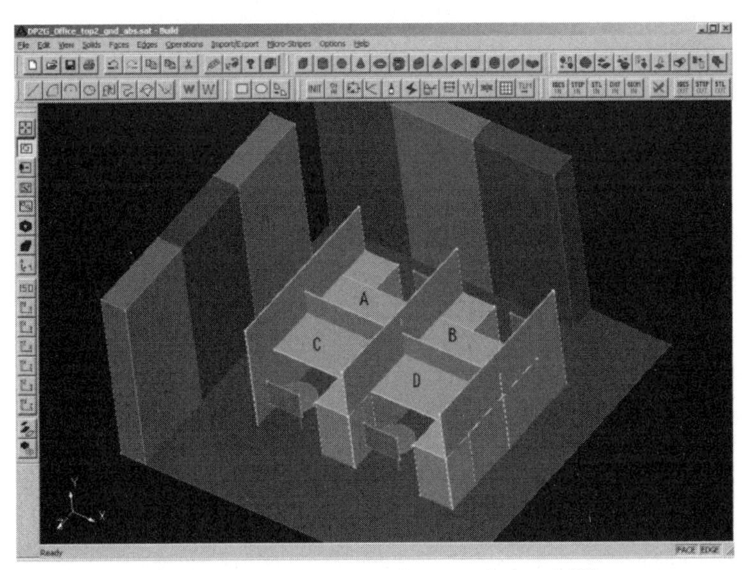

Figure 7.43 Simulation model of an office to evaluate a wireless LAN system.

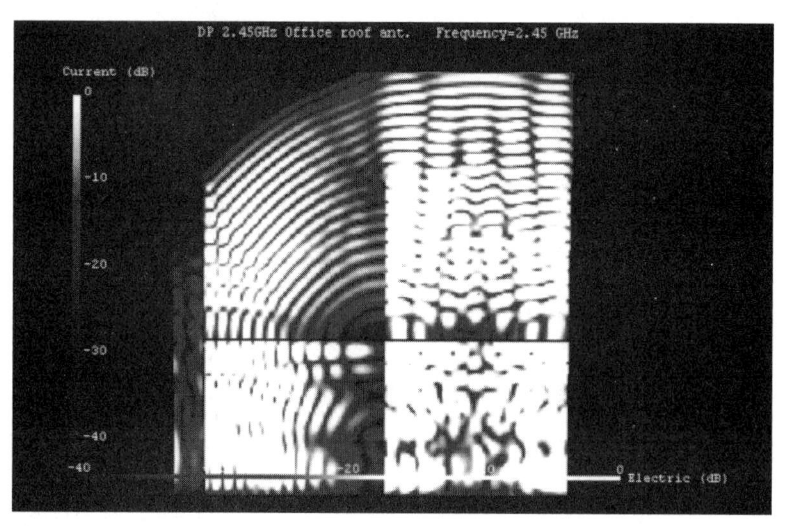

Figure 7.44 Root mean square value of the electric field in the office.

a noise suppression sheet on these surfaces. A typical product, shown in Figure 7.45, is manufactured by Nitta Corporation. Figure 7.46 is the result of a model using absorbing surfaces of –15 dB on the desks and partitions. It shows that the standing waves are gone, yielding a homogeneous field distribution. Thus, it is possible to evaluate the effects of a radio wave absorber using electromagnetic field simulations.

Figure 7.45 Radio wave absorber manufactured by Nitta Corporation.

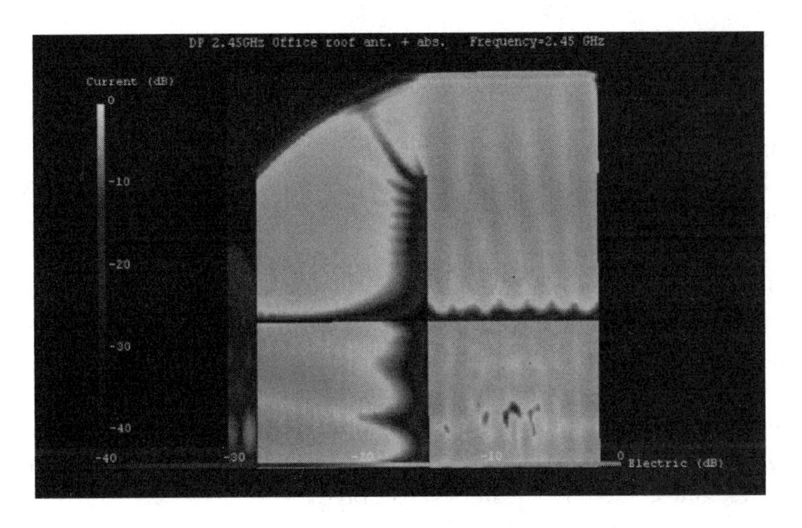

Figure 7.46 Model including absorbing surfaces of –15 dB.

7.8 Waveguide Mode Tutorial

Electromagnetic waves travel along various kinds of transmission lines (see Chapter 1), including parallel wires, microstrip lines, and waveguide tubes. These can be analyzed by numerical electromagnetic field analysis software programs based on Maxwell's equations. The transmission line is solved under boundary conditions determined by the various materials and shapes. At a given frequency, there are a number of independent solutions for the field configuration. These solutions are called propagation modes.

The typical propagation modes we might see are as follows:

1. *TEM mode:* The transverse electric magnetic mode has the electric and magnetic fields both completely perpendicular to the wave's direction of travel. This is the typical mode of propagation in free space.
2. *TE mode:* The transverse electric mode has all of the electric field perpendicular to the direction of travel of the electromagnetic wave. There is some magnetic field pointing in the direction of travel.
3. *TM mode:* The transverse magnetic mode has all of the magnetic field perpendicular to the direction of travel of the electromagnetic wave. There is a portion of the electric field that points in the direction of travel.
4. *Hybrid mode:* The electric and the magnetic fields both have components in the direction of travel of the electromagnetic wave.

The notational convention is, for example, TE_{mn} mode, where the two integer subscripts indicate the number of half wavelengths along the *x* and *y* axes in the cross section of the *z* directed transmission line. Examples of rectangular and circular waveguide are shown in Figures 7.47–7.50.

The mode that can propagate at the lowest possible frequency is called the dominant mode. The dominant mode of a rectangular waveguide is TE_{10} mode, shown in Figure 7.47(a).

In addition, the lowest resonant frequency of a resonant cavity is also called the dominant mode, with the response of that and several higher order modes shown in Figure 7.10. For example, there are three orientations for a given resonant mode in a rectangular parallelepiped (a "rectangular cavity"), shown in Figure 7.51. In this case, these three are functionally identical. This is because there is no rule for which side is longitudinal. We arbitrarily set *x*, *y*, and *z* axes, and set the mode names accordingly.

7.9 Confirming This Chapter by Simulation

As Sonnet Lite is based on the method of moments applied to a shielded (boxed) circuit, we can use Sonnet to simulate rectangular cavity resonances. Here, we

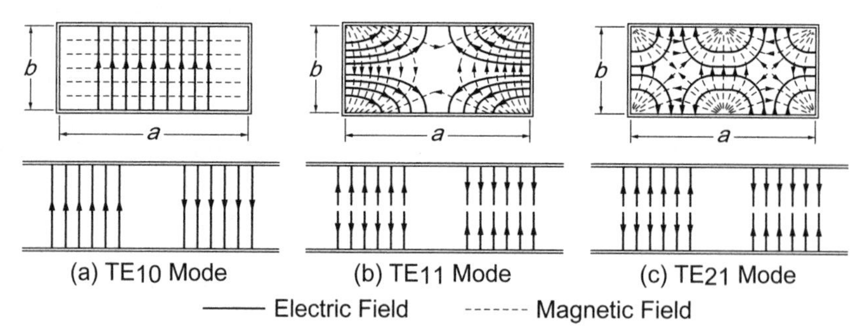

Figure 7.47 (a–c) Transverse electric modes of a rectangular waveguide.

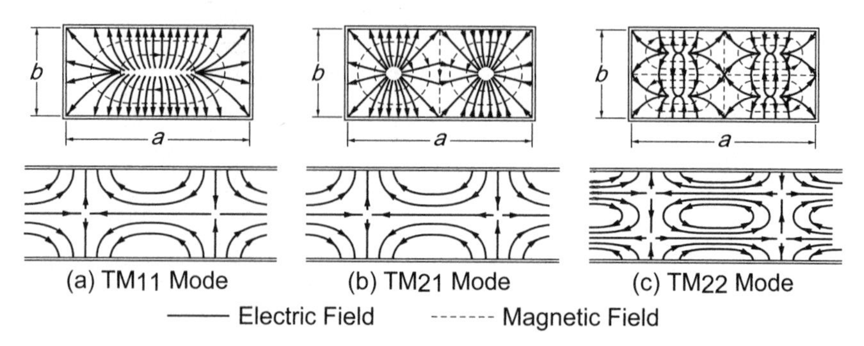

Figure 7.48 (a–c) Transverse magnetic modes of a rectangular waveguide.

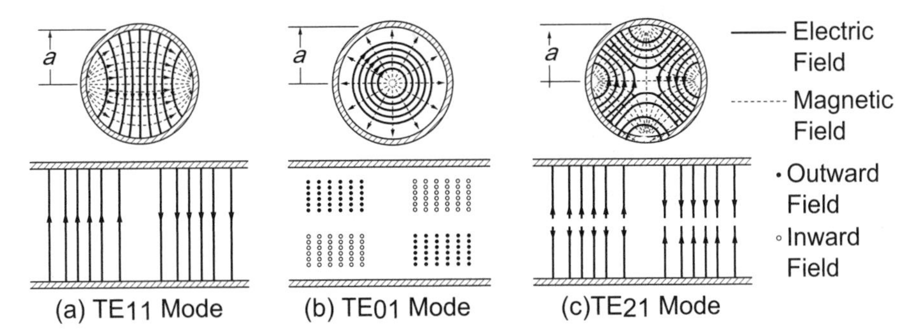

Figure 7.49 (a–c) TE modes of a circular waveguide.

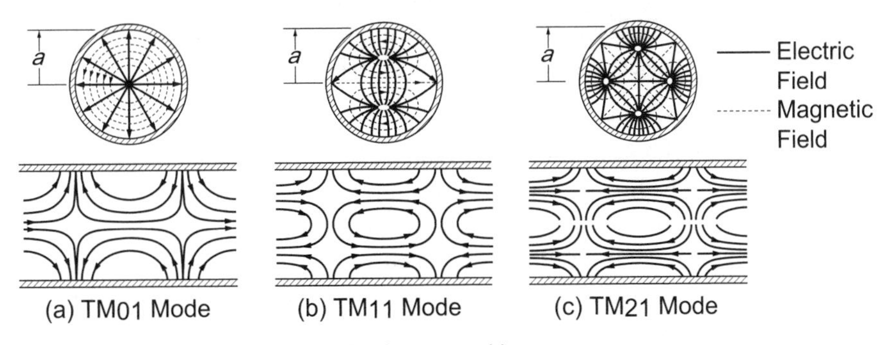

Figure 7.50 (a–c) TM modes of a circular waveguide.

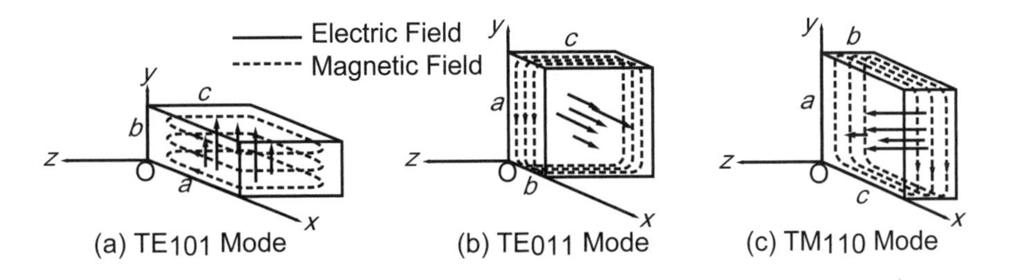

Figure 7.51 (a–c) Three possible orientations for a resonant mode in a rectangular waveguide.

use this ability to verify the effect of a noise suppression sheet on a microstrip line.

7.9.1 Examining a Cavity Resonance

Sonnet analyzes a circuit enclosed by a shielding, conducting box. After the box size and analysis frequencies are set, selecting Analysis > Estimate Box Resonances… displays the possible resonant frequencies, shown in Figure 7.52.

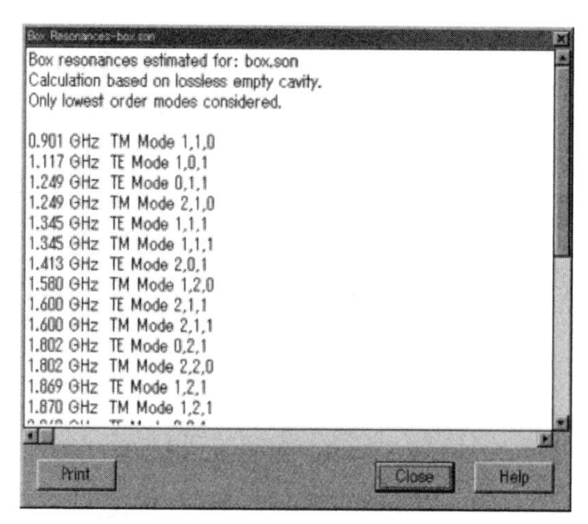

Figure 7.52 Box (rectangular cavity) resonant frequencies.

As shown in the 3-D display in Figure 7.53, we set $x = 300$ mm, $y = 200$ mm for Box size, The box size in the z (vertical) direction is set by the thickness of the dielectric layers. We set the bottom dielectric layer to 10 mm. This sets the distance from level 0, the level on which we will draw our circuit, to the Box Bottom. We set the upper dielectric layer to 140 mm. This sets the distance to the Box Top. This makes the Box a total of 150 mm tall in the z direction.

The short lines coming from Port 1 and Port 2 are connected to the box bottom with vias. These small loops act as micro loop probes.

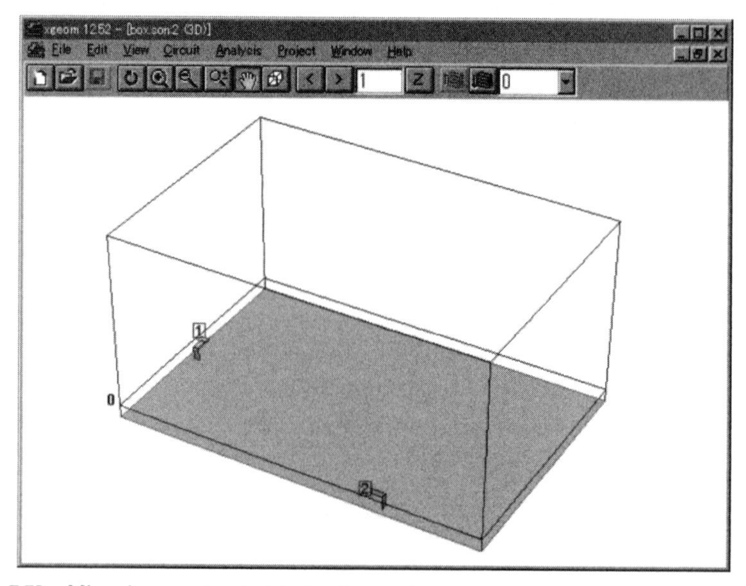

Figure 7.53 Micro loop probes inside the Sonnet box.

There is a peak in S_{21} (transmission coefficient between the two ports), seen in Figure 7.54, when the electromagnetic energy inside the box is at a box resonant frequency. The frequencies where peaks appear in the lower level curve correspond to each frequency of Figure 7.52. The curve due to coupling between the two micro loop probes is the lower curve. The higher curve is the result of a model that excites the box by means of a slot in the top cover of the housing, shown in Figure 7.55. The frequencies where peaks appear are almost the same. The levels of the two curves are different because we are using two different mechanisms for exciting the resonances. In both cases, the resonant frequencies are determined by the size of the box, and they are the same.

When exciting resonances in a housing, like that shown in Figure 7.53, note that any specific excitation mechanism (e.g., micro loop probes or plane wave illuminated slot) there might be possible resonant frequencies that are not excited and thus are not seen in the result. A different excitation mechanism (e.g., changing the location of the probes or slots) might make the resonances visible.

Figures 7.56 and 7.57 show the surface current distributions on the top cover. We can confirm the modes inside the housing just by viewing the current distributions.

7.9.2 Simulation of a Noise Suppression Sheet

Figure 7.58 is a model of the structure of Figure 7.42 using Sonnet Lite. It uses autoground ports so that we can place ports away from the Sonnet box side-

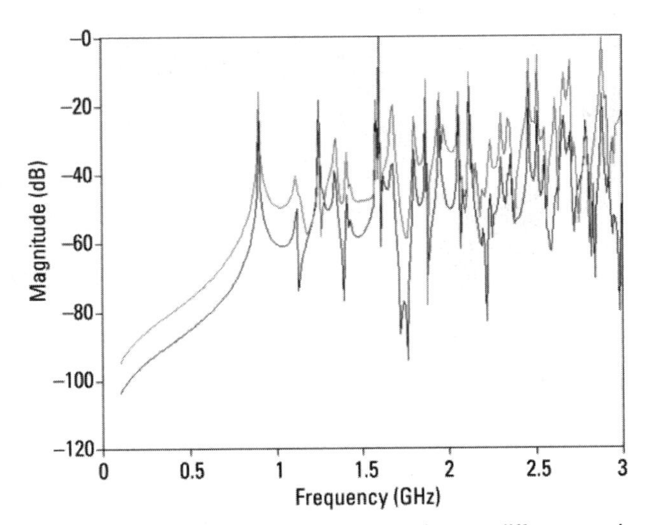

Figure 7.54 Result of S_{21} showing the box resonances for two different excitation mechanisms.

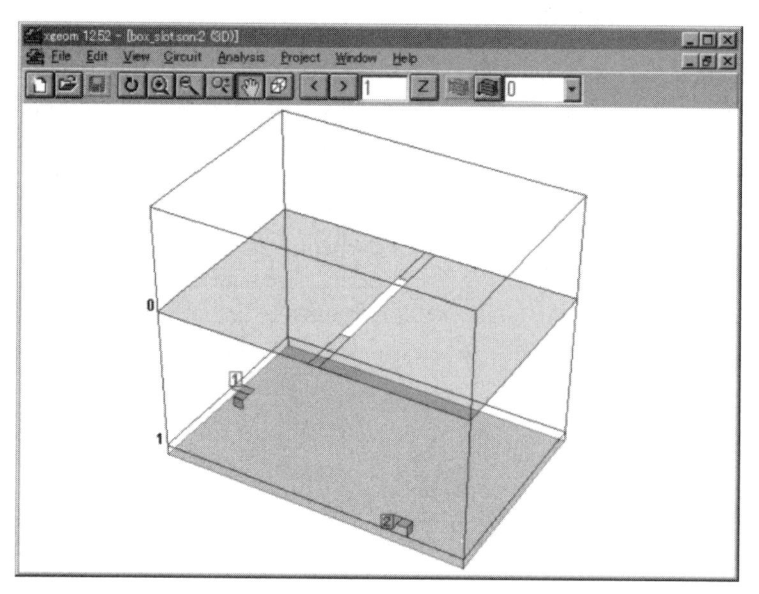

Figure 7.55 Housing model with a slot in the top cover. The slot is illuminated with a plane wave to excite resonances.

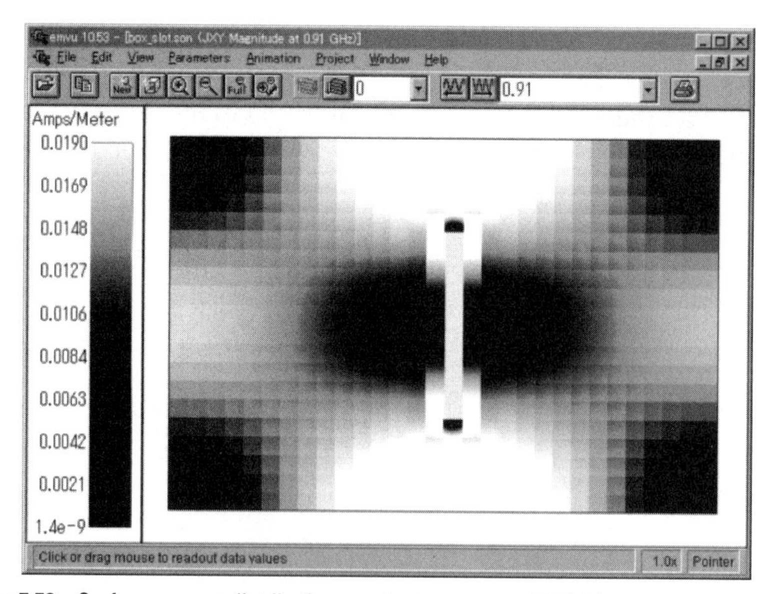

Figure 7.56 Surface current distributions on the top cover at 900 MHz.

wall, at the position of the SMA connectors. To set this port, after drawing the transmission line, select Tools > Add Port and click the edge of the line where we want a port. Then, double click the port number and change the port Type to Autognd, as shown in Figure 7.59.

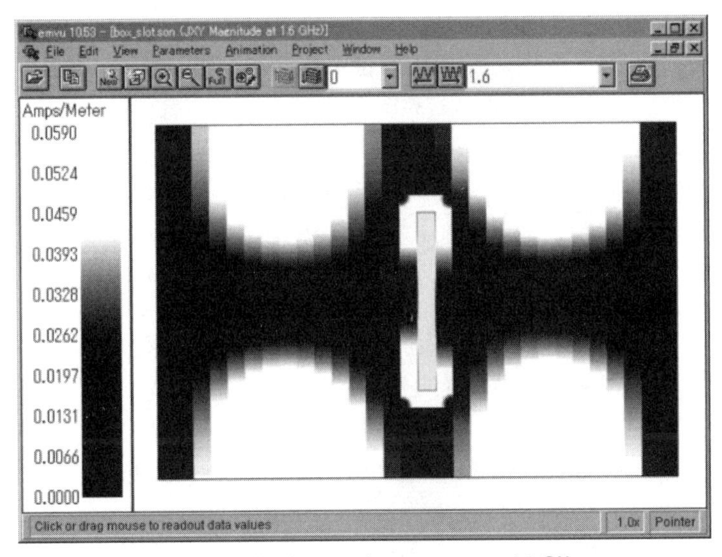

Figure 7.57 Surface current distributions on the top cover at 1.6 GHz.

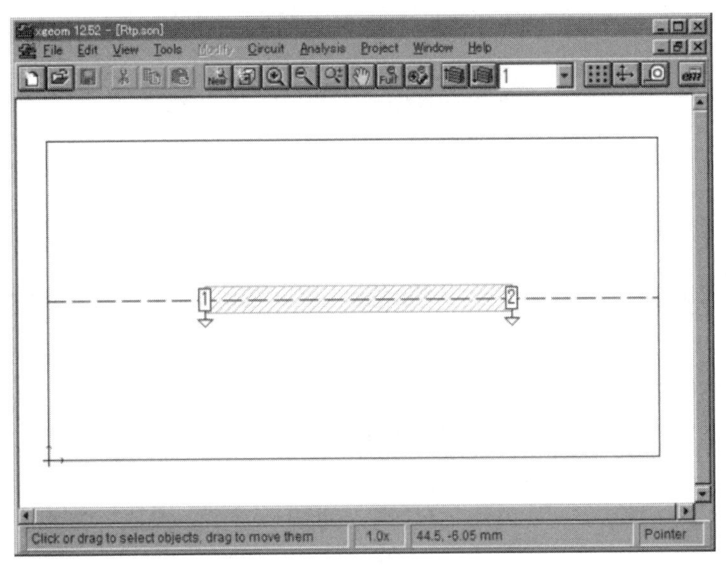

Figure 7.58 Sonnet model with a noise suppression sheet on a microstrip line.

As for the substrate dielectric, assign Teflon (PTFE), which is registered in the Sonnet library. Set the cell size as shown in Figure 7.60. We reduce the required memory by half by checking Symmetry because this circuit is perfectly symmetrical about the horizontal center line.

Figure 7.61 shows the settings for dielectric layers to add a noise suppression sheet. Because the free version of Sonnet Lite can use up to three dielectric

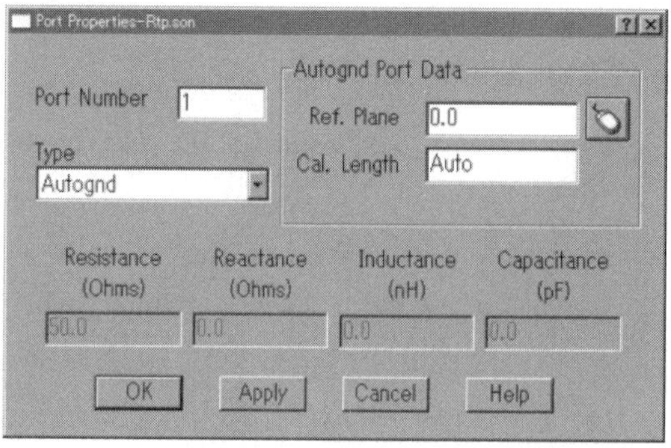

Figure 7.59 Dialog box for setting port options.

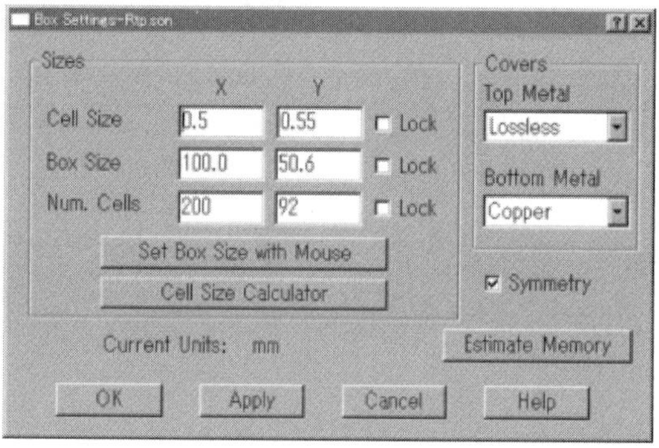

Figure 7.60 Dialog box to set the Sonnet Box parameters.

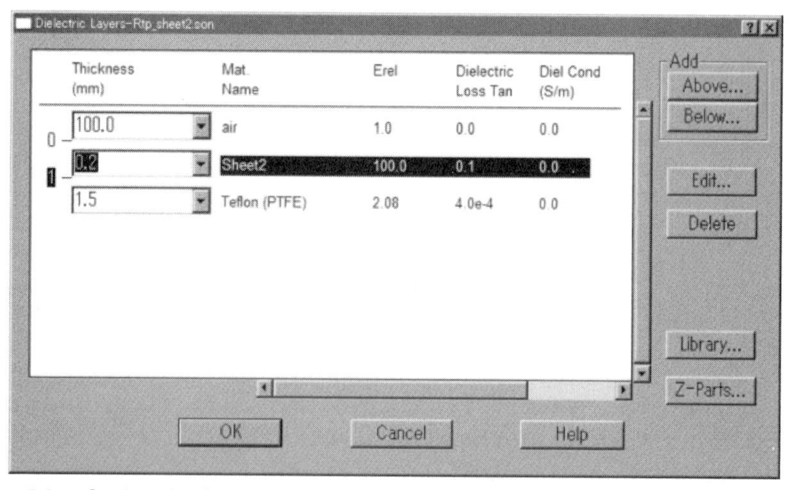

Figure 7.61 Settings for dielectric layers to add a noise suppression sheet.

layers, including the air layer, the absorbing sheet must be perfectly adhered (no air gap) to the transmission line and the top of the Teflon substrate. Figure 7.62 shows the material constants. These values were arbitrarily selected for illustration; they are not based on actual measurement.

In general, the relative permittivity (which affects electric field) and permeability (which affects magnetic field) are dependent on frequency. Here, we just use one value for all frequencies. If you have an equation that models the frequency dependency, you can type the equation into Sonnet, even into the free Sonnet Lite, for higher accuracy.

Figure 7.63 shows the result of a simulation from 1 MHz to 1 GHz after setting the plot Type to Zin (input impedance at a port). It is 50Ω at 1 MHz with or without a noise suppression sheet. Without the noise suppression sheet at 1 GHz, it is 58Ω. With the noise suppression sheet, it decreases to 37Ω at around 700 MHz.

The S-parameters, shown in Figure 7.64, show that S_{21} decreases with increasing frequency. This indicates that the sheet absorbs more power at the higher frequencies. From the plot of S_{11}, we see that the sheet causes the reflection to increase.

Next, we save the data in a spreadsheet .csv file. First, select Output > S, Y, Z-Parameter Files... and change the Format to Spreadsheet.... Next, click the Save button, and all the data is saved in csv format for easy use in a spreadsheet.

For this example, we read the data into Microsoft Excel and input the formula for R_{tp} provided earlier in this chapter. (This could also be done entirely in Sonnet; we are using Excel here to show how to use Excel with Sonnet.) We now create a plot of R_{tp} (transmission attenuation power ratio). Figure 7.65 is

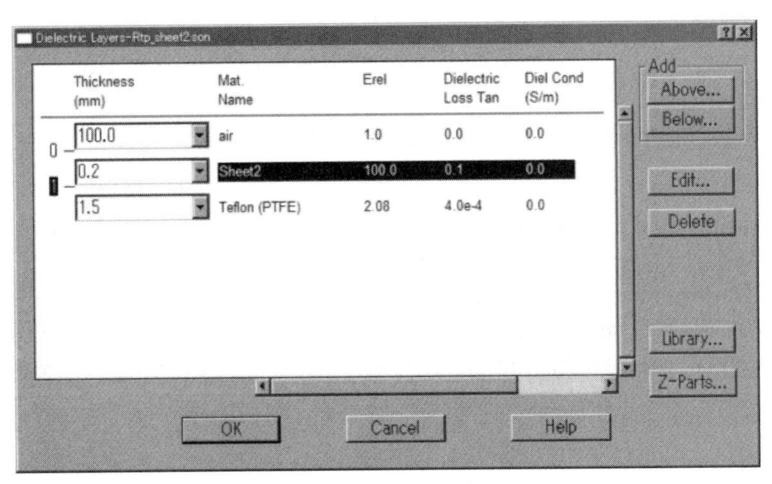

Figure 7.62 Material constants for a noise suppression sheet.

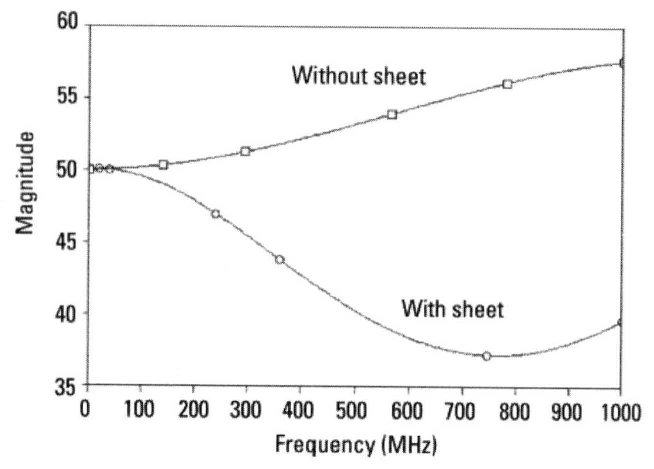

Figure 7.63 Input impedance seen at port 1.

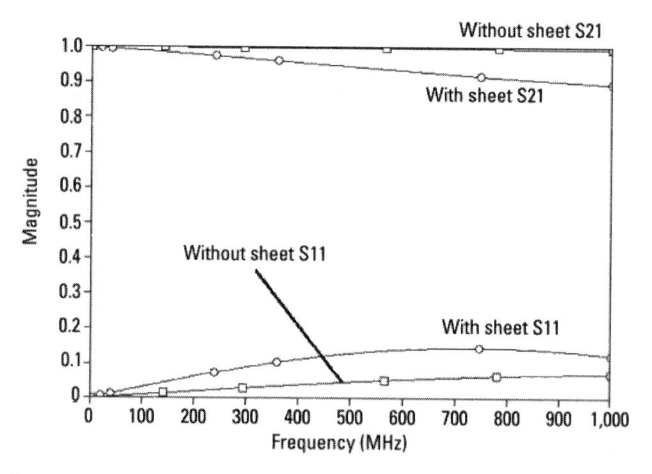

Figure 7.64 S-parameter results showing the result of adding the absorption sheet.

an example of the transmission attenuation power ratio of the noise suppression sheet that we investigated in this example.

If the target system is lossless and there is no radiation, then, by the law of conservation of energy, the sum of the reflection energy and the transmission energy equals the incident energy. As S_{21} and S_{11} are derived from voltage ratios, these values are squared for power, and we have

$$|S_{11}|^2 + |S_{21}|^2 = 1$$

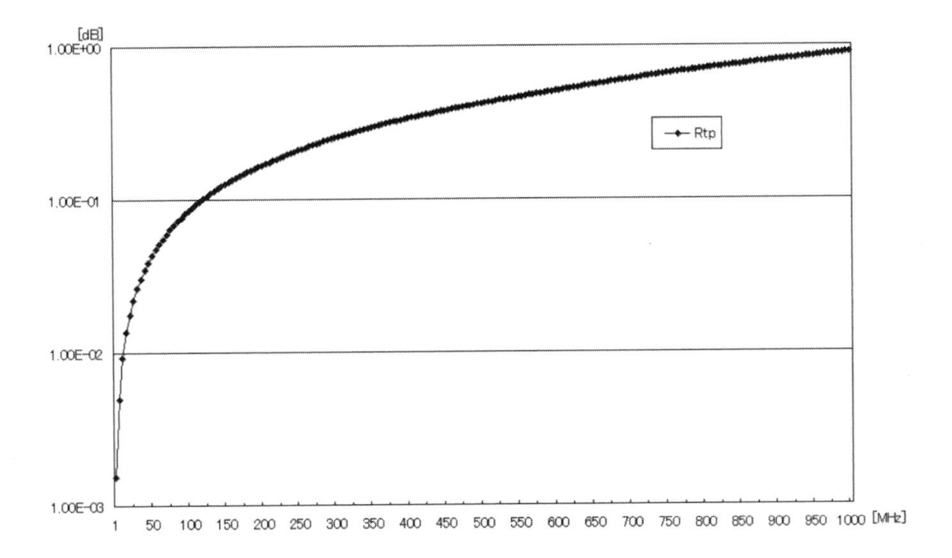

Figure 7.65 Transmission attenuation power ratio of a noise suppression sheet.

When energy is lost only to the noise suppression sheet, subtracting the reflected and transmitted power from the incident power (100 percent or 1.0) tells us how much power is changed to heat. This is the absorption capability of the noise suppression sheet.

We have studied a system to measure the effectiveness of an absorber of electromagnetic waves. This is important so that we can limit the amount of absorber required and apply it only to areas where it is effective.

7.10 Summary

1. The behavior of artificial systems in the electromagnetic environment is called electromagnetic compatibility.
2. A sealed metallic perfectly conducting housing (or case) has an infinite number of discrete resonant frequencies.
3. When there is a printed circuit in a sealed case, there is a possibility that the unwanted electromagnetic energy couples to the transmission lines and other components. This happens most strongly at resonant frequencies and can cause system failure.

4. Sometimes strong current flows along a joint with a resonant slit in the top cover. Secondary radiation can then couple to internal circuitry, and electromagnetic waves can leak outside.
5. Noise suppression sheets and radio wave absorbers can be simulated and used to reduce EMC problems.

8

All Roads Lead to Antennas

8.1 Antennas Where We Least Expect Them

In Chapter 7, we learned about designing for electromagnetic compatibility and electromagnetic interference. We demonstrated how problems can be identified and solved using electromagnetic field simulation.

When we search for the origin of electromagnetic noise, we are actually searching for an unintended antenna. Any such antenna transmits electromagnetic noise, and at the same time it receives any external noise and couples the noise into our circuits.

As for intended antennas, we might imagine the Yagi antenna for television and FM reception. In addition, the rod-like dipole and the monopole antennas used for wireless terminals are everywhere.

In this chapter, we learn that the "antenna" can be viewed as the ultimate electromagnetic field problem as we approach the heart of "the world of high frequency."

8.1.1 A Transmission Line Named Space

The first transmission line we discuss in this chapter does not appear to be a transmission line at all. In fact, this transmission line is nothing, literally. The technical term for this is free space. Developing mathematical models for how an electromagnetic wave can travel in a medium that is nothing is one of the great achievements of nineteenth-century physics.

8.1.2 Discovery of Displacement Current

The Scottish physicist James Clerk Maxwell (1831–1879) wondered what happens if we cut a wire where the electric current generates a magnetic field around the wire and connect a flat plate capacitor (with two conductive plates) across the gap. The strange thing is that RF current (which varies sinusoidally with time) can flow right through the capacitor, even if there is absolutely nothing between the plates. How could that be?

As described in Chapter 2, he carefully considered the circumference of the capacitor when current is flowing. At that time, no one knew what current was; they thought it might be some kind of strange fluid, or perhaps even two different kinds of fluids (positive and negative). It seems like current cannot possibly flow because this fluid (which today we know to be electrons) cannot possibly move in the empty space between two plates. This would then force the magnetic field to break off around the capacitor. This did not seem right. So Maxwell just simply assumed that the magnetic field would somehow be generated between the plates just like it was around the actual physical current flowing in the wire, as shown in Figure 8.1.

While the current is flowing (and changing sinusoidally with time), the charge on the capacitor plates is also changing. As the charge changes, the magnitude of the electric field changes. Maxwell made the amazing deduction that the changing electric field must generate a magnetic field. This effect, a changing electric field acting just like a real current, is called the displacement current. (The exact details of how Maxwell actually discovered the displacement current are lost to history.)

If we include both displacement current and the more easily understood conduction current (i.e., moving electrons), then this "total" current always flows in continuous loops no matter where it goes.

8.1.3 Prediction of Electromagnetic Waves

When Maxwell included his displacement current along with all the other known laws of electromagnetics, he was able to mathematically predict the existence of electromagnetic waves. Maxwell's close friend, Michael Faraday, had many years earlier discovered that that a changing magnetic field generates an electric field. This is Faraday's law of induction and comes into play whenever a generator makes electricity. Faraday tells us that a changing magnetic field generates an electric field, and Maxwell's displacement current tells us that a changing electric field makes a magnetic field. So each field can keep going back and forth re-generating the other. Maxwell then mathematically calculated the speed of the resulting electromagnetic wave in free space. It is approximately 3×10^8 m/s. Maxwell noticed that this is essentially the same as the mechanically measured speed of light. Thus he proposed that light is an electromagnetic

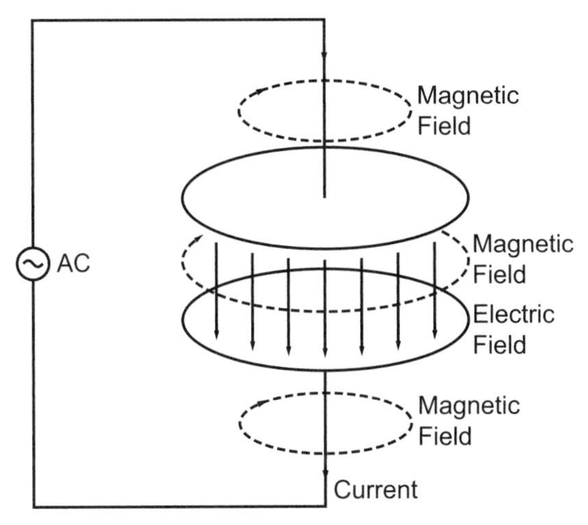

Figure 8.1 Magnetic field in a capacitor.

wave, and the electromagnetic theory of light was born in 1861, with the full formulation first published in 1865.

In classical physics, gravity can only attract; it does not repel. Because there are two kinds of charge, positive and negative, we can get both repulsion and attraction in electricity. Electric field is defined as the direction and strength with which a positive charge is pulled. The electric field is represented by a vector defined over a volume (which is called a vector field) that has a magnitude and direction, just like a gravitational field. Electric lines of force can be visualized as the electric field vectors at many points that are connected head to tail.

Figure 8.2 is an illustration taken from Maxwell's treatise, the founding document of the field of electromagnetics. This is a figure drawn by Maxwell showing the electric lines of force surrounding a flat plate capacitor. The electric lines of force are perpendicular to the capacitor plates where they touch the capacitor plates. The other lines, which are perpendicular to the lines of force, are constant voltage (potential) lines. If the electric lines of force at the edge of capacitor extend out into space, an electromagnetic wave is radiated by traveling on a transmission line called space.

8.2 The Hertzian Dipole, the Very First Antenna

The very first intentional antenna is the Hertzian dipole created by the physicist Heinrich Rudolf Hertz (1857–1894) of Germany and first published in 1888. The basic structure is shown in the Figure 8.3. Two soccer ball–sized metallic balls are attached at the ends of two thin metallic rods. There are two small me-

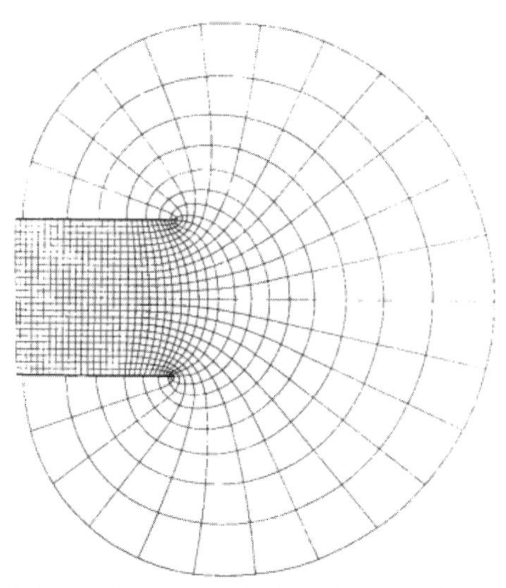

Figure 8.2 The electric lines of force around a flat plate capacitor as drawn by Maxwell in his *Treatise on Electricity and Magnetism*, the founding document of the field of electromagnetics.

Figure 8.3 Original Hertz's dipole at Deutsches Museum in Munich. (Photo by H. Kogure at Deutsches Museum in Munich.)

tallic balls in the middle with a narrow gap between them. The power source at the time was Leiden jars or Volta's batteries (or Bunsen batteries). The antenna was excited with high frequencies generated by a high-voltage spark discharge from an induction coil. This is called a Hertzian oscillator.

Hertz also used metallic plates, shown in Figure 8.4, in addition to metallic spheres. These plates hold positive and negative charges isolated in space like

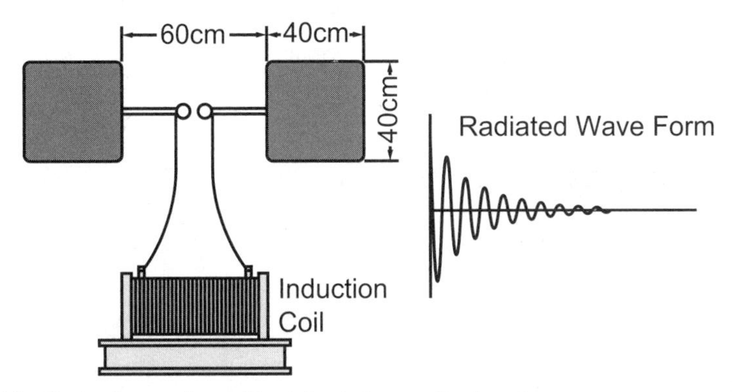

Figure 8.4 Hertz also used metallic plates to form a dipole antenna.

the plates of a capacitor, only now the electric field between them is spread out into space. A dipole distributes this electric field between the two (di) poles.

In addition, because a strong magnetic field is generated around a long current-carrying metallic rod, an LC (inductor-capacitor) resonance will occur, and it will form a resonant type antenna.

Figure 8.5 shows a metallic loop made by Hertz. There are two small metallic balls with a gap in the loop. A spark discharge is generated in the gap when an induction coil forms enough voltage around the loop. One day Hertz noticed that it is possible to observe a spark discharge in one loop when an induction coil is making sparks in another nearby and identical coil. Today, we call the loop being driven by an induction coil a transmitter. We call the other loop a receiver.

Receivers of various sizes are shown in Figure 8.3. If the loop length is short compared to the wavelength, they effectively detect the magnetic field of any electromagnetic wave traveling through them.

In addition, Hertz observed that the size of the spark discharge changes in accordance with the loop length. His experimental setup is shown in Figure 8.6. Hertz instantly recognized an electromagnetic resonance due to electromagnetic waves going from one loop to the other. The specific frequency depends on the size of metallic plates, spheres, or loops. The strength of the spark depends on the distance between them.

As for the resonance of a wind instrument (like a clarinet or flute), the size of the interior cavity determines the frequency of the sound that resonates inside the instrument. The frequency of an electrical resonance is set by the electrical energy being stored alternately in voltage across the capacitor and then in the current and magnetic field of the inductance. The repetition rate at which the energy surges back and forth between these two forms determines the resonant frequency.

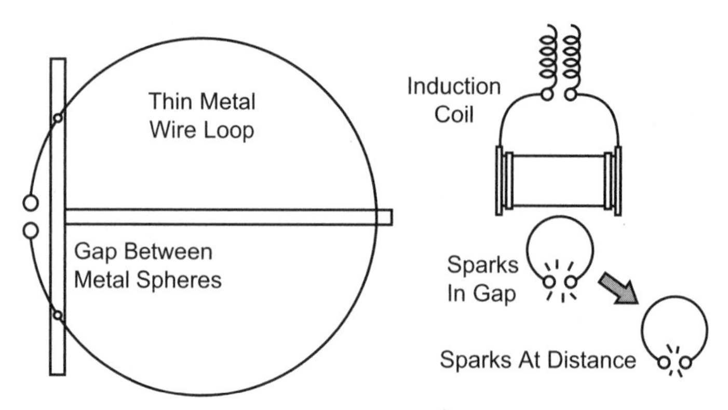

Figure 8.5 Receiving antenna formed from a metallic loop made by Hertz.

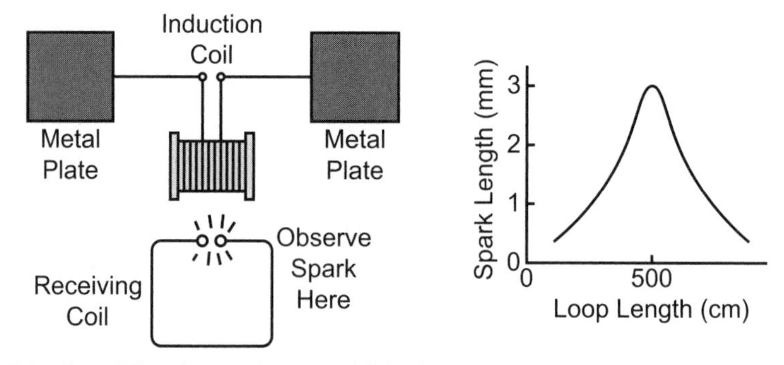

Figure 8.6 One of Hertz's experiments published in 1888.

Because large electromagnetic energy is easily obtained at the resonant frequency of a resonant circuit, antennas related to Hertz's original dipole are commonly designed to resonate at the desired radiation frequency.

8.3 Patch Antenna for Global Positioning Service

A global positioning service (GPS) antenna for a car navigation system is embedded in a plastic case, as seen in Figure 8.7 (Sony Corporation). As installed on a metallic car body, the patch antenna can work well even when mounted right on the chassis metal.

The patch antenna is also called a microstrip antenna. We investigate a rectangular patch antenna held a short distance above a wide metallic surface, which we call ground (GND), in Figure 8.8. The size of the patch is one half wavelength. However, if a high-permittivity material is used between the patch and ground, one half wavelength (and, thus, the size of the patch) becomes shorter due to the shortening effect described in Chapters 4 and 5. Be careful,

Figure 8.7 GPS antenna for a car navigation system.

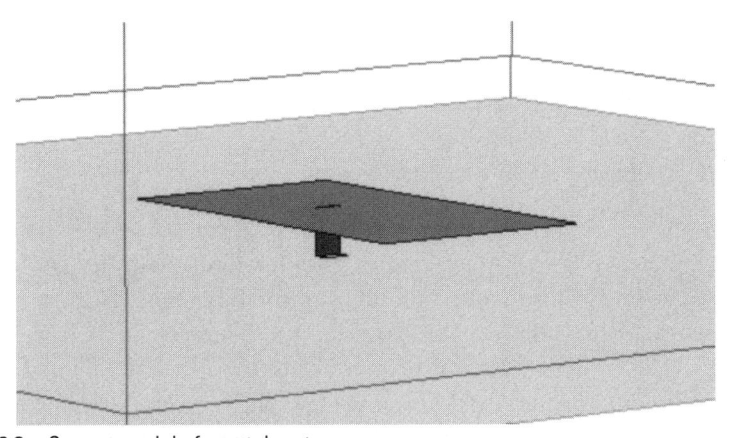

Figure 8.8 Sonnet model of a patch antenna.

however. If too high a dielectric constant is used, the antenna will indeed be small, but it will be lossy and will not radiate well.

Figure 8.9 shows small GPS patch antennas produced by Yokowo Co. Ltd. Outside dimensions are 25 mm × 25 mm × 4 mm, 20 mm × 20 mm × 4 mm, and 18 mm × 18 mm × 4 mm.

One of the frequency bands used by GPS satellites is 1.575 GHz. Thus, its wavelength is

$$\lambda = \frac{3\times10^{8}}{1.575\times10^{9}} = 0.19\ \text{m} = 19\ \text{cm}$$

So the length of a half wave dipole is 9.5 cm, as in Figure 8.10. In this equation, 3×10^8 is the speed of light in free space. Its exact value is 2.99792458×10^8 m/s. To get a feel for how fast this is, this speed would take an electromagnetic wave completely around the earth about seven and one half times per second.

Figure 8.9 GPS antennas produced by Yokowo Co. Ltd.

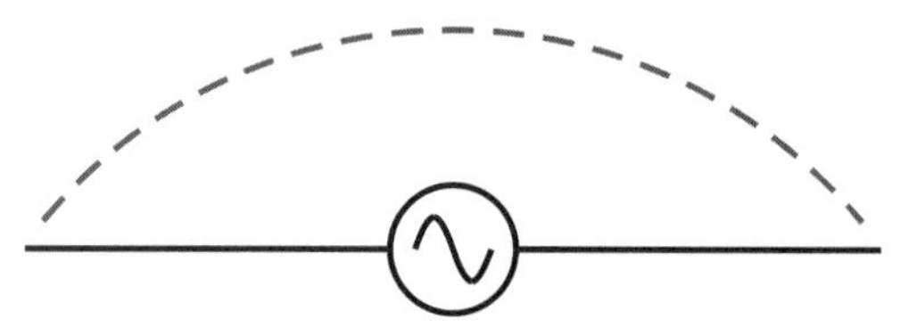

Figure 8.10 Current distribution on a dipole antenna.

The electromagnetic wave from a GPS satellite enters the two metallic plates (the patch and its mating ground) of a patch antenna. Because the electromagnetic energy is alternately contained in the plates like in a capacitor, then transfers to the inductance that is present due to the size of the plates (one half wavelength), we have a resonance, and the antenna receives the signal.

The structure of a patch antenna is like a very wide half-wavelength-long MSL with open ends. Thus, it can be called a microstrip antenna, and it can be viewed as a half wavelength microstrip resonator.

When we transmit using this antenna, we launch an electromagnetic wave onto this MSL. It is totally reflected at both ends. Thus, we have two waves traveling in opposite directions, and this gives us a standing wave. When the standing wave is at a resonant frequency of the antenna, we obtain very strong electric and magnetic fields. The electric field extending from both ends (edges) of the patch reminds us of the capacitor Maxwell used to illustrate displacement current.

8.4 A Patch Antenna Fed by an MSL

A patch antenna fabricated on a double-sided substrate is fed with an SMA connector, as shown in Figure 8.11. This method is convenient because it is easy to measure the input impedance at the feed point using a vector network analyzer. In contrast, Figure 8.12 shows a patch antenna fed by an MSL. When simulating it with Sonnet Lite, S_{11} (reflection coefficient) is calculated as shown in Figure 8.9, and we see that it resonates at 3.6 GHz.

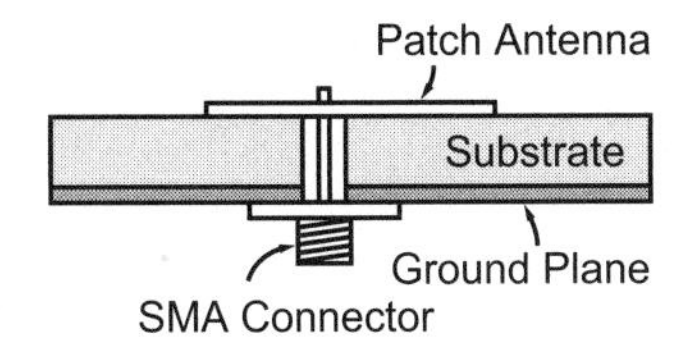

Figure 8.11 Patch antenna fed by a coaxial cable.

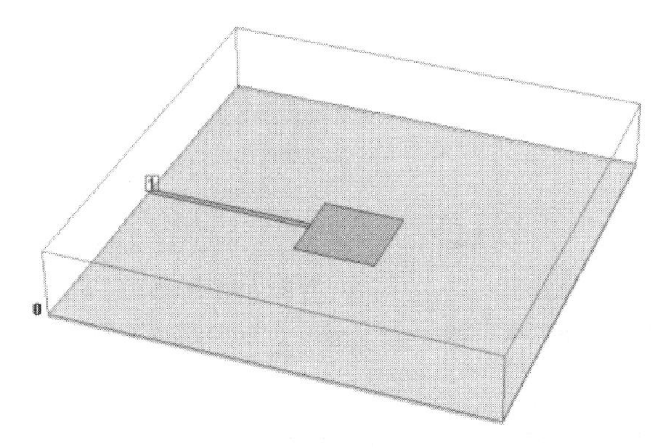

Figure 8.12 Patch antenna fed by an MSL. The substrate is 128 mm × 128 mm, 0.8 mm thick. The dielectric constant is 3.0, and line width is 2 mm. File: pat0.son.

A single antenna with a single connector means these antennas are one-ports. Thus, we measure or simulate only S_{11}, as there is no S_{21} (see Figure 8.13). Reading the vertical axis of the graph, we see that the minimum reflection coefficient is 0.78. This means that a wave with a voltage amplitude of 0.78V is reflected from the antenna when there is a 1V amplitude wave input into the antenna. This is a large reflection and cannot be easily used if nothing is done.

The dielectric of the substrate is 0.8 mm thick and the relative permittivity is 3.0. For this situation, if we set the width of the line to 2 mm, the characteristic impedance of the line is close to 50Ω. However, because the matching between the feed point input impedance and the transmission line are important, we need to determine the feed point input impedance. We must connect a feed line of some kind to the antenna, and then we'll measure the input impedance of the feed line. Remember, the characteristic impedance of a transmission line depends only on the dimensions of the line. The input impedance that we see at one end of the line depends on what we connect to the other end, how long the line is, and the characteristic impedance of the line. The length of feed line transforms the input impedance of the antenna so that a different value is seen at the input to the feed line. It would be nice if there were some way to remove the effect of the feed line.

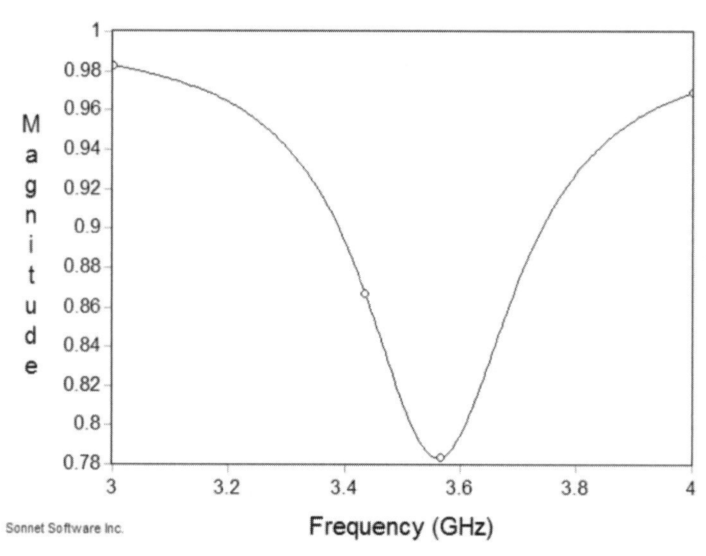

Figure 8.13 S_{11} (reflection coefficient) of the patch antenna fed by MSL.

And, of course, there is a way to do this. Figure 8.14 shows how to calculate the input impedance right at the feed point of a patch antenna by shifting the measurement reference plane in Sonnet. The thick arrow extended from the left port to the feed point indicates that Sonnet will remove the influence of the MSL. This is called the de-embedding.

The simulation result shows that the real part of the input impedance, *R*, is approximately 480Ω at 3.6 GHz, as shown in Figure 8.15. The imaginary part, *X*, is zero, so it is pure resistance at resonance. The impedance at the resonant frequency of an antenna is pure resistance, as we have here. The problem now becomes how to connect a 50Ω line to an input impedance of 480Ω.

To understand that a connection to the edge of a patch has high impedance, like the 480Ω we see here, we can imagine the distribution of current and voltage on a patch, as in Figure 8.16. We see that the current at the edge of a patch is low, and the voltage is high. As the electric field is the gradient (rate of change with distance) of voltage, we see that there are strong electric fields at the edges of a patch. A strong electric field vector field spreading to space from the opposite edges of a patch can be imagined. Remember the electric lines of force drawn by Maxwell, seen in Figure 8.2.

As for the patch antenna fed by a coaxial cable, in Figure 8.11, we can shift the feed point within the patch. Judging from Figure 8.16, the center of the patch is zero voltage, so the feed point impedance there would be 0Ω. In addition, the edge of the patch exhibits the high impedance of 480Ω. If we just move the feed point to the appropriate point, it will be 50Ω This is sometimes called offset feeding.

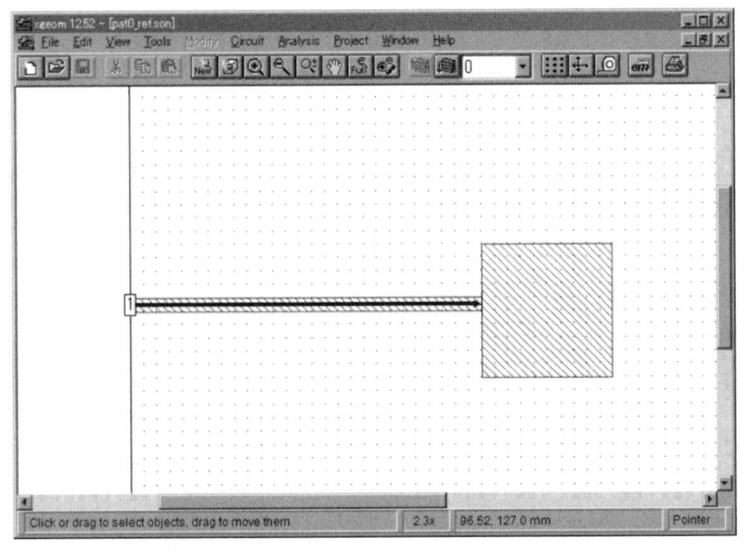

Figure 8.14 De-embedding the patch antenna fed by an MSL. File: pat0_ref.son.

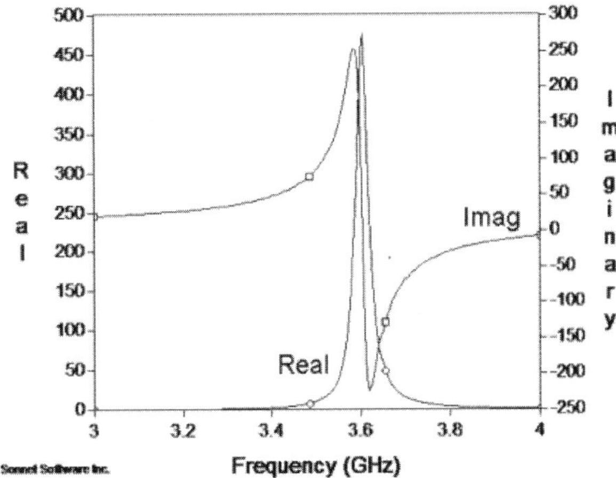

Figure 8.15 Zin (input impedance) of the patch antenna fed by MSL.

8.4.1 Patch Antenna with an Offset Feed

Figure 8.17 is a model of an offset fed patch in Sonnet Lite. The port is set after drawing a via at the feed point. The port location is set by trial and error. You can shift it to the left when R is smaller than 50Ω or to the right when R is larger than 50Ω. The procedure is described in more detail near the end of this chapter.

Figure 8.18 shows the electric field distribution around a patch antenna on a surface passing through the center of the patch. We see that a strong

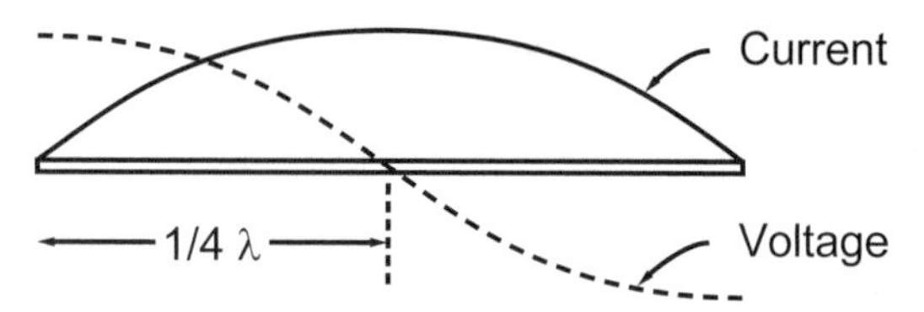

Figure 8.16 Current and voltage distribution on a patch.

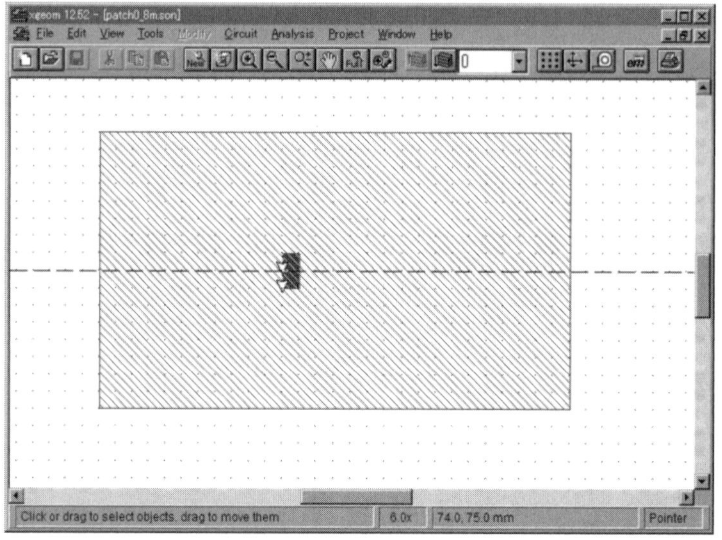

Figure 8.17 Offset fed patch antenna. File: patch0_8m.son.

electric field extends out from both edges of the patch into space, as simulated by XFdtd.

8.5 Vehicle Mounted Antennas

Motor vehicles have a variety of antennas for different frequencies, including digital terrestrial television, FM and AM radio, GPS, the electronic toll collection system, and keyless entry systems. Such antennas pose special problems.

8.5.1 The Influence of a Car Body

Figure 8.19 is a model of a motor vehicle with a digital terrestrial television receiving antenna on a windowpane. The L-shaped antenna, indicated by the arrow, is an inverted L antenna on both sides. The distance between them is 2m, so it is equivalent to approximately 3.3 wavelengths (one wavelength is 60

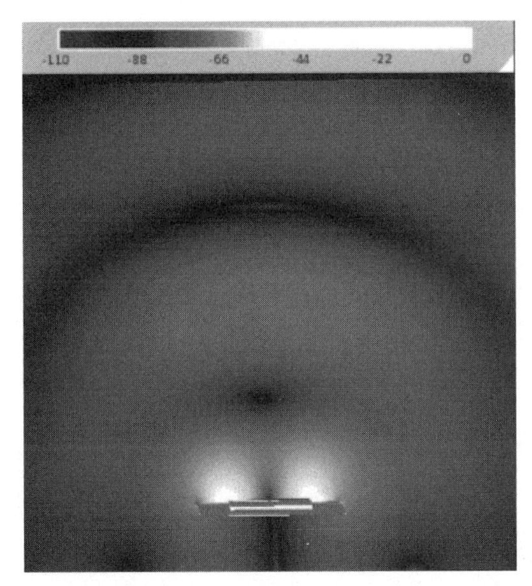

Figure 8.18 Electric field distribution around a patch antenna.

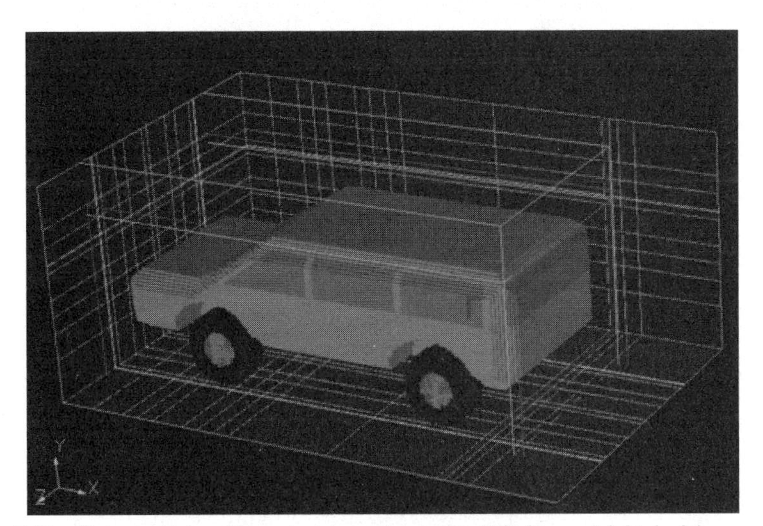

Figure 8.19 Motor vehicle with a receiving antenna for digital terrestrial television on a windowpane.

cm at 500 MHz) at the center frequency of the digital terrestrial television band in Japan.

As shown in Figure 8.20, two half-wave dipoles a quarter wavelength apart realize directivity from A to B, when the phase of the current in element A leads B by 90 degrees. The resulting antenna pattern is called a Cardioid pattern. The

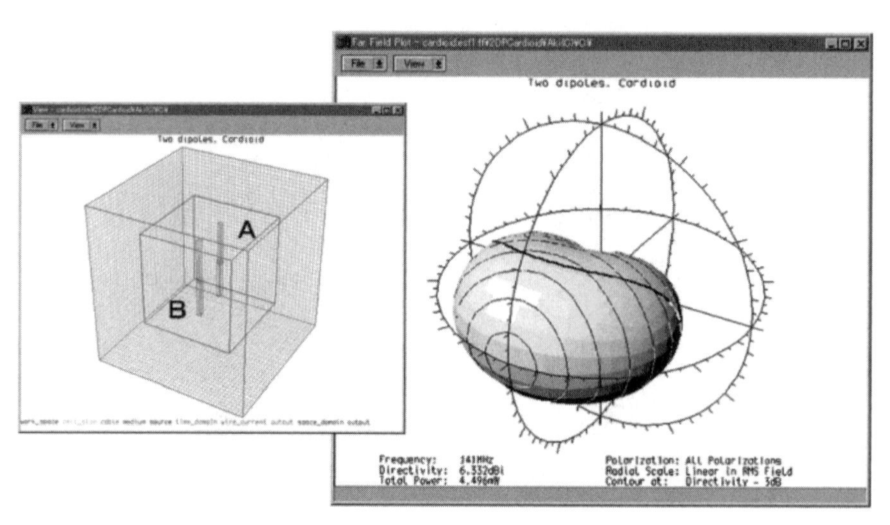

Figure 8.20 Two half-wave dipoles a quarter wavelength apart.

phase difference is achieved by designing an appropriate feed network. When we combine two electromagnetic waves transmitted from antenna A and B, the in-phase waves reinforce and out-of-phase waves cancel each other. This gives us stronger radiation in a specific direction.

Figure 8.21 shows the direction of transmission for these antennas on a vehicle. When the phase difference is set to 135 degrees at 500 MHz, the

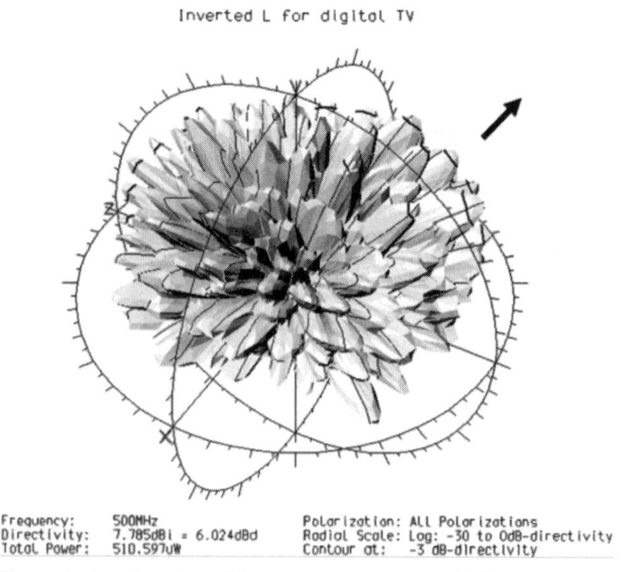

Figure 8.21 Transmission direction of these antennas on a vehicle.

strongest radiation is in the direction of the arrow. The magnetic field vectors in the vicinity of the antenna become parallel to the metallic surface of the car body. This is because the field must be parallel to the induction current (eddy current) flowing on the surface of a car body. This current also contributes to the total antenna radiation. This is also called secondary radiation. We see that the radiation pattern becomes complicated because the secondary radiation combines in and out of phase with the direct wave from the antenna.

Antennas that transmit well (or poorly) in a particular direction also receive well (or poorly) in that direction. This characteristic is called reciprocity. Thus, we can also control the receiving direction by changing the relative phase of the antennas, just as we did for transmitting in Figure 8.21.

8.6 Electromagnetic Susceptibility and Electromagnetic Interference

EMI and EMC are described in Chapter 7. Because undesired radiation has a close link with antennas, here we further discuss the idea of the antenna.

EMC problems can also be viewed as electromagnetic environmental issues. Electromagnetic susceptibility describes sensitivity to the surrounding electromagnetic environment. Even though this term is widely used, it is effectively only a conceptual, subjective term. There is no physical quantitative definition. Figure 8.22 shows a possible physical definition of electromagnetic susceptibility (EMS) based on equipment inside a housing with an aperture.

Like Figure 8.22, we assume a plane wave of x directed electric field illuminates the housing from the y direction. The illuminating electromagnetic wave is partially reflected on the top plate of the housing. Some of the electromagnetic energy is coupled through the aperture. The electromagnetic energy

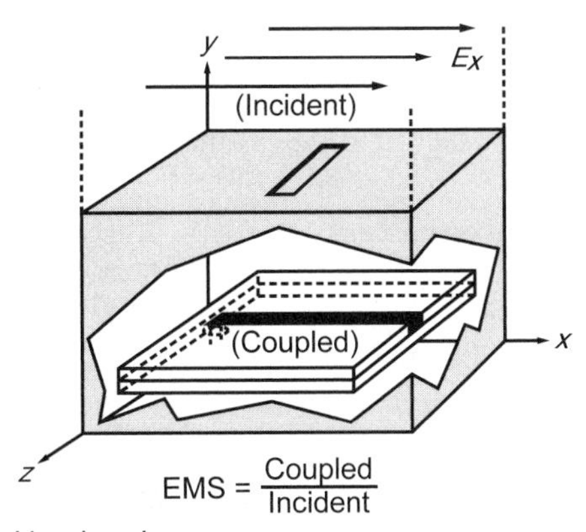

Figure 8.22 Circuit inside a housing.

that makes it through the aperture then couples to the circuit inside the equipment housing.

Considering this example, electromagnetic susceptibility can be defined as the ratio of the electromagnetic energy coupled to the circuit inside the housing to the electromagnetic energy that is available to couple into the housing through an aperture.

The origin of the electromagnetic energy inside the housing is an external uniform plane wave. We also assume that the incident wave is perpendicular to the aperture. Then, the total power available to be coupled inside the housing is the total power in the incident plane wave over the area of the housing.

Once some of the incident energy has coupled inside the housing, some of this energy is lost in the resistance of the housing walls and some is absorbed by coupling to the circuit inside the housing. Only the energy coupled to the circuit is of interest for EMS. If there is more energy coupled to the circuit, susceptibility is higher. If there is more energy coupled to lossy walls of the housing, then susceptibility is lower. Thus, we define electromagnetic susceptibility as the ratio of the energy coupled to the circuit to the total energy illuminating the housing.

Though various directions are possible for an incidence wave, as for our definition of EMS, we assume that the wave is incidence perpendicular to the housing.

The factors that dominate EMS are as follows:

1. Location of the aperture on the plane of incidence and the relative dimension of the aperture;
2. Orientation of the incident electric field to the aperture;
3. The shapes and locations of circuits inside the housing, including multilayered substrates;
4. Frequency.

It is most confusing if we consider all these variables. Thus, we treat factors 1 to 3 as constant values. Then, EMS is calculated as a function of frequency.

8.6.1 Calculation of the EMS

Using the electromagnetic simulation result for the circuit shown in Figure 8.23, we obtain the EMS by application of the recipe in the previous section. The electric field of the illuminating plane wave is x directed, E_x, with a magnitude of 1 V/m, and the z directed component of magnetic field, H_z, is $1/120\pi$ A/m; the power is their product, 2.65×10^{-3} W/m^2. This is the magnitude of the Poynting vector (see Section 6.2.3). We multiply this value by the area of the top of the housing to give us the total power available, 1.6×10^{-4} Ω.

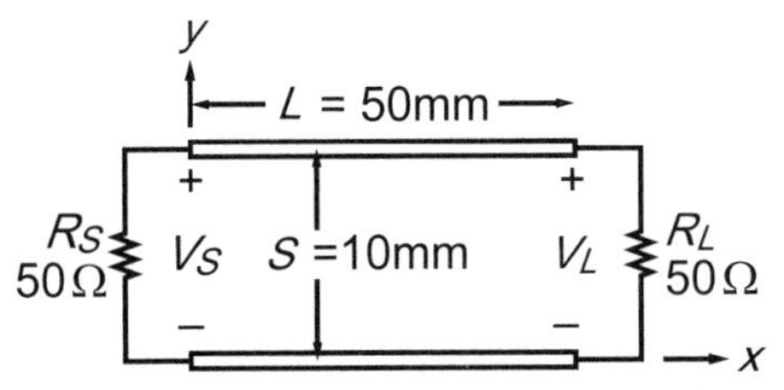

Figure 8.23 Circuit model for which we calculate electromagnetic susceptibility.

Next, by using the housing model of Figure 8.24 to get the value of induced current in the circuit inside the housing, we take the electric power consumption of the termination as the energy coupled into the circuit.

The EMS characteristic calculated using our approach is shown in Figure 8.25. The thick line shows the result with the circuit located at the center of the housing: for (a), y = 75 mm. The thin line shows what happens as the circuit approaches the aperture: for (b), y = 115 mm. The dotted line shows the circuit just above the bottom of the housing: for (c), y = 35 mm.

According to this result, at the frequency of 1.15 GHz where the slot works as a resonant slot antenna, EMS is large, and we see that the closer the circuit is to the aperture, the stronger it receives the electromagnetic wave. Thus, the value at location B (near the aperture) is 20 dB larger than that at C (near the bottom).

In addition, at the housing resonant frequencies of 1.35 GHz (TM_{111}) or 1.60 GHz (TM_{211}), we see that the difference in EMS due to the location is not as much as at the frequency where the aperture works as a slot antenna.

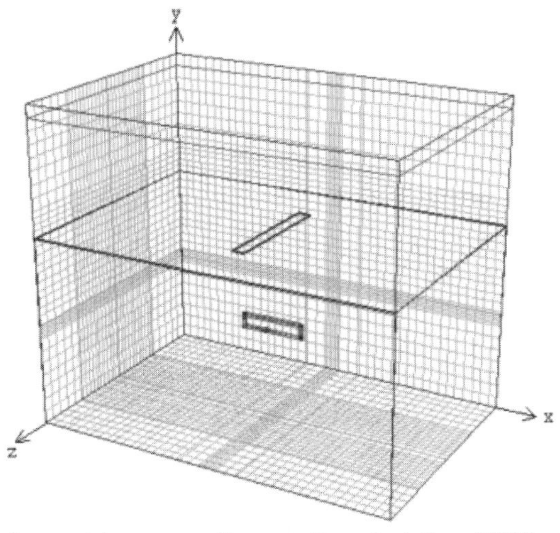

Figure 8.24 Housing model we use to illustrate the calculation of EMS.

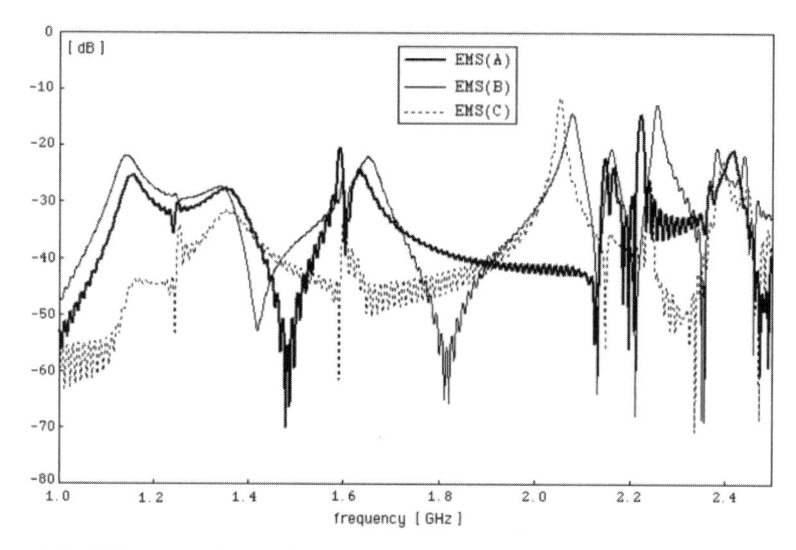

Figure 8.25 EMS as calculated using our approach.

At the higher-order housing resonant modes of 2.24 GHz (TM_{410}) and 2.36 GHz (TM_{411}), we see that the difference in EMS due to location also becomes small. At higher frequencies, in addition to the resonant frequencies of the housing, we see that there are specific frequencies where EMS becomes large due to resonances in the transmission lines composing the circuit.

It is possible to think of the housing as if it is a receiving antenna and to repeat this simulation. Due to reciprocity, we will get exactly the same result if we input a signal into the circuit and measure the power transmitted by the housing. The role of receiver and transmitter can be swapped, and exactly the same result is obtained.

8.7 EMI and Antennas

The purpose of the transmission line is to deliver electromagnetic energy to the load without leaking into space, even at high frequency. However, this ideal is often not realized because of discontinuities like sharp corners and slits. In addition, a standing wave can be generated along the edges of finite ground planes, and we see unwanted radiation occurring from the apertures inherent in multilayered substrates. We design the microstrip line to minimize radiation. But the patch antenna, which can be viewed as an expanded width microstrip line, can radiate strongly.

To eliminate EMI, we must know when a structure will be an antenna and when it will not. Some rules we might use are as follows:

- "The edge of a finite ground is an antenna!"
- "The slot in a metallic box is a slot antenna!"
- "The power distribution cable is an antenna!"

These rules of thumb alone are not enough. When we suspect an antenna in a circuit that we do not want to radiate, we need to perform the required EM analyses to be sure.

In the age of gigahertz electronics, substrates and circuits are easily one half wavelength and more in size. There can be many unintended antennas scattered over a substrate. These potential antennas are made even smaller because of the wavelength shortening effect of the dielectric.

After we make the long journey to solving an EMI or EMC problem, we might conclude that, indeed, "All roads lead to antennas!"

8.8 Confirming This Chapter by Simulation

Sonnet Lite, which provides 3-D CAD drawing for planar multilayered structures, is suitable for the simulation of MSLs and patch antennas. Here, we simulate the offset fed patch antenna described earlier.

8.8.1 Modeling of a Patch

Cell size and box size are set as shown in Figure 8.26. As this is the simulation of an antenna, Top Metal is set to Free Space and Bottom Metal is used as a ground plane. Keep the default of Lossless.

Next, the dielectric layer is set to 1.6 mm thick, relative permittivity, ε_r, is 4.8, $\tan\delta$ is 0.008, and the air layer (above the substrate) is set to 60 mm thick (Figure 8.27). This antenna is designed to resonate at around 2.4 GHz. As required for radiation to be properly analyzed, the 60-mm distance to the Top Cover is approximately one half of a wavelength.

Figure 8.28 shows the patch antenna layout, a rectangle 28 mm × 16 mm. The resonant frequency, f_0, of the patch antenna is determined by the horizontal length d of the patch, and it is calculated as follows:

$$f_0 = \frac{3\times10^8}{2d\sqrt{\varepsilon_r}}$$

$$= \frac{3\times10^8}{2\times28\times10^{-3}\sqrt{4.8}} = 2.45\text{ GHz}$$

where $1/\sqrt{\varepsilon_r}$ represents the wavelength shortening coefficient.

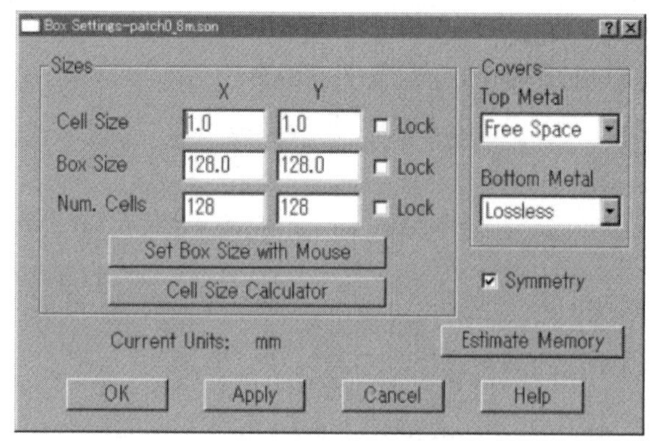

Figure 8.26 Box and cell size settings for this example.

8.8.2 Setting a Via Port

As this antenna uses an offset feed (Figure 8.27), we must draw a via for the feed point and set a port on the via. First, draw a small rectangle for the base of the via on the bottom level (GND) of the Box. Next, select Tools > Add Via > Up One Level. This tells Sonnet Lite that we are now adding vias that go up from the level we are viewing. Now, select Tools > Add Via > Edge Via, and click the metal edge where there is a small triangle mark, as in Figure 8.28. Next, add a port in the same location by first selecting Tools > Add Port.

Set the port to a suitable location, as shown in Figure 8.29. After placing it a little to the left of the patch center, we investigate the resulting input impedance.

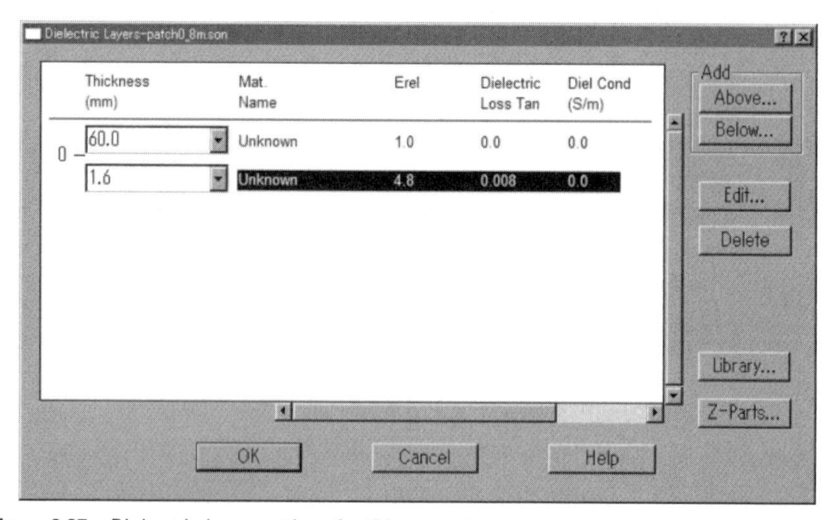

Figure 8.27 Dielectric layer settings for this example.

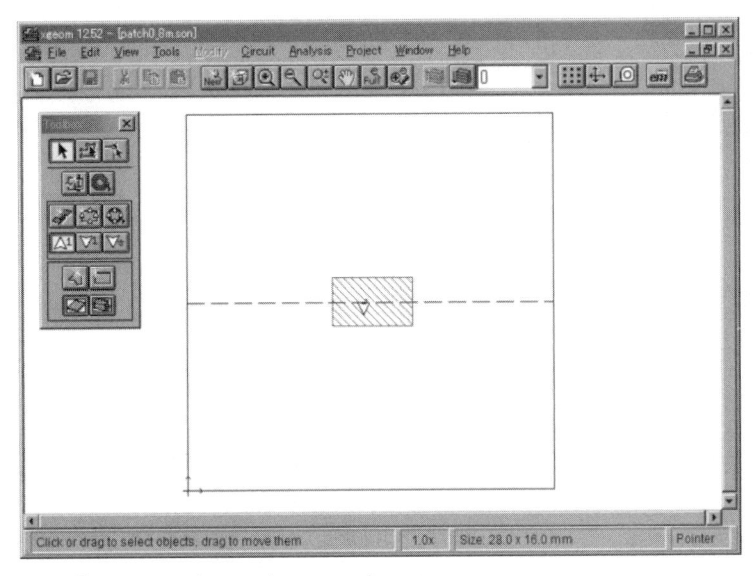

Figure 8.28 The rectangular patch antenna layout.

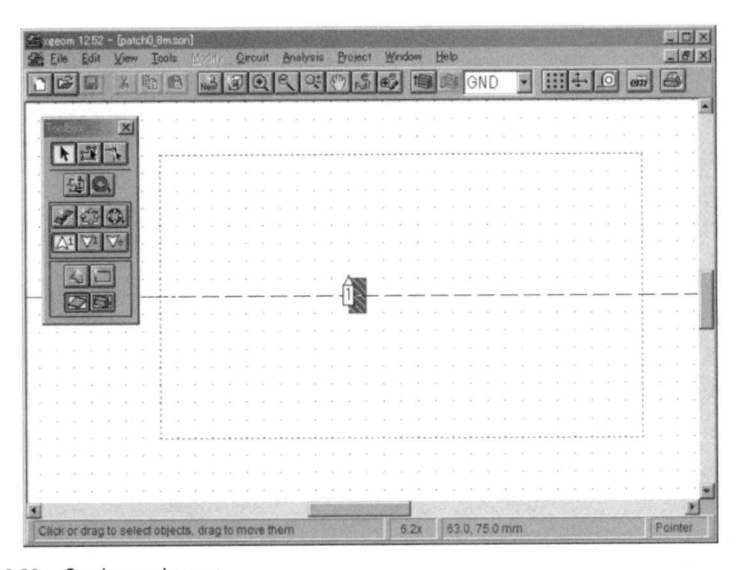

Figure 8.29 Setting a via port.

8.8.3 Interpreting the Result

Figure 8.30 shows the S_{11} (reflection coefficient) result. With the vertical axis in decibels, we often call this return loss. Strictly speaking, return loss for a passive circuit should be positive decibels, but we often plot the negative decibels of the magnitude of the reflection coefficient and call it return loss anyway. The reflec-

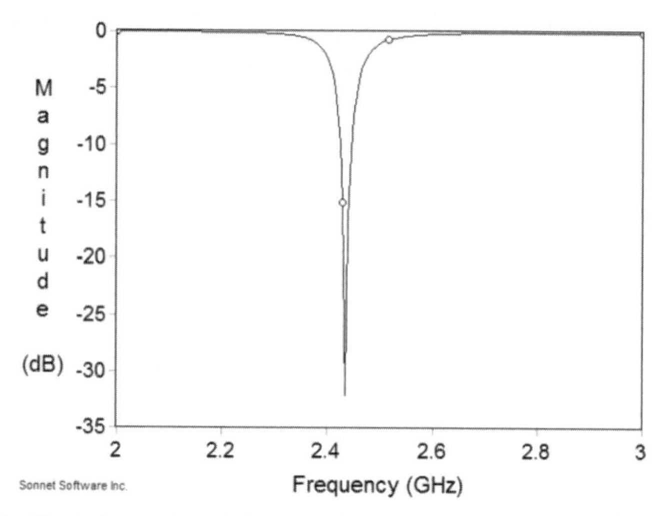

Figure 8.30 Simulation result of S_{11} (reflection coefficient).

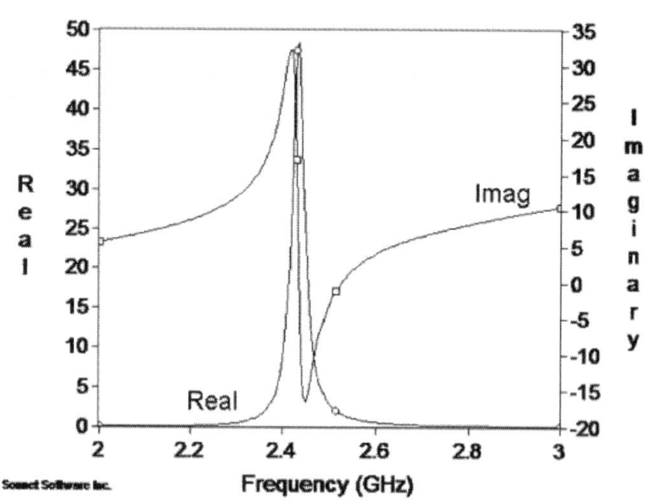

Figure 8.31 Plot of the simulated input impedance.

tion is very low at around 2.45 GHz, and it does in fact radiate electromagnetic energy as an antenna. As this graph is normalized to the usual 50Ω, the small reflection coefficient indicates that the input impedance, *R*, should be close to 50Ω.

As we see in Figure 8.31, *R* is 49Ω at 2.45 GHz, which means that this port location is excellent. If we move the port to the right from the location of Figure 8.29, *R* will be less than 50Ω. And if we move it to the left, it will be greater than 50Ω. Try it and see.

The right vertical axis shows the Imaginary (imaginary part of the impedance), which is also called the reactance, *X*. We see that a resonant type antenna has zero *X* at the resonant frequency.

8.8.4 Investigating the Surface Current Distribution

Select Analysis > Setup… and check the Compute Current Density box in the top left corner, as shown in Figure 8.32, and reanalyze the circuit. Now, we can display the surface current distribution, Project > View Current, as in Figure 8.33.

Figure 8.33 is the default display at the lowest frequency of 2 GHz. We see that the strongest currents concentrate around the via.

In Sonnet we use planar 3-D CAD to draw metallic conductors. These conductors are zero thickness by default. While often useful for analysis, this is actually physically impossible. When we need a more realistic model, we make an equally spaced stack of these infinitely thin conductors to model thick conductors.

Another common model used by 3-D CAD electromagnetic simulators appears to include the full effect of thick metal. However, some of these models restrict all current to the infinitely thin surface of the thick metal. There is no current allowed inside the metal. This is acceptable at high frequency, where skin depth keeps all current on the surface of the metal anyway. However, it does not work at lower frequency, where current flows inside the thick metal.

The Sonnet multisheet model does allow current to flow inside the thick metal, at least when there are three or more sheets used to model the thick metal. Keep in mind that whatever model is used for thick metal in whatever electromagnetic simulator, the analysis will take substantially more time. Thus, thick metal models should be avoided unless you actually need the increased

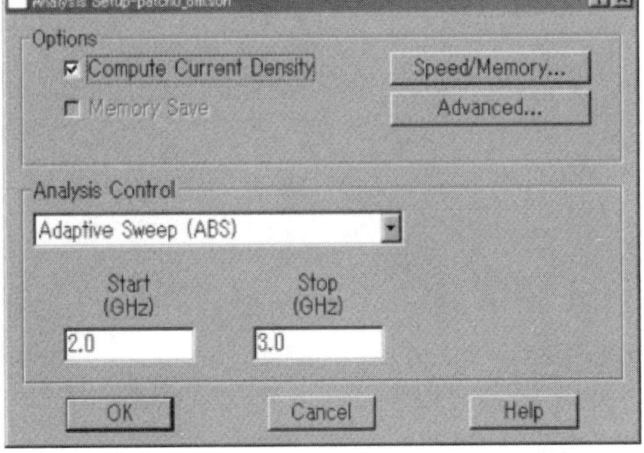

Figure 8.32 Check the Compute Current Density box in the top left corner.

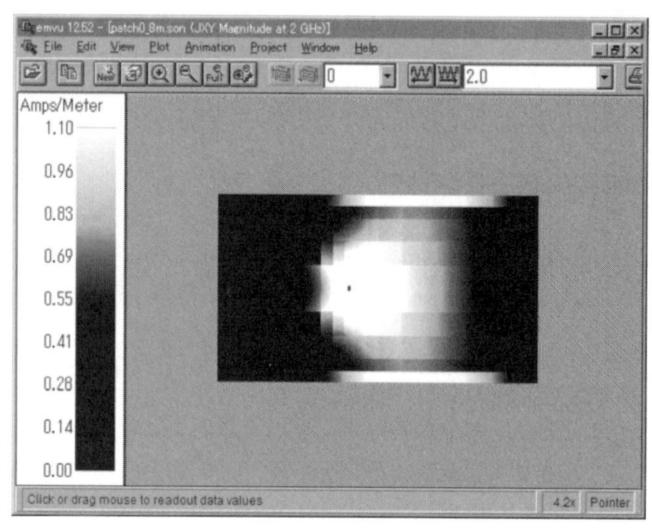

Figure 8.33 Surface current distribution at the lowest frequency, 2 GHz.

accuracy. To determine if you need the increased accuracy, just analyze a simple circuit with, and then without, thick metal. Then make a judgment as to whether the increased analysis time is worth the difference you see between the two results.

Figure 8.34 is the surface current distribution at 2.43 GHz. You can choose the frequency settings at the top right. In this case, we simulated with ABS Sweep, so we can only display the frequencies that were automatically selected during analysis.

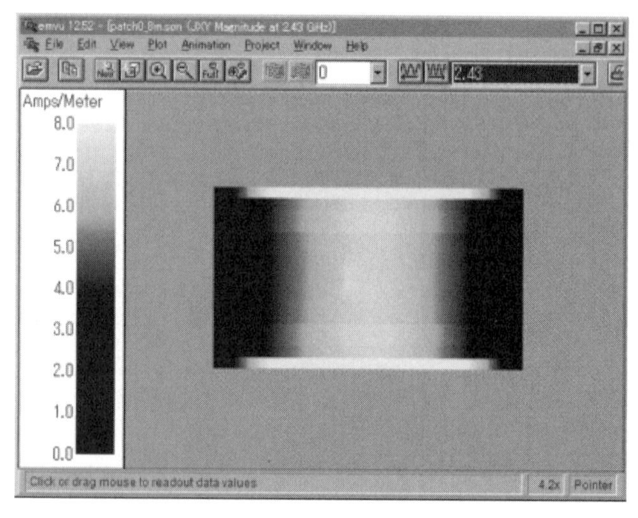

Figure 8.34 Surface current distribution at 2.43 GHz.

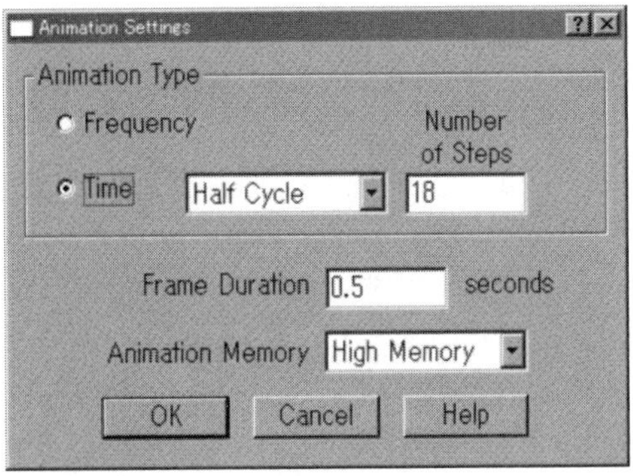

Figure 8.35 Select animation.

Figure 8.36 Play animation.

Compared with Figure 8.33, the current flows beautifully and symmetrically on the upper and lower edges. We see that the patch edges are one half wavelength long. As this is a standing wave, the distribution does not change with time. Let's set up a time animation (sweep the current distribution through one half cycle) at this frequency. Select Animation > Settings…, shown in Figure 8.35, to animate the current distribution. Select Time under Animation Type.

Next, select Animation > Animate View, as shown in Figure 8.36, and use the controls to start and stop the animation. We see that the standing waves along both edges stay in place through the entire cycle.

An interesting exercise is to set the analysis for a linear sweep that covers a few higher-order resonances of the patch antenna. Then set up a frequency animation over all the analyzed frequencies. The resulting modal patterns, as they melt from one into another, are beautiful.

8.9 Summary

1. Maxwell predicted electromagnetic waves could travel in a completely empty, "transmission line named space," and Hertz proved the existence of electromagnetic waves experimentally.

2. Electromagnetic field simulators are useful for the design of antennas.
3. Strong electric currents flow when an antenna resonates, and a standing wave is generated.
4. For antennas, "the object is to receive and transmit electromagnetic waves efficiently and accurately" and "an excellent receiving antenna is an excellent transmission antenna."
5. For EMI, EMC, and EMS, circuits must be viewed as antennas, only now we wish to suppress the radiation, rather than enhance it.
6. "All roads lead to antennas!"

9

Fundamentals and Utilization of Electromagnetic Simulators

9.1 Many Different Electromagnetic Simulators

The commercial electromagnetic field simulator first appeared in the latter half of the 1980s. Countless numbers of electromagnetic simulators are being developed in the world today. Needless to say, because humans make software, each product is characterized by them. The authors of this book have become intimately familiar with many of the developers and many of the electromagnetic tools that they have developed at each company since the 1980s. We have learned about the successes and the troubles that arise during their development. In this book, we bring together a number of those experiences to help others climb the same learning curve, only faster.

Next, we describe the theory (from a conceptual point of view) and utilization of a few of the electromagnetic field simulators we have used in this book. We also compare the relative advantages and disadvantages. Keep in mind that this is only a high-level comparison. As you become more familiar with the details of each tool that you use, you will find a wealth of information. The most important point is that different electromagnetic simulators have different strengths and weaknesses. No single simulator is appropriate for all tasks.

9.1.1 Capabilities of Electromagnetic Simulators

The electromagnetic field simulator solves for the electromagnetic fields as determined by Maxwell's equations. For planar, layered circuits, the main input is the layout and the substrate stackup. As shown in the examples described in this book, the principal outputs are S-parameters, current distributions, and models, summarized here.

- A. Visualization:

  1. Current distributions on conductor surfaces (animation is useful).
  2. Electric and magnetic field distributions in the near field (vector representation, root mean square value, and so on), electric power distributions, energy distributions.
  3. Far-field radiation patterns.

- B. High-accuracy analytic data:

  1. S-, Z-, and Y-parameters (e.g., Cartesian coordinate system, Smith chart).
  2. Characteristic impedance of a line, effective relative permittivity.
  3. SPICE lumped model file extraction.
  4. Electric field and magnetic field at an observation point (Cartesian coordinate system, display in decibels, evaluation of electric field strength for EMC problems).
  5. Current or voltage on a line.

9.2 The Method of Moments and Its Friends

The Method of Moments is a technique that is performed in the frequency domain. By applying a sine wave as a signal source, current distributions and S-parameters are obtained one frequency at a time. Because the same simulation is repeated for each frequency, if we are satisfied with data at fewer frequencies, we can get results faster.

The Method of Moments (in the formulation that is usually used in electromagnetics) starts with the integral form of Maxwell's equations. In general, the Method of Moments can solve any linear integral-differential equation by converting it to a set of simultaneous linear equations that can be solved by inverting a matrix on a computer. It can be used in areas other than electromagnetics (e.g., in mathematical statistics).

9.2.1 Implementation of Shielded Method of Moments

For planar circuits, electromagnetic field simulation by the Method of Moments starts by dividing the metal of a circuit into small subsections, as in Figure 9.1. It is important to realize that only the circuit metal is meshed. The

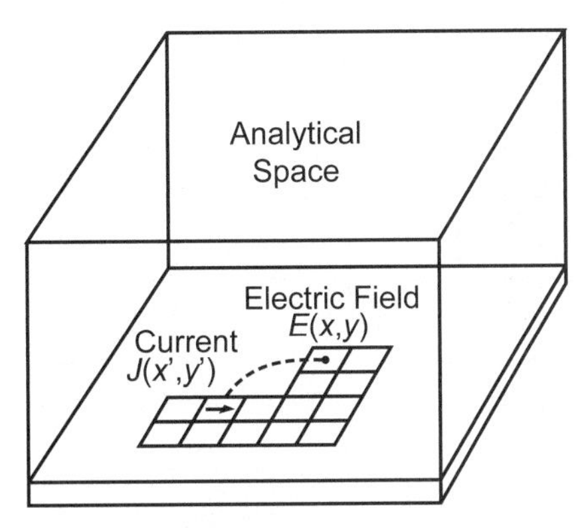

Figure 9.1 Analytical space of the Method of Moments.

volume of the substrate is not meshed. The technique then solves for the surface current flowing in each subsection of a multilayered substrate.

In Figure 9.1, $J(x', y')$ is a small current element. The central problem is to calculate the fields due to this current induced on all other subsections in the circuit. The electric field, E, is at the observation point of another cell and is given by the following equation:

$$E(x,y) = \int_{x'}\int_{y'} G(x,y,x',y')J(x',y')dx'dy'$$

Here, $G(x, y, x', y')$ is the Green's function. It was invented by George Green of Britain (1793–1841). A bit of trivia is that he lived in a windmill that is preserved today as a tourist site. Very little is known about his life. The Green's function represents the fields due to an infinitesimal dipole at (x', y'), and it exactly includes the effect of the layered substrate. It is like an impulse function, only it is a function of location, not of time. To get the electric fields due to a patch of current, we simply convolve the Green's function with the current distribution assumed on the patch. Application of the previous equation to every subsection, one at a time, allows calculation of the coupling between every pair of subsections. The Method of Moments is then used to solve for the current distribution, S-parameters, and so on.

As shown in Figure 9.2, the surface of the conductor is divided into N subsections. First, we assume current on subsection i. The current distribution usually assumed is the rooftop function. The sloping sides of each rooftop function overlap with the next adjacent rooftops, allowing a smooth variation of current from one subsection to the next. Thus, one rooftop current distribution actually extends over two of the adjacent rectangles shown in Figure 9.2. The current ramps up on one rectangle and then ramps down on the next rectangle.

Next, the electric field at subsection j is determined from the known current at subsection i, as shown in Figure 9.3, using the previous equation. The

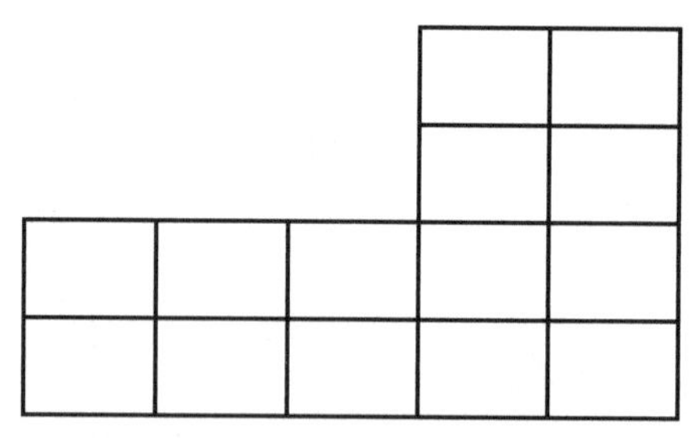

Figure 9.2 The surface of the circuit metal (only) is divided into subsections.

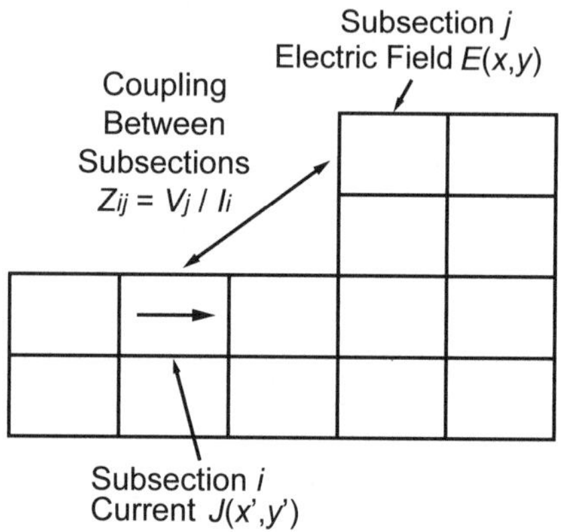

Figure 9.3 The voltage induced in each subsection due to current on any other subsection is calculated.

voltage induced in subsection i due to the current on subsection *j* is calculated by the integrating the electric field.

$$V_j = \int_x \int_y \int_{x'} \int_{y'} G(x, y, x', y')J(x', y')dxdydx'dy'$$

As shown in Figure 9.3, the voltage induced in all subsections is obtained by putting a known current (which is called a basis function) on only one subsection, and the current on all others is set to zero. By iterating this for the current on every subsection, one at a time, we can fill a matrix with the couplings (the current on one subsection generates voltage on another). A matrix inversion then determines the actual surface current distributions, shown in Figure 9.4, by using the boundary condition that the total voltage, due to current on all subsections simultaneously, is zero.

Sonnet solves for a circuit inside a shielding conducting box. The Green's function, which determines the electric field generated by an infinitesimal dipole, is given next:

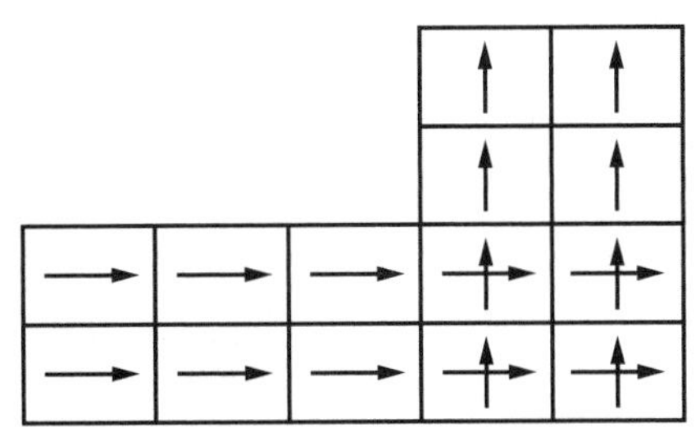

Figure 9.4 Surface current distributions are determined.

$$G(x,y,x',y') = \sum_{m=0}^{\infty}\sum_{n=1}^{\infty} C_{mn} \cos\left(\frac{m\pi x}{A}\right)\sin\left(\frac{n\pi y}{B}\right)\cos\left(\frac{m\pi x'}{A}\right)\sin\left(\frac{n\pi y'}{B}\right)$$

Because the Green's function is represented by a sum of rectangular waveguide modes (the cosine and sine functions of the previous equation; also see Chapter 7), we can perform the integrations indicated by the equation for subsection voltage listed earlier and we have

$$V_j = \sum_{m=0}^{\infty}\sum_{n=1}^{\infty} D_{mn} \cos\left(\frac{m\pi x_1}{A}\right)\sin\left(\frac{n\pi y_1}{B}\right)\cos\left(\frac{m\pi x_0}{A}\right)\sin\left(\frac{n\pi y_0}{B}\right)$$

The only difference is the constants, C_{mn}, are changed to different constants, D_{mn}. Performing the summation is extremely fast and accurate using a 2-D FFT.

9.2.2 Other Method of Moments Implementations

Sonnet is a well-developed solution that represents the coupling between subsections as a sum of waveguide modes. In general, problems involving finite volume regions result in a summation for their solution. On the other hand, unshielded tools, like Agilent Momentum, EM Software & Systems FEKO, and Zeland (Mentor) IE3D, are solved in an unshielded, infinite volume environment. Infinite volume problems usually involve integrals, rather than summations, for their solution.

FEKO is based on the Method of Moments. It uses multilevel fast multipole method (MLFMM) for very large-scale problems. In addition, FEKO incorporates several other EM analysis techniques, like finite element method (FEM), physical optics (PO), geometric optics (GO), and uniform theory of diffraction (UTD).

It has been twenty years since the first commercial electromagnetic field solvers were developed. With the evolution of fast and powerful PCs, the scale of the problems that can be solved has increased substantially. For one particular problem that is central to EM analysis, matrix inversion, J. Rautio inverted his first matrix for 100 subsections on a 4.77-MHz IBM PC in 1985 in about 1 hour using hand-coded assembly language. More recently, in 2010, he has inverted a matrix for 100,000 subsections on a dual hexacore 2.67-GHz computer in about three hours. Since matrix solve time increases with the cube of the number of subsections, computers and algorithms have seen a combined increase in speed of 300 million times since 1985.

The situation is even more dramatic for electrically large problems at high frequencies. The electromagnetic wave wavelength is short in the gigahertz range. For instance, analysis of a wireless LAN in an office needs a huge memory. The car antenna now must handle frequencies higher than the usual TV and FM radio bands (tens to hundreds of megahertz), like GPS (1.5 GHz) and electronic toll collection system (5.8 GHz). Problem size is measured in terms of wavelengths. The automobiles are still about the same size, but the wavelength is now much shorter. We need new ways to solve problems that extend over such a large number of wavelengths. A number of speed-enhanced techniques have been developed to meet the need.

Figure 9.5 shows a simulation to determine the level of electromagnetic radiation from wireless equipment inside a car. A human Phantom (model) is modeled by FEM and an antenna on a car body is simulated by a speed-enhanced Method of Moments (courtesy of Farad Corp. and EM Software & Systems).

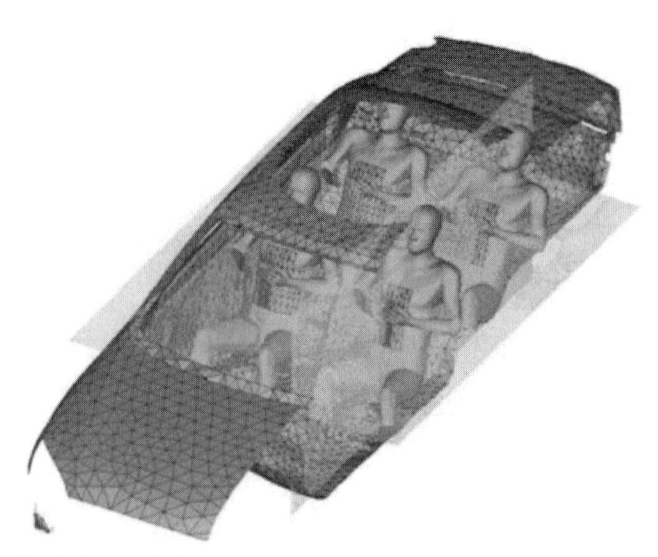

Figure 9.5 Simulation model inside a car.

9.3 FDTD and Its Friends

FDTD and TLM are techniques that perform their analyses in the time domain. This technique inserts an initial (broadband) pulse into a system and evaluates the electromagnetic fields as the pulse propagates through the system, as shown in Figure 9.6. This is instead of solving a system of equations like the Method of Moments.

These methods discretize all space into a small mesh and simulate the propagating electromagnetic field along the mesh using a difference formula based on the differential form of Maxwell's equations.

Since these techniques mesh the entire volume, we use 3-D CAD tools to set up models for arbitrarily shaped objects. The detailed human model (from Medical Virtual Reality Studio, http://www.mavrstudio.de/) was developed to simulate, among other things, the influence of portable electronic devices on the human body.

A signal source of one Gaussian pulse is inserted into the system. Such a pulse (given that it is sufficiently narrow) includes wideband frequencies. As the entire space, including the conductors, the dielectric substrates, and space, are discretized with meshes, the pulse propagates to all cells of the mesh.

Observation points (for viewing fields) can be set in space or inside a dielectric. The simulation stops when the time response data converges to almost zero. A Fourier transform (one form of which is the FFT) of the time response data is taken yielding a wideband frequency response. Figure 9.7 shows an example time response (transient response) by MicroStripes of CST, which uses the TLM method.

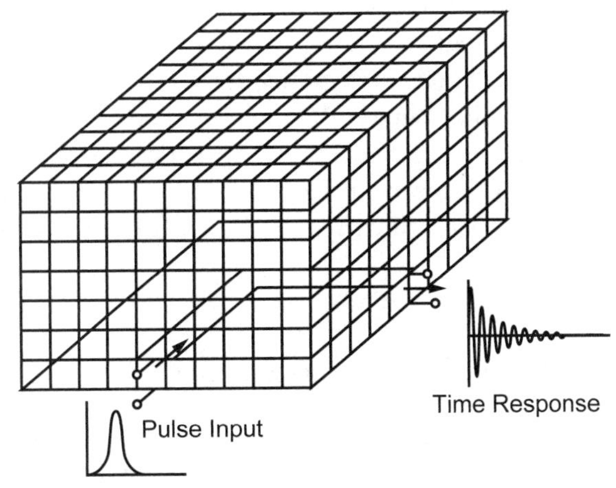

Figure 9.6 Analytical space of FDTD and TLM.

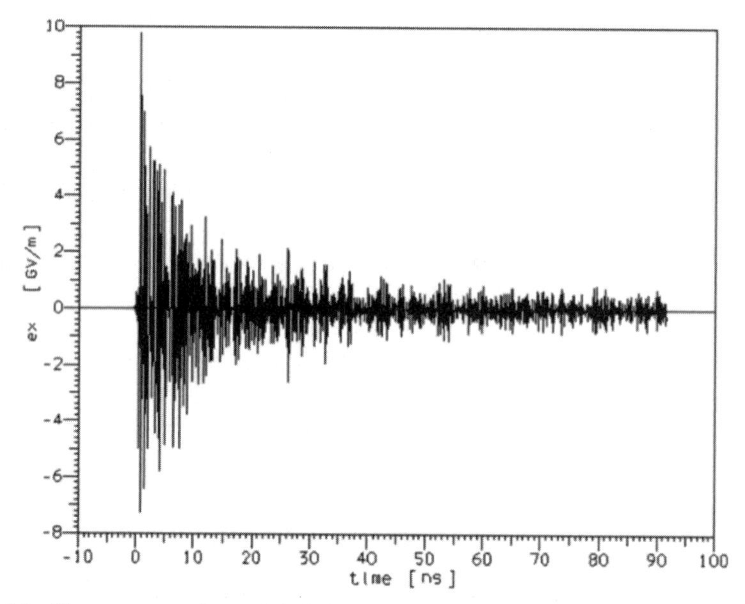

Figure 9.7 Time response data example (transient response).

Even though the amplitude of the electric field right at the edge of the graph does not quite reach zero, if we perform an FFT on this data, the data in the frequency domain still looks good, as seen in Figure 9.8.

The FDTD method directly simulates the electromagnetic field propagating in space by modeling the differential form of Maxwell's equations as a set of difference equations.

On the other hand, the TLM method models discrete points in space as a mesh of transmission lines. It discretizes the entire space by cells composed of TLM meshes, as seen in Figure 9.9.

An impulse excited at one node propagates to the next, one after another. This operates very much like Huygens' principle (all points on one wave surface generate secondary waves to create the next wave surface). Because this transient response is calculated by a computer iteratively, this is a unique method in that it first solves for the voltage and current and then converts to the electric and magnetic fields.

9.4 Frequency Domain Versus Time Domain

Numerical electromagnetic analysis is roughly divided into two kinds of methods. Frequency-domain techniques analyze frequency by frequency. Time-domain techniques analyze timestep by timestep and then use an FFT to obtain a broadband frequency response.

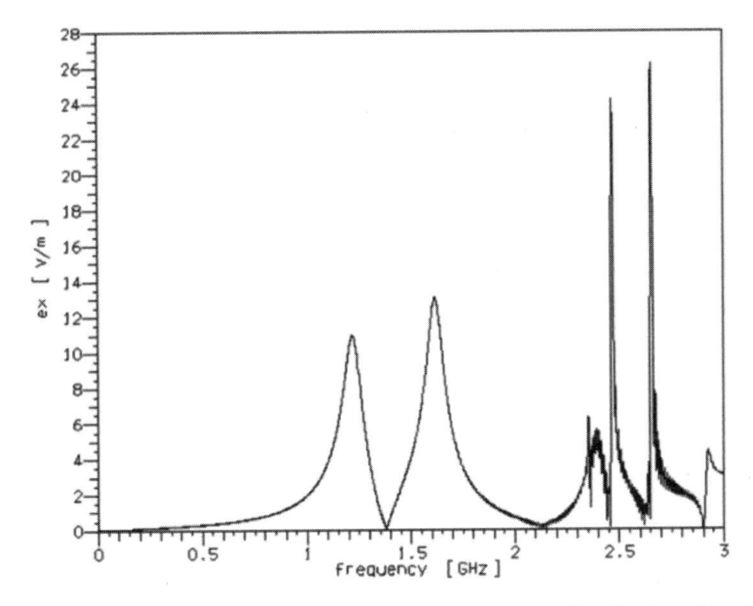

Figure 9.8 Frequency-domain data obtained from by Fourier transforming time-domain data.

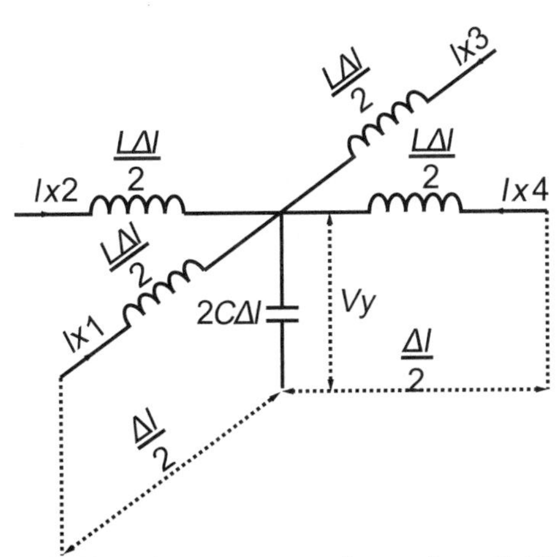

Figure 9.9 One cell in a TLM mesh represents a volume using a distributed circuit.

Frequency-domain methods are best when we only need a few frequencies. Interpolation algorithms (taking a few carefully chosen frequencies and accurately generating many frequencies) are very advanced today. However, extremely wide bandwidth results are still difficult.

On the contrary, time-domain methods excel in generating wideband data responses but have difficulty when there are resonances. The time-domain response tends to keep ringing for a long time. There are, however, well-developed techniques to extrapolate the ringing and obtain good responses even for resonant structures.

A few of the more important commercially developed general-purpose methods are as follows:

1. Frequency domain:
   - Method of Moments;
   - Finite element method (FEM);
   - Boundary element method (BEM).
2. Time domain:
   - TLM;
   - FDTD;
   - Spatial network method.

We see that these techniques are largely divided into the two groups described here. The electromagnetic simulators used by the authors are easily classified. Most frequency-domain simulators are equipped with a 2-D multilayer CAD drawing capability. The time-domain simulators and the frequency-domain FEM simulators come with a 3-D CAD capture tool for arbitrary shapes.

9.5 Accuracy of Electromagnetic Simulators

The authors of this book have been concerned with accuracy since we first started using electromagnetic field simulators. Simulation is always an abstraction of reality. We digitize the analog world on a computer. No matter how much we refine the cells or mesh, we cannot completely eliminate error. It is important to gain an understanding of the relation between discretization and error, especially in view of memory usage and analysis time.

There are a great variety of problems to be analyzed. Thus, it is difficult to make absolute statements about error. However, we can certainly find ways to quantify, estimate, and bound the error.

9.5.1 Discretization Error in Method of Moments

Here we discuss discretization error in the Method of Moments based on an evaluation method that Sonnet Software (J Rautio) introduced. This error quantification technique is based on the strip line. The characteristic impedance of a lossless, zero-thickness strip line can be calculated exactly by conformal mapping. Sonnet uses such a stripline to precisely measure analysis error.

Figure 9.10 is a stripline whose characteristic impedance is exactly 50Ω. By modeling a quarter-wavelength line at 15 GHz, we compare the result to the exact answer. If the line is terminated, as usual, by a 50Ω load, the exact correct magnitude of S_{11} is 0.0 (which equals minus infinity decibels) and the correct phase angle of S_{21} is exactly –90 degrees.

However, the simulation, gives us an S_{11}magnitude of = 0.01026 and an S_{21} phase angle of –89.999 degrees. Error in the analysis of the characteristic impedance mainly appears as error in the magnitude of S_{11}, and here it is approximately 1.026 percent. Error in the velocity of propagation appears mainly as error in the phase of S_{21}. We see that it is approximately 0.001 percent.

The error is determined by the degree of discretization. For example, the error in characteristic impedance is approximately 1 percent when a line is discretized into 16 cells across the line width. In fact, the percent error is bounded by a simple formula that is a function of the discretization numbers.

$$E_T \le \frac{16}{N_W} + 2\left(\frac{16}{N_L}\right)^2 \qquad N_W > 3 \qquad N_L > 15$$

where $N_W$is the number of cell across a line width, N_L is the number of cells per wavelength, and $E_T$is the error in percent.

The stripline in Figure 9.10 was analyzed with N_W = 16 and N_L = 128. The expected error due to the discretization is 1.03 percent by the formula, which agrees almost exactly with our result.

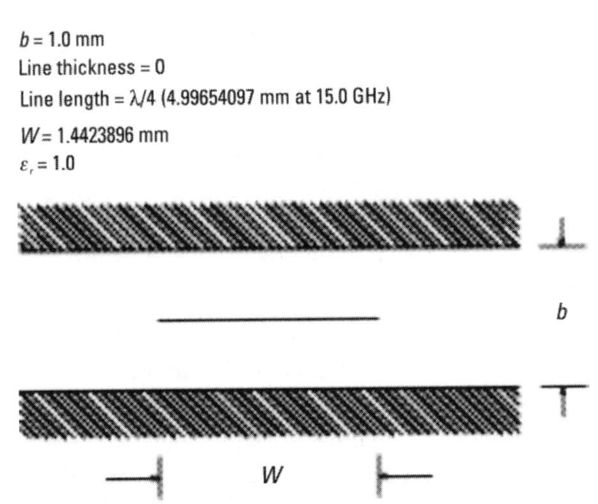

Figure 9.10 Stripline has an exact solution and so is easily used to quantify electromagnetic analysis error.

In addition, we see that the number of cells per wavelength, N_L, is related to the error by the inverse square in the second term of the formula. In most practical situations, we see that the number of cells across line width is dominant in determining error. Contrary to popular wisdom (see the following discussion of urban legends), subsections per wavelength typically are of secondary concern. If you would like to conduct your own tests using single or coupled stripline, the Excel file, "Z0.xls," on the DVD can be used.

9.6 Utilization of Electromagnetic Simulators

Based on the authors' experience with electromagnetic simulators, the following points are important for their most effective usage.

1. Do not discretize a problem too finely at an early stage. We tend to worry about error, and, as a result, we might tend to discretize a problem too finely at an early stage. A rough rule of thumb for early-stage discretization is to keep the subsections one-tenth to one-twentieth of wavelength long and have about five or so subsections across the width of a line.
2. A portion of the circuit can benefit from being snapped. The FFT used in shielded Method of Moments (Sonnet) requires a very fine underlying FFT mesh for our circuit layout. The FFT provides extremely high robustness. After setting the FFT mesh size (Circuit > Box) we might find our desired geometry is not exactly aligned to that mesh. A finer mesh provides better alignment but also requires a longer analysis time. Alternatively, we can snap portions of the circuit to the FFT mesh, or simply tolerate the offset until we are ready for a final, high-accuracy verification analysis.
3. Use a finer discretization where strong current flows. We need to set finer meshes where the edge singularity (high edge current) is found for both edges of all transmission lines. How this is set depends on the specific electromagnetic simulator you use. In the case of Sonnet, it is set automatically.
4. Error might be large with a zero metal thickness model. When we model an MSL conductor with zero thickness, we can get a good result when we take care of items 1–3. For all electromagnetic simulators, the total memory usage grows rapidly when we discretize it to a very fine mesh. Usually, we analyze with zero metal thickness. However, in some cases, thickness might be important. For example, a coupler with a gap between lines that is on the order of the metal thickness

has especially strong (odd mode) electric field between thick line edges and should typically be modeled with thick metal, at least as a last step before deciding the design is complete. If your accuracy requirements are extremely high, you might need to include thickness even in other situations as well. If in doubt, do an EM analysis with, and then without, thickness. If the difference is large (you are the judge of what large is), then include thickness. Otherwise, enjoy the faster analysis of not including thickness.

5. The analytical space for a radiating structure is not large enough. Analysis of an antenna needs enough space around the antenna in the simulator for the near field to set up and be correctly analyzed. Otherwise, the far field and input impedance will be incorrect. In fact, this is true (but not quite as critical) even for nonradiating objects and circuits. We do not want the edge of the analytical space to modify the response of our circuit.

   As a rule of thumb, set the edge of the analytical space to about one wavelength from the object in all directions. To be sure, we need to apply convergence techniques as described next.

6. Confirm results using convergence techniques. Convergence techniques allow us to confirm the quality of the discretization of a model. For best efficiency, we start with a coarse, fast discretization. Then we analyze finer and finer discretizations and evaluate the difference between subsequent results. We can then decide when we have a fine enough discretization. It can be tedious to do this on every design. However, after doing this a few times, a designer becomes experienced and can reduce the amount of time spent confirming that EM analysis error is not going to jeopardize a design.

7. Do a reality check. When we get a simulation result, we should always do a reality check. For example, sometimes smaller parts might not be connected. If we notice the results look a little odd, when we have a careful look we might find a tiny gap that should not be there. Sometimes it is good to have an experienced colleague look at the data. It might be something they have seen before.

8. Compare to measured data. If possible, when the design has been fabricated, compare the EM analysis result to measured data. In the old days, we would just build a circuit and tune it up, maybe with a razor knife. Electromagnetic software was not needed. Those days are gone.

However, EM software does not cure all problems. Dielectric constants might be wrong, there might be critical etch tolerances in a circuit, transistor models might not be valid. So, to be safe, we must be skeptical; we must doubt everything.

One thing we must doubt is our measurements. One very useful trick we have found is to include a simple circuit as part of any major fabrication. A good example is a through line. If you see a difference between measured and calculated for a simple through line, the difference might be due to the measurement. There is very little that can go wrong with the EM analysis of a through line. If measured versus calculated does not agree for a through line, the problem is most likely with the measurement or with the parameters (dimensions, dielectric constants, and so on) that you used in the analysis model.

We have described many examples drawn from the applied experiences of the authors of this book. But we cannot teach everything. Now, you will proceed into the world of microwave design to learn even more.

9.6.1 The Impact of Electromagnetic Simulation

For a finale, we describe the impact of the simulators that the authors have used for more than 20 years.

1. *Eliminate the urban legends.* Urban legends is a term that had not yet been invented when we first started working with EM simulators. It refers to the false illusions that tend to spread unchecked through a believing, unskeptical community. Look at things skeptically. For example, we can calculate all the resonant modes that might occur in a housing. Do not believe that a particular aperture will excite all the modes without testing that possibility with EM analysis. Likewise, do not believe that a mode will not be excited without testing that idea, too. When you see modes that are or are not excited, try to understand why that happens. The wisdom you acquire by doing these experiments is priceless.
2. *Fresh discoveries.* When you quietly look at a simulation result, pay close attention to any unexpected results. This might be the most important advantage in using EM simulators. For example, when we investigate unexpected loss in a transmission line, we find strong radiation in the horizontal direction from between the layers of a multilayer substrate at a specific frequency.
3. *Grow your knowledge base.* When we first start simulating various circuits, we start gathering many unrelated bits and pieces of information. As we continue gathering experience, the bits and pieces start to connect together and we see patterns forming. Without simulation, we would have to acquire new experience at great expense in time and effort by actually building and troubleshooting our mistakes. The electromagnetic field simulator lets us make these mistakes and gain

experience efficiently. Then, before we realize it, we have become experts.

4. *Clues to improvement.* When we find a trouble spot using EM simulation, we can continue performing various numerical experiments to isolate the cause of the problem. Then we can use our creativity to promptly and quantitatively realize improvements.

9.7 Summary

1. Techniques based on frequency-domain and time-domain analysis are used in electromagnetic field simulators.
2. It is dangerous and incorrect to assume that EM simulators give the right answer. Also be skeptical and look for things that might have gone wrong. For example, all EM simulators have discretization error, which should be understood. No EM simulator can give us the right answer if we give it the wrong input.
3. For most efficient EM simulator usage, start with a coarse but fast discretization. As a final step, go to a slow, but more accurate, fine discretization.
4. With EM simulators, we can make our mistakes and gain experience quickly, without costly failed fabrications.

Appendix: Sonnet Lite Installation

You can install Sonnet Lite from the DVD that accompanies this book, or you can go to http://sonnetsoftware.com/products/lite/ and download the latest version.

Before installing Sonnet Lite, please read the License Agreement. An important term in the license agreement is that Sonnet Lite cannot be used for publishing competitive comparisons; doing so violates the License Agreement.

Follow these steps to install Sonnet Lite:

1. Download or copy from the DVD the following file to any directory. sl1253.exe (50 Mbytes)
2. Execute (double click) the file (sl1253.exe).
3. Sonnet Lite can be run by selecting Start > Programs > Sonnet 12.53 > Sonnet. We highly recommend that you go through the tutorials, which can be found by selecting Getting Started from the Help menu. A little effort spent in the tutorials will save you time and make you productive with Sonnet Lite in the quickest way possible. There are also numerous tutorials and videos on the Sonnet Web site.

   You should download Abobe Acrobat Reader if you don't already have it. You need Adobe Acrobat Reader 7.0.8 or above to read the manuals and tutorials. You will need the manuals to get the most out of Sonnet Lite. If you already have Adobe Acrobat Reader installed on your machine, you may skip this step. To download Adobe Acrobat Reader, go to http://get.adobe.com/reader/.

4. Register Sonnet Lite by selecting Start > Programs > Sonnet 12.53 > Register. You may use Sonnet Lite without the requirement to register the software and solve problems that require up to 1 MB of RAM to analyze. However, if you register your copy of Sonnet Lite, we'll send you a license that will enable you to solve problems requiring up to 16 MB of RAM. It only takes a few minutes to register your software.

If you experience any trouble installing or registering, please check the Sonnet Lite Troubleshooting Guide.

About the Authors

Hiroaki Kogure received a BSEE from Tokyo University of Science in 1977. In 1977, he joined Hitachi Engineering Co. Ltd., and worked with the development of electric power control systems and in the test production of the pioneer object-oriented precompiler language "objC" in 1985. Since 1992, he has been a representative of Kogure Consulting Engineers. He completed his doctorate (Dr. Eng.) in 1998 in electromagnetic field analyses at Tokyo University of Science. He has also been a part-time lecturer at the Tokyo University of Science since 2004 and Tokyo City University (formerly the Musashi Institute of Technology) since 2006. He primarily works as a registered professional engineer of information technologies to support research and development for leading companies in Japan. Dr. Kogure has written many technical books in the fields of electromagnetics, high frequencies, EMC problems, and antennas.

Yoshie Kogure received a B.A. in Chinese literature from Waseda University in 1983. Since 1992, she has worked with Kogure Consulting Engineers as a technical writer and a translator. She has written many technical books in collaboration with Dr. Kogure.

James C. Rautio received a BSEE from Cornell in 1978, an M.S. in systems engineering from the University of Pennsylvania in 1982, and a Ph.D. in electrical engineering from Syracuse University in 1986. From 1978 to 1986, he worked for General Electric, first at the Valley Forge Space Division, then at the Syracuse Electronics Laboratory. At this time, he developed microwave design and measurement software, and he designed microwave circuits on alumina and on GaAs. From 1986 to 1988, he was a visiting professor at Syracuse University and Cornell. In 1988, he went full time with Sonnet Software, a company he had founded in 1983. In 1995, Sonnet was listed on the *Inc.* 500 list of the fastest-growing privately held U.S. companies, the first microwave software company ever to be so listed. Dr. Rautio was elected a fellow of the IEEE in 2000 and received the IEEE MTT Microwave Application Award in 2001. He has lectured on the life of James Clerk Maxwell more than 100 times.

Index

Recent Titles in the Artech House Microwave Library

Active Filters for Integrated-Circuit Applications, Fred H. Irons

Advanced Techniques in RF Power Amplifier Design, Steve C. Cripps

Automated Smith Chart, Version 4.0: Software and User's Manual, Leonard M. Schwab

Behavioral Modeling of Nonlinear RF and Microwave Devices, Thomas R. Turlington

Broadband Microwave Amplifiers, Bal S. Virdee, Avtar S. Virdee, and Ben Y. Banyamin

Computer-Aided Analysis of Nonlinear Microwave Circuits, Paulo J. C. Rodrigues

Designing Bipolar Transistor Radio Frequency Integrated Circuits, Allen A. Sweet

Design of FET Frequency Multipliers and Harmonic Oscillators, Edmar Camargo

Design of Linear RF Outphasing Power Amplifiers, Xuejun Zhang, Lawrence E. Larson, and Peter M. Asbeck

Design Methodology for RF CMOS Phase Locked Loops, Carlos Quemada, Guillermo Bistué, and Iñigo Adin

Design of RF and Microwave Amplifiers and Oscillators, Second Edition, Pieter L. D. Abrie

Digital Filter Design Solutions, Jolyon M. De Freitas

Discrete Oscillator Design Linear, Nonlinear, Transient, and Noise Domains, Randall W. Rhea

Distortion in RF Power Amplifiers, Joel Vuolevi and Timo Rahkonen

EMPLAN: Electromagnetic Analysis of Printed Structures in Planarly Layered Media, Software and User's Manual, Noyan Kinayman and M. I. Aksun

An Engineer's Guide to Automated Testing of High-Speed Interfaces, José Moreira and Hubert Werkmann

Essentials of RF and Microwave Grounding, Eric Holzman

FAST: Fast Amplifier Synthesis Tool—Software and User's Guide, Dale D. Henkes

Feedforward Linear Power Amplifiers, Nick Pothecary

Foundations of Oscillator Circuit Design, Guillermo Gonzalez

Frequency Synthesizers: Concept to Product, Alexander Chenakin

Fundamentals of Nonlinear Behavioral Modeling for RF and Microwave Design, John Wood and David E. Root, editors

Generalized Filter Design by Computer Optimization, Djuradj Budimir

High-Linearity RF Amplifier Design, Peter B. Kenington

High-Speed Circuit Board Signal Integrity, Stephen C. Thierauf

Intermodulation Distortion in Microwave and Wireless Circuits, José Carlos Pedro and Nuno Borges Carvalho

Introduction to Modeling HBTs, Matthias Rudolph

Introduction to RF Design Using EM Simulators, Hiroaki Kogure, Yoshie Kogure, and James C. Rautio

Klystrons, Traveling Wave Tubes, Magnetrons, Crossed-Field Amplifiers, and Gyrotrons, A. S. Gilmour, Jr.

Lumped Elements for RF and Microwave Circuits, Inder Bahl

Lumped Element Quadrature Hybrids, David Andrews

Microwave Circuit Modeling Using Electromagnetic Field Simulation, Daniel G. Swanson, Jr. and Wolfgang J. R. Hoefer

Microwave Component Mechanics, Harri Eskelinen and Pekka Eskelinen

Microwave Differential Circuit Design Using Mixed-Mode S-Parameters, William R. Eisenstadt, Robert Stengel, and Bruce M. Thompson

Microwave Engineers' Handbook, Two Volumes, Theodore Saad, editor

Microwave Filters, Impedance-Matching Networks, and Coupling Structures, George L. Matthaei, Leo Young, and E.M.T. Jones

Microwave Materials and Fabrication Techniques, Second Edition, Thomas S. Laverghetta

Microwave Mixers, Second Edition, Stephen A. Maas

Microwave Network Design Using the Scattering Matrix, Janusz A. Dobrowolski

Microwave Radio Transmission Design Guide, Second Edition, Trevor Manning

Microwaves and Wireless Simplified, Third Edition, Thomas S. Laverghetta

Modern Microwave Circuits, Noyan Kinayman and M. I. Aksun

Modern Microwave Measurements and Techniques, Second Edition, Thomas S. Laverghetta

Neural Networks for RF and Microwave Design, Q. J. Zhang and K. C. Gupta

Noise in Linear and Nonlinear Circuits, Stephen A. Maas

Nonlinear Microwave and RF Circuits, Second Edition, Stephen A. Maas

Q Factor Measurements Using MATLAB®, Darko Kajfez

QMATCH: Lumped-Element Impedance Matching, Software and User's Guide, Pieter L. D. Abrie

Practical Analog and Digital Filter Design, Les Thede

Practical Microstrip Design and Applications, Günter Kompa

Practical RF Circuit Design for Modern Wireless Systems, Volume I: Passive Circuits and Systems, Les Besser and Rowan Gilmore

Practical RF Circuit Design for Modern Wireless Systems, Volume II: Active Circuits and Systems, Rowan Gilmore and Les Besser

Production Testing of RF and System-on-a-Chip Devices for Wireless Communications, Keith B. Schaub and Joe Kelly

Radio Frequency Integrated Circuit Design, Second Edition, John W. M. Rogers and Calvin Plett

RF Bulk Acoustic Wave Filters for Communications, Ken-ya Hashimoto

RF Design Guide: Systems, Circuits, and Equations, Peter Vizmuller

RF Measurements of Die and Packages, Scott A. Wartenberg

The RF and Microwave Circuit Design Handbook, Stephen A. Maas

RF and Microwave Coupled-Line Circuits, Rajesh Mongia, Inder Bahl, and Prakash Bhartia

RF and Microwave Oscillator Design, Michal Odyniec, editor

RF Power Amplifiers for Wireless Communications, Second Edition, Steve C. Cripps

RF Systems, Components, and Circuits Handbook, Ferril A. Losee

The Six-Port Technique with Microwave and Wireless Applications, Fadhel M. Ghannouchi and Abbas Mohammadi

Solid-State Microwave High-Power Amplifiers, Franco Sechi and Marina Bujatti

Stability Analysis of Nonlinear Microwave Circuits, Almudena Suárez and Raymond Quéré

Substrate Noise Coupling in Analog/RF Circuits, Stephane Bronckers, Geert Van der Plas, Gerd Vandersteen, and Yves Rolain

System-in-Package RF Design and Applications, Michael P. Gaynor

TRAVIS 2.0: Transmission Line Visualization Software and User's Guide, Version 2.0, Robert G. Kaires and Barton T. Hickman

Understanding Microwave Heating Cavities, Tse V. Chow Ting Chan and Howard C. Reader

For further information on these and other Artech House titles, including previously considered out-of-print books now available through our In-Print- Forever® (IPF®) program, contact:

Artech House Publishers
685 Canton Street
Norwood, MA 02062
Phone: 781-769-9750
Fax: 781-769-6334
e-mail: artech@artechhouse.com

Artech House Books
16 Sussex Street
London SW1V 4RW UK
Phone: +44 (0)20 7596 8750
Fax: +44 (0)20 7630 0166
e-mail:
artech-uk@artechhouse.com

Find us on the World Wide Web at: www.artechhouse.com